High Performance Programming
for
Soft Computing

High Performance Programming for Soft Computing

Editors

Oscar Montiel Ross and Roberto Sepúlveda Cruz
Instituto Politécnico Nacional-Centro de Investigación y
Desarrollo de Tecnología Digital
Tijuana, México

CRC Press
Taylor & Francis Group
Boca Raton   London   New York

CRC Press is an imprint of the
Taylor & Francis Group, an **informa** business

A SCIENCE PUBLISHERS BOOK

Preface

Soft Computing (SC) refers to a collection of several modern computational techniques in computer science that includes fuzzy systems, neural networks, machine learning, evolutionary and natural computing, etc. Soft Computing differs from conventional (hard) computing, mainly in that hard computing methods are developed using traditional mathematical concepts based on rigorous concepts that lead us to the use of rigid equations that are sometimes out of reality, specially, for complex systems. On the other hand, inspired on the ability of the human mind to solve complex problems, SC exploits the tolerance to imprecision, uncertainty and partial truth, these techniques are for the most part complementary more than competitive, that works synergistically with other traditional techniques. Currently, SC techniques are the principal components of any intelligent system. It has gained acceptability in science and engineering for several reasons, for example in the creation of mathematical models of complex systems with huge amounts of uncertainty and lack of knowledge in applications related to biology, medicine, humanities and engineering, where scientific methods of previous centuries inexorably failed.

The aim of this book is to expose to the readers the present and future of SC techniques, and provide them with the latest technological tools, i.e., multicore processors and graphics processing units (GPUs), to implement highly efficient intelligent system methods using a general purpose computer.

We want to thank all the people who encouraged us in the effort of making this book. We also want to thank our colleagues and students of the Intelligent System area of our research center Centro de Investigación y Desarrollo de Tecnología Digital del Instituto Politécnico Nacional (IPN), as well as to our colleagues from Instituto Tecnológico de Tijuana for their contributions. It is important to mention and thanks our institution IPN, and the CONACYT for supporting our research work.

September 2013

Prof. Dr. Oscar Montiel Ross
Prof. Dr. Roberto Sepúlveda Cruz

Contents

Preface   v

1. Fundamentals of Soft Computing   **1**
  1.1 Introduction   1
  1.2 Type-1 Fuzzy Sets   4
  1.3 Fuzzy Logic Systems   6
  1.4 Type 2 Fuzzy Sets   9
  1.5 Type-2 Fuzzy Inference Systems   11
       1.5.1 Interval Type-2 fuzzification stage   12
       1.5.2 Interval Type-2 inference stage   12
       1.5.3 Type-reduction and defuzzification in an interval   13
              Type 2 FIS
  1.6 Artificial Neural Networks   15
       1.6.1 A single neuron   16
       1.6.2 ANN architectures   20
       1.6.3 The learning process   22
       1.6.4 The back-propagation algorithm   26
  1.7 Adaptive Neuro-Fuzzy Inference System   27
       1.7.1 ANFIS   27
       1.7.2 ANFIS architecture   28
       1.7.3 Hybrid learning algorithm   29
  References   30

2. Introduction to Novel Microprocessor Architectures   **33**
  2.1 Introduction   33
  2.2 History of Computer Processors   35
       2.2.1 Beginning of the computer processors   36
       2.2.2 Modern commercial multicore processors   39
  2.3 Basic Components of a Computer   42
  2.4 Characteristics of Modern Processors   48
  2.5 Classification of Computer Architectures   52
       2.5.1 SISD   53
       2.5.2 SIMD   53
       2.5.3 MISD   55
       2.5.4 MIMD   56
  2.6 Parallel Computer Memory Architectures   56
  2.7 Measuring Performance   58
       2.7.1 Computer performance   59

  2.7.2 Analyzing the performance in multicore/   63
       multiprocessor systems
  2.8 Conclusions   69
  References   70

3. **High-Performance Optimization Methods**   **72**
  3.1 Introduction   73
  3.2 Unconstrained Iterative Optimization Methods   75
      3.2.1 Gradient descent method   76
      3.2.2 Newton's method   77
      3.2.3 Conjugate gradient method   78
  3.3 Unconstrained Optimization Using Parallel Computing   79
      3.3.1 MATLAB mex-file implementation   81
      3.3.2 Testing algorithm performance   83
  3.4 Global Optimization Using Memetic Algorithms   83
      3.4.1 Set up of the implemented MA   84
      3.4.2 Experiments and results   86
  3.5 Concluding Remarks   90
  3.6 Apendix   91
  References   93

4. **Graphics Processing Unit Programming and Applications**   **94**
  4.1 Introduction   94
  4.2 CUDA: Compute Unified Device Architecture   96
      4.2.1 Programming model   96
      4.2.2 Kernel functions   97
      4.2.3 Thread hierarchy   97
      4.2.4 Memory spaces and hierarchy   99
  4.3 CUDA Programming: Fractal Generation and Display   100
      4.3.1 Code analysis   103
      4.3.2 Benchmarking GPU vs. CPU implementation   108
  4.4 Conclusions   108
  References   109

5. **GPU-based Conjugated Gradient Solution for Phase**   **110**
   Field Models
  5.1 Introduction   110
  5.2 Conjugate Gradient Method   111
      5.2.1 Solution of a linear system as an optimization   112
           problem
      5.2.2 Steepest descent (SD) method   113
      5.2.3 Conjugate directions method   116
      5.2.4 Linear conjugate gradient method   120
      5.2.5 Non-linear conjugate gradient method   122
  5.3 Phase Field Model   123

5.3.1	Discretization of the Allen-Cahn equation	125
5.3.2	CUSPARSE and CUBLAS	127
5.3.3	GPU-based solution for the A-C equation	127
5.4	Results	128
5.4.1	1-D case	129
5.4.2	2-D case	129
5.5	Conclusions and Future Work	130
References		131

6. Parallel Computing Applied to a Diffusive Model — **133**
6.1	Introduction	133
6.2	Random Walk	134
6.3	Continuous Distributions	137
6.4	Solution via Finite Difference Approach	139
6.5	High Performance Computing Resources	142
6.6	Conclusions	147
References		147

7. Ant Colony Optimization (Past, Present and Future) — **148**
7.1	Introduction	148
7.1.1	Introduction to swarm intelligence	149
7.1.2	Biological inspiration	150
7.2	The Simple ACO Algorithm	152
7.3	Ant System	153
7.3.1	Tour construction in ant system	154
7.3.2	Update of pheromone trails	154
7.3.3	ACO for the Traveling Salesman Problem	155
7.4	Elitist AS: An Extension of AS	158
7.5	Ant-Q: Introduction to Q-Learning	158
7.6	Ant Colony System (ACS)	160
7.7	$\mathcal{MAX}$-$\mathcal{MIN}$ AS	161
7.8	Rank-based AS	162
7.9	ANTS	163
7.10	BWAS: 2000	163
7.11	10 Hyper–Cube AS (HC–AS): 2001	165
7.12	The Present of ACO: High Performance Computing	166
7.12.1	Introduction to parallel ant colony optimization	166
7.13	Application of SACO in Mobile Robotics	167
7.14	The Future of ACO: A Final Conclusion	168
References		169

8. Tool Path Optimization Based on Ant Colony Optimization for CNC Machining Operations — **172**
| 8.1 | Introduction | 172 |
| 8.2 | Introduction to CNC Programming using CAD/CAM | 173 |

8.2.1 Numerical control programming   174
8.2.2 The use of Computer Aided Manufacturing (CAM)   177
8.2.3 Drilling holes   177
8.2.4 Absolute and incremental programming   178
8.2.5 Canned cycles   179
8.3 Ant Colony Optimization for Hole Making NC Sequences: 182
    A Special Case of the Traveling Salesman Problem (TSP)
    8.3.1 Tour construction in ant system   182
    8.3.2 Update of pheromone trails   183
    8.3.3 NC for drilling operations using ant colony   183
          optimization: Case Study
8.4 Parallel Implementation of Ant Colony Optimization   186
    8.4.1 Parallel Implementation: Problem Formulation   188
    8.4.2 Performance of Parallel Computers   191
8.5 Experimental Results   191
    8.5.1 Analysis of the parallel implementation of ACO   192
    8.5.2 Tool path optimization analysis   193
8.6 Conclusions   195
References   195

9. **A Compendium of Artificial Immune Systems**   **197**
   9.1 Introduction   197
   9.2 A Brief History of Natural and Artificial Immune Systems 198
   9.3 Interesting Properties of the Immune System   200
   9.4 What Exactly is an Artificial Immune System?   201
   9.5 Main Algorithms   202
       9.5.1 Negative selection algorithm   202
       9.5.2 Clonal selection   204
   9.6 Immune Networks   207
   9.7 Other Sources of Inspiration   210
   9.8 Are Artificial Immune Systems Worth It?   211
   9.9 Conclusions   212
   References   212

10. **Applications of Artificial Immune Algorithms**   **215**
    10.1 Introduction   215
    10.2 Step 1. Reduction   217
         10.2.1 City representation using a Class   217
         10.2.2 Tour representation using a Class   218
         10.2.3 Artificial Vaccines representation using the 219
                Vaccine Class
         10.2.4 Vaccine Generation by Random Selector (VRS) 222
         10.2.5 Vaccine Generation by Elitist Selector (VES) 223

10.3 Step 2: Optimization   224
10.4 Step 3: Expansion   225
10.5 Example of Vaccinated TSP   226
10.6 Example: Application of the ROE Method and a GA   229
10.7 Conclusions   230
     References   231

11. **A Parallel Implementation of the NSGA-II**   **232**
11.1 Introduction   233
11.2 Basic Concepts   234
11.3 Genetic Algorithms   237
     11.3.1 Crossover operators   238
     11.3.2 Mutation operators   242
11.4 NSGA-II   244
11.5 Parallel Implementation of NSGA-II (PNSGA-II)   246
11.6 Results   248
11.7 Conclusions   254
     References   256

12. **High-performance Navigation System for Mobile Robots**   **258**
12.1 Introduction   258
12.2 Artificial Potential Field   259
     12.2.1 Attraction potential   262
     12.2.2 Repulsive potential   263
     12.2.3 Limitations of the artificial potential field   265
12.3 Genetic Algorithms   266
     12.3.1 How does a genetic algorithm work?   267
     12.3.2 Genetic algorithm operators   268
12.4 High-performance Implementation   270
     12.4.1 Phase 1: Simple navigation system with artificial   271
             potential field
     12.4.2 Phase 2: Complete navigation system with   273
             artificial potential field and genetic algorithms
     12.4.3 Phase 3: High-performance navigation system   275
             with artificial potential field and parallel genetic
             algorithms
12.5 Results and Conclusions   277
     References   280

13. **A Method Using a Combination of Ant Colony Optimization**   **282**
    Variants with Ant Set Partitioning
13.1 Introduction   283
     13.1.1 Ant Colony Optimization (ACO)   284
     13.1.2 ACO variations   285

13.2 Proposed Method   285
        13.2.1 Methodology   286
13.3 Experiments   287
13.4 Simulation Results   287
        13.4.1 Berlin 52 cities   288
        13.4.2 Bier 127 (127 cities)   288
13.5 Conclusions   289
References   290

14. **Variants of Ant Colony Optimization: A Metaheuristic for**   **291**
     Solving the Traveling Salesman Problem
     14.1 Introduction   291
     14.2 ACO Variants   292
             14.2.1 Traveling Salesman Problem (TSP)   292
             14.2.2 Elitist Ant System   293
             14.2.3 Rank based ant system   293
             14.2.4 Max-Min ant system   293
             14.2.5 Ant Colony System (ACS)   293
     14.3 Graphical Interface in Matlab   294
     14.4 Sequential Processing   295
     14.5 Parallel Processing   296
     14.6 Simulation Results   297
             14.6.1 Speedup   301
     14.7 Conclusions   302
     References   305

15. **Quantum Computing**   **306**
     15.1 Introduction   306
     15.2 Classic Computation   308
     15.3 Basic Mathematics Used in Quantum Computing   311
     15.4 Quantum Mechanic: basic Principles   317
     15.5 Elements of Quantum Computing   320
             15.5.1 The Bloch sphere   321
             15.5.2 Quantum registers   324
             15.5.3 Quantum measurements   325
             15.5.4 Quantum Gates   326
             15.5.5 Quantum circuits   328
     15.6 Concluding Remarks   329
     References   329

Index   **331**
Color Plate Section   **339**

Why this Book and Target Audience of this Book

Soft Computing (SC) refers to a collection of several modern computational techniques that includes fuzzy systems, neural networks, evolutionary and natural computing, and others. Machine learning uses SC techniques to make computers improve their performance over time using learning methods, which are highly demanding of computational resources; therefore, there are emerging novel algorithmic and hardware proposals that help to overcome such requirements. Many of the new algorithms are focus in taking advantage of multicore processors and General Purpose Graphics Units (GPU), with the aim of reducing execution time; on the other hand, computer scientist are also working in new proposals that help to alleviate computational algorithmic complexity, independently of the hardware platform.

The target audience of this book is graduate and undergraduate students from all science and engineering disciplines that are working in any of the fields of artificial intelligence, such as intelligent systems where massively computational processing are needed.

1

Fundamentals of Soft Computing

Roberto Sepúlveda* and Oscar Montiel

ABSTRACT

This is an introduction to the most common soft computing (SC) techniques. It provides to the novice reader interested in this field of knowledge, the fundamentals of this technology, and for the experimented practitioner it can be used as a practical quick reference. In this chapter, the main topics discussed are the type-1 and type-2 fuzzy logic systems, neural networks and artificial neuro-fuzzy inference system (ANFIS).

1.1 Introduction

In the 1940s and 50s, several researchers proved that many dynamic systems can be mathematically modeled using differential equations. These previous works, in addition to the Transform theory, represent the foundation of the Control theory which provides an extremely powerful means of analyzing and designing control systems (Cordón et al. 2001). Many developments were achieved until the 70s, when the area was known as System Theory to indicate its definitiveness (Mamdani 1993). Its principles have been used to control a very large amount of systems using mathematics as the main tool, for several years. Unfortunately, in too many cases this approach could not

Instituto Politécnico Nacional, CITEDI, Tijuana, México.
  Email: rsepulvedac@ipn.mx
* Corresponding author

be sustained because many systems have unknown parameters or highly complex and nonlinear characteristics that make them not amenable to the full force of mathematical analysis as dictated by the Control Theory.

Soft Computing techniques such as fuzzy logic, neural networks, evolutionary and natural computing, etc. have now become a research topic which is applied, among others, in the design of controllers (Jang et al. 1997). These techniques as far as possible avoid the above-mentioned drawbacks, and allow us to obtain efficient controllers, which utilize the human experience in a more related form rather than the conventional mathematical approach (Zadeh 1971, Zadeh 1973, Zadeh 1975c). In cases where a mathematical representation of the controlled systems cannot be obtained, it is possible to express the relationship existing in them, i.e., the process behavior.

Uncertainty affects decision-making and appears in a number of different forms. The concept of information is fully connected to the concept of uncertainty. The most fundamental aspect of this connection is that the uncertainty involved in any problem-solving situation is a result of some information deficiency, which may be incomplete, imprecise, fragmentary, not fully reliable, vague, contradictory, or deficient in some way or the other (Klir and Yuan 1995). The general framework of fuzzy reasoning allows handling much of this uncertainty, fuzzy systems employ type-1 fuzzy sets, which represents uncertainty by numbers in the range [0, 1]. When something is uncertain, like a measurement, it is difficult to determine its exact value, and of course type-1 fuzzy sets make more sense than using crisp sets (Zadeh 1975a, Zadeh 1975b, Zadeh 1975c).

A Fuzzy Logic System (FLS) is composed of a knowledge base that comprises the information given by the process operator in the form of linguistic control rules; a fuzzification stage, which has the effect of transforming crisp data into fuzzy sets; an inference system stage, that uses them in conjunction with the knowledge base to make inference by means of a reasoning method; and a defuzzification stage, which translates the resulting fuzzy set to a real control action (Cordón et al. 2001).

Type-2 fuzzy sets deal with uncertain information (Karnik and Mendel 1998, Mendel 1998, Karnik and Mendel 2001). It is not reasonable to use an accurate Membership Function (MF) for something uncertain; so, in this case what we need is another type of fuzzy sets, those which are able to handle these uncertainties, the so-called type-2 fuzzy sets (Mendel 2001). This is possible by using type-2 fuzzy logic because it offers better capabilities to cope with linguistic uncertainties by modeling vagueness and unreliability of information (Mendel 1999, Liang and Mendel 2000, Karnik and Mendel 2001, Wagenknech and Harmann 1988, Mendel 2003).

The quest for new controllers which provide functionality and guarantee a precise performance has led to the search of new technologies, such as

Fuzzy Logic. We can find some papers emphasizing on the implementation of a type-2 fuzzy logic system (FLS) (Karnik et al. 1999); in others, it is explained how type-2 fuzzy sets let us model and minimize the effects of uncertainties in rule-based FLSs (Mendel and Bob John 2002). Some research works are devoted to solving real-world applications in different areas; for example, in signal processing, type-2 fuzzy logic is applied in the prediction in Mackey-Glass chaotic time-series with uniform noise presence (Mendel 2000, Karnik and Mendel 1999). New ways to control systems have been developed based mostly in this area of knowledge and there has been an increasing interest in the research of type-2 fuzzy systems because they offer more advantages in handling uncertainty with respect to type-1 fuzzy systems (Melin et al. 2013, Sepúlveda et al. 2012, Hagras and Wagner 2012, Sahab and Hagras 2011).

New developments in science and engineering require processing data at high speed, therefore the interest in creating new hardware platforms and algorithms that can fulfill such requirement is increasing. Intelligent systems that use soft computing techniques need to use learning algorithms and process information with uncertainty, the application of such methods requires huge quantities of processing capabilities, hence computer architects are designing new processing units with several cores to reduce execution time. The idea is to share the workload by the use of the processors parallely. This technology needs modern algorithms that take advantage of the novel processors' architectures such as the graphics processors units. Other efforts have been devoted to the design of specific application hardware that make use of parallelism at the circuit level, for instance, in the case of the implementation of fuzzy systems, neural networks, genetic algorithms, etc., in Field Programmable Gate Arrays (FPGA) and VLSI technology. Some of these efforts have been reported in Ngo et al. 2012 where an implementation of a high performance IT2FS on a GPU is presented without the use of a graphic API, which can be in other research works with knowledge of C/C++. The authors demonstrate that use of the computer CPU outperforms the GPU for small systems and conclude that as the number of rules and sample rate grow, the GPU outperforms the CPU. They show that there is a switch point in the performance ratio which indicates when the GPU is more efficient than the CPU. GPU runs approximately 30 times faster. They tested the software in a robot navigation system; in (Montiel et al. 2012) is explained a methodology for integrating a fuzzy coprocessor described in VHSIC Hardware Description Language (VHDL) to a soft processor embedded into an FPGA; the same co-processor with small changes to the circuitry could be used by a PC to improve performance in applications that requires high speed processing. The objective of the system is to increase the throughput of the whole system; since the controller uses parallelism at the circuitry level for high-speed-demanding applications, the rest of the application can be written in C/C++. The authors tested the design using an inverted pendulum system.

1.2 Type-1 Fuzzy Sets

This section is dedicated to providing definitions of set theory, starting with "crisp sets", in order to introduce the basic concepts of type-1 and type-2 fuzzy set theory.

In the classical set theory (Jech 2006), a set A comprises elements x with membership into a universe of discourse X; i.e., $x \in X$. The membership of an element can be expressed using an MF that is also known as the characteristic function or discrimination function (Liang and Mendel 2000). This definition is valid to represent an MF in the classical bivalent set (crisp set) theory. One of the most common methods of representing the crisp set A is,

$$A = \{x | x \in X\} \tag{1.2.1}$$

For a crisp set A, MF μ_A is defined as

$$\forall x \in X, \ \mu_A(x) = \begin{cases} 1 & \text{if} \quad x \in A \\ 0 & \text{if} \quad x \notin A \end{cases} \tag{1.2.2}$$

In other words, the function μ_A maps the elements x, in the universal set X to the set $\{0,1\}$,

$$\mu_A(x) : X \to \{0,1\}. \tag{1.2.3}$$

Ordinary fuzzy sets, also known as type-1 fuzzy sets (T1FS) were developed by Lotfi Zadeh in 1965 (Zadeh 1965). They are an extension of the classical set theory where the concept of membership was extended to have various grades of membership on the real continuous interval [0,1]. The original idea was to use a fuzzy set (FS), i.e., a linguistic term to model a word. T1FSs have demonstrated to work efficiently in many applications. Most of them use the mathematics of fuzzy sets but losing the focus on words, that are mainly used in the context to represent a function that is more mathematical than linguistic (Mendel 2003). MFs are used to characterize T1FSs.

Earlier, we had mentioned that FS and its MF is an extension of a crisp set defined in (1.2.1) and in (1.2.3), respectively. Therefore, that definition has been modified for FS and MF. Type-1 FS is a set of ordered pairs expressed by (1.2.4) (Liang and Mendel 2000),

$$A = \{(x, \mu_A(x)) | x \in X\} \tag{1.2.4}$$

So, for T1FS A, each element is mapped to [0,1] by its μ_A, where closed intervals [0,1] mean real numbers between 0 and 1, including the values 0 and 1,

$$\mu_A(x) : X \to [0, 1] \tag{1.2.5}$$

For a T1FS A, see Fig. 1.1, each generic element x is mapped to [0,1] by its $\mu_A(x)$.

MFs can represent discrete or continual universes. If a fuzzy set has a finite number of elements $\{x_1, x_2, ..., x_n\}$, then (1.2.4) can be represented as:

$$A = \mu_1/x_1 + \mu_2/x_2 + ... + \mu_n/x_n \tag{1.2.6}$$

or

$$A = \sum_{i=1}^{n} \mu_i/x_i \tag{1.2.7}$$

where the summation sign indicates the union of the individual values in a discrete universe of discourse.

In the case of a continuous universe of discourse, fuzzy set A can be represented as:

$$A = \int_{x \in X} \mu_A(x)/x \tag{1.2.8}$$

where the integral sign indicates the union of the individual values. In (1.2.7) and (1.2.8), the symbol "/" is just a separator; it does not imply a division.

The characterization of a fuzzy set depends on: the selection of the appropriate universe of discourse, and the definition of the MF which represents as much as possible the notion of a concept.

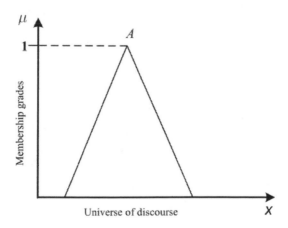

Fig. 1.1 Type-1 Membership function.

1.3 Fuzzy Logic Systems

Fuzzy logic is a branch of artificial intelligence that is based on the concept of perception; it gives the opportunity to obtain the approximate reasoning emulating the human brain mechanism, from facts described using natural language with uncertainty, it allows handling vague information or of difficult specification. Fuzziness concept is born when one is left to think that all phenomena founded everyday may qualify under crisp sets, when it is admitted the necessity of expressing mathematically vague concepts that cannot be represented in an adequate way using crisp values (Klir and Yuan 1995).

Fuzzy Logic Systems are being increasingly used because they tolerate inaccurate information, and can be used to model nonlinear functions of arbitrary complexity to build a system based on the knowledge of experts using natural language (Zadeh 1975c).

Generally speaking, this kind of system implements a non-linear mapping of its input to the output space. This mapping is achieved through the IF—THEN fuzzy rules, where each rule describes the local behavior of the mapping.

In the design of a FLS, after the analysis of a problem to be solved, it is helpful to describe, using the natural language, the necessary steps in the approach to solving the problem, then to translate each step to the IF - THEN fuzzy rule format:

$$\text{IF } x \text{ is } A \text{ THEN } y \text{ is } B \tag{1.3.1}$$

which is abbreviated as

$$A \rightarrow B \tag{1.3.2}$$

In this case, A represents the antecedent of the rule, and B the consequent; so, the expression describes a relation between them for the two variables x and y.

For a two-input, single-output FIS, the IF-THEN rule is written as:

$$\text{IF } x \text{ is } A \text{ AND } y \text{ is } B \text{ THEN } z \text{ is } C \tag{1.3.3}$$

Here, x and A are defined in the universe of discourse A^1, y and B are defined in the input universe of discourse B^1, while z and C are defined in the output universe of discourse C^1. This fuzzy rule can be put into the simpler form

$$A \times B \rightarrow C \tag{1.3.4}$$

which can be transformed into a ternary fuzzy relation R_m based on Mamdani's fuzzy implication function (Wagenknech and Harmann 1988):

$$R_m(A,B,C) = (A \times B) \times C$$
$$= \int\limits_{X \times Y \times Z} \mu_A(x) \wedge \mu_B(x) \wedge \mu_C(x)/(x,y,z)$$

(1.3.5)

For the two-input simple-output, in case we have the inputs x in A', y in B' and the output C', can be expressed as

$$C' = (A' \times B') \circ (A \times B \to C)$$

(1.3.6)

which can be decomposed as (Jang et al. 1997):

$$C' = [A' \circ (A \to C)] \cap [B' \circ (B \to C)]$$

(1.3.7)

The preceding expression states that the resulting consequence C' can be expressed as the intersection of

$$C'_1 = A' \circ (A \to C)$$

(1.3.8)

and

$$C'_2 = B' \circ (B \to C')$$

(1.3.9)

each one corresponding to a single-input single-output fuzzy rule as (1.3.1).

Figure 1.2 shows the three stages of FLS: Fuzzification, Inference and Defuzzification.

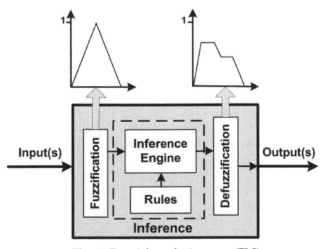

Fig. 1.2 Type-1 fuzzy logic system (FLS).

Fuzzification, whose function is to transform the crisp values to the entry system to fuzzy values for each label linguistic defined in the database, is defined as "one that turns data into appropriate linguistic values, which can be seen as labels on fuzzy sets" (Klir and Yuan 1995, Mendel 1999, Karnik and Mendel 2001). *Inference* is the core driver diffuse: inferred control actions simulating the decision-making process using a human diffuse involvement and the rules of inference logic blurred. It uses the techniques of systems based on rules for the inference of the results (Liang and Mendel 2000, Wagenknech and Harmann 1988). *Defuzzification* involves extracting a numerical value of a diffuse collection, to select a point that is the most representative of the action that will be taken (Mendel 2003).

FLS takes the input crisp values to convert them to fuzzy values, although the output is a fuzzy set, through the defuzzification stage, it is possible to obtain a crisp value at the output.

FLS can have different inference models, one of the best known is the *Mamdani model*, which historically was the first one used to control a steam machine through a set of linguistic fuzzy rules designed based on the operator's experience. In Fig. 1.3 is illustrated this kind of system, where there are two rules, with two antecedents and one consequent in each rule.

Rule 1: IF x is A_1 AND y is B_1 THEN z es C_1

Rule 2: IF x is A_2 AND y is B_2 THEN z es C_2

The Mamdani model operates repeating a cycle with the following four steps (Mendel 1999):

1. Measurements of the x, y variables are taken; they represent the relevant conditions of the process to be controlled.
2. In the fuzzification process, the measurements are used to represent the intersection of their projection to the appropriate fuzzy sets, i.e.: x in A_1 and in A_2; y in B_1 and B_2 for a two input system.
3. The fuzzified measurements are used by the inference stage to evaluate the fuzzy rules stored in the fuzzy rules base. The result of this evaluation is a fuzzy set, which is obtained with the min and max operations, in Fig. 1.3, the min of the first rule "min (.5, .8)" is obtained; similarly, for the second rule the "min (.8, .3)" is also obtained. Then, a max operation is performed to the output fuzzy set (consequents) of each active rule in the FLS. This stage is called the inference process.
4. The resulting fuzzy set of the inference process is converted to a crisp value using a defuzzification method. The defuzzified values represent the actions to be taken by the FLS.

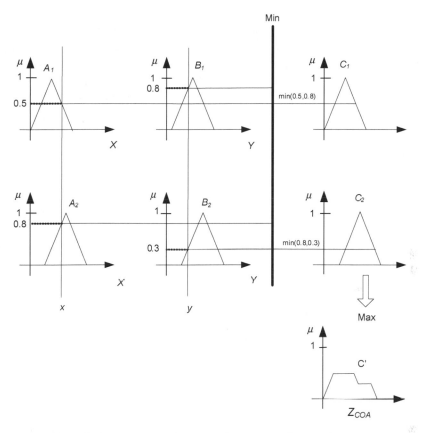

Fig. 1.3 A Mamdani Type-1 Fuzzy System with two antecedents and one consequent.

1.4 Type 2 Fuzzy Sets

The concept of a type-2 Fuzzy Set (T2FS) was introduced by Zadeh in 1975 (Zadeh 1975b), as an extension of the concept of a fuzzy set, also known as Type-1 Fuzzy Set (T1FS) in 1965 (Zadeh 1965), to cope with the uncertainty from words, because "words mean different things for different people, and are therefore uncertain" (Mendel 2003). Type-1 FISs are unable to handle uncertainties directly, because they use T1FSs that are certain (Mendel 2001). On the other hand, type-2 FISs, are very useful in situations where it is difficult to determine an exact MF because there is uncertainty from many sources. When the membership grade of each of the elements of a T2FS is indeed a fuzzy set in [0,1] it is called a Generalized T2FS (GT2FS), unlike a T1FS whose membership is a crisp number in [0,1] as it was explained in Section 3.2.

If we have a T1FS represented as in Fig. 1.1, and we blur its MF to the left and to the right then, at a specific value $x = x'$, the MF value u', takes on different values which will not all be weighted the same, so we can assign an amplitude distribution to all of those points (see Fig. 1.4). By doing this for all $x \in X$, we create a three-dimensional membership function that characterizes a T2FS (Mendel 2001, Mendel and Mounzouris 1999). The definition of a T2FS $\tilde{A}$, is given by

$$\tilde{A} = \{(x, u), \mu_{\tilde{A}}(x, u) | \, \forall \, x \, \in \, X, \forall \, u \, \in \, J_x \, \subseteq [0, 1]\} \tag{1.4.1}$$

where $\mu_{\tilde{A}}(x, u)$ is a type-2 MF whose value is in the range of $0 \leq \mu_{\tilde{A}}(x, u) \leq 1$.

Another way to express $\tilde{A}$ is,

$$\tilde{A} = \int_{x \in X} \int_{u \in J_x} \mu_{\tilde{A}}(x, u)/(x, u) \qquad J_x \subseteq [0, 1] \tag{1.4.2}$$

where $\int \int$ denote union over all admissible input variable x', and u'. For discrete universes of discourse the symbol $\int$ is replaced by $\sum$ (Jang et al. 1997). In fact $J_x \subseteq [0, 1]$ represents the primary membership of x and $\mu_{\tilde{A}}(x', u)$ is a T1FS known as the secondary set. Hence, a T2MF can be any subset in [0,1], the primary membership, and corresponding to each primary membership, there is a secondary membership (which can also be in [0,1]) that defines the uncertainty for the primary membership. This kind of fuzzy set is known as GT2FS as illustrated in Fig. 1.4, and the uncertainty is represented by a region called Footprint Of Uncertainty (FOU).

When $\mu_{\tilde{A}}(x, u) = 1$ where $x \in X$ and $u \in J_x \subseteq [0, 1]$, we have an Interval Type-2 Membership Function (IT2MF), as shown in Fig. 1.5.

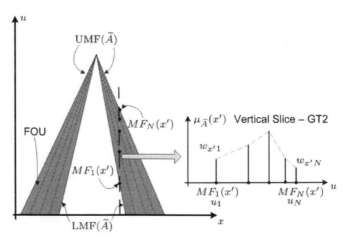

Fig. 1.4 Generalized Type-2 Membership Function.

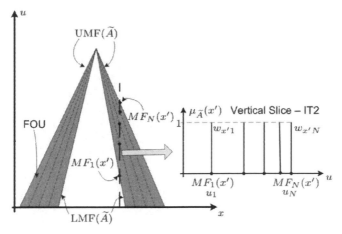

Fig. 1.5 Interval Type-2 Membership Function.

The uniform shading for the FOU represents the entire Interval Type-2 Fuzzy Set (IT2FS) and it can be described in terms of an upper membership function and a lower membership function:

$$\bar{\mu}_{\tilde{A}}(x) = \overline{FOU(\tilde{A})} \quad \forall x \in X. \tag{1.4.3}$$

$$\underline{\mu}_{\tilde{A}}(x) = \underline{FOU(\tilde{A})} \quad \forall x \in X. \tag{1.4.4}$$

In Fig. 1.5 the shadow region is the FOU of the IT2MF.

1.5 Type-2 Fuzzy Inference Systems

The basics and principles of fuzzy logic do not change from T1FS to T2FS (Mendel 2007, Mendel and Bob John 2002), they are independent of the nature of membership functions, and in general, will not change for any type-n. When an FIS uses at least one T2FS, it is a T2FIS (Mendel 2001, Liang and Mendel 2000). The structure of the type-2 fuzzy rules, is the same as for the type-1 case because the distinction between them is associated with the nature of the membership functions. Hence, the only difference is that now some or all the sets involved in the rules are of type-2. The most popular type-2 fuzzy inference system is based on T2FS, which we name IT2FIS. It contains four components: fuzzification, inference engine, type-reducer, and defuzzification, which are connected as shown in Fig. 1.6.

The IT2FIS can be seen as a mapping from the inputs to the output and it can be interpreted quantitatively as $Y = f(X)$, where $X = x_1, x_2, \cdots, x_n$ are the inputs to the IT2FIS, and $Y = y_1, y_2, \cdots, y_n$ are the defuzzified outputs. In a T1FIS, where the output sets are type-1 fuzzy sets, the

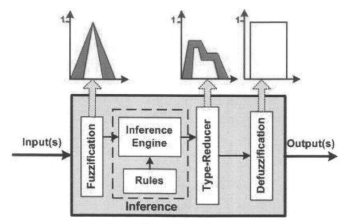

Fig. 1.6 Type-2 Fuzzy System.

defuzzification is performed to get a number, which is in some sense a crisp representation of the combined output sets. In the type-2 case, the output sets are type-2, so it is necessary for the extended defuzzification operation to get T1FS at the output. Since this operation converts type-2 output sets to a type-1 fuzzy set, it is called type reduction, and the T1FS obtained is called a type-reduced set. The type-reduced fuzzy set may then be defuzzified to obtain a single crisp number.

1.5.1 Interval Type-2 Fuzzification Stage

The task of the fuzzifier is to map a crisp value $x = (x_1, \ldots, x_p)^T \in X_1 \times X_2 \times \ldots \times X_p \equiv \mathbf{X}$ into a T2FS $\widetilde{A_x}$ in $\mathbf{X}$. For each singleton input, there are two membership grades values obtained in the fuzzification stage, $\bar{\mu}_N$ for the upper bound of the MF (UMF), and $\underline{\mu}_N$ for the lower bound of the MF (LMF), both of them associated respectively with the upper and lower bound of their FOU (Mendel and Bob John 2002), see Fig. 1.7.

1.5.2 Interval Type-2 Inference Stage

A type-2 fuzzy inference engine combines the if-then rules, according to Liang and Mendel (Liang and Mendel 2000), for an interval type-2 FIS the results of the input and antecedent operations are contained in the interval type-1firing set $[\underline{f}(x'_1), \bar{f}(x'_1)]$, computed using the minimum *t*-norm, as follows:

Lower firing level: $[\underline{f}(\mathbf{x}') = min[\underline{\mu}_{F_1}(x'_1), \underline{\mu}_{F_2}(x'_2)]$

Upper firing level: $[\bar{f}(\mathbf{x}') = min[\bar{\mu}_{F_1}(x'_1), \bar{\mu}_{F_2}(x'_2)]$

To obtain the output rule calculation, the firing interval $F(\mathbf{x}')$ is *t*-normed with the consequent $\widetilde{G}$; i.e., $\underline{f}(\mathbf{x}')$ is *t*-normed with the lower bound of the FOU of $\widetilde{G}$, and $\bar{f}(\mathbf{x}')$ is *t*-normed with the upper bound of the FOU of $\widetilde{G}$ (see Fig. 1.7).

1.5.3 Type-reduction and Defuzzification in an Interval Type-2 FIS

Some type-reduction (TR) methods like center-of-sets, centroid, center-of-sums, height type-reduction and modified height type reductionare described in (Mendel 2001, Karnik et al. 1999, Liang and Mendel 2000), that can all be expressed as:

$$Y_{TR}(\mathbf{x'}) = [y_l(\mathbf{x'}), y_r(\mathbf{x'})] \equiv [y_l, y_r] \qquad (1.5.1)$$

$$[y_l, y_r] = \int_{y^1 \in [y_l^1, y_r^1]} \cdots \int_{y^M \in [y_l^M, y_r^M]} \int_{f^1 \in [\underline{f}^1, \bar{f}^1]} \cdots \int_{f^M \in [\underline{f}^M, \bar{f}^M]} 1 / \frac{\sum_{i=1}^M f^i y^i}{\sum_{i=1}^M f^i}$$

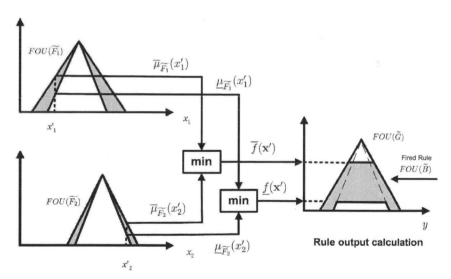

Fig. 1.7 A Mamdani Type-2 FLS, with two antecedents and one consequent system.

the multiple integral signs denote the union operation (Mendel 2001, Mendel et al.), where

Y_{TR} is the interval T1 FS called the type reduced set.
y_l is the left end point of the interval.
y_r is the right end point of the interval.
y_l^i is the left end point of the centroid of the consequent of the i-th rule.
y_r^i is the right end point of the centroid of the consequent of the i-th rule.
$\underline{f}^i$ is the lower firing degree of the i-th rule.
$\overline{f}^i$ is the upper firing degree of the i-th rule.

M is the number of fired rules.

In general, there are no closed-form formula for y_l and y_r; however, Karnik and Mendel (Karnik and Mendel 2001) have developed two iterative algorithms (known as the Karnik-Mendel or KM algorithms) for computing these end-points exactly, see Algorithm 1.1, and they can be run in parallel (Mendel et al.). The KM algorithm to calculate y_r is as shown in Algorithm 1.1.

Algorithm 1.1 KM algorithm.

---

1. It is assumed that the pre-computed y_r^i are arranged in ascending order, $y_r^1 \leq y_r^2 \leq \cdots y_r^M$.
2. Compute $y_r = \sum_{i=1}^{M} f_r^i y_r^i / \sum_{i=1}^{M} f_r^i$ by initially setting $f_r^i = (\underline{f}^i + \overline{f}^i)/2$ for $i = 1, \ldots, M$ and let $y_r'' \equiv y_r$.
3. Find R $(1 \leq R \leq M - 1)$ such that $y_r^R \leq y_r' \leq y_r^{R+1}$.
4. Compute $y_r = \sum_{i=1}^{M} f_r^i y_r^i / \sum_{i=1}^{M} f_r^i$ with $f_r^i = \underline{f}^i$ for $i \leq R$ and $f_r^i = \overline{f}^i$ for $i > R$, and let $y_r'' \equiv y_r$.
5. If $y_r'' \neq y_r'$, then go to step 6. If $y_r'' = y_r'$, then stop and set $y_r'' \equiv y_r$.
6. Set y_r' equal to y_r'', and return to step 3.

Similarly, it is possible to compute y_l, instead of using y_r^i use y_l^i, and in step 3 find $L (1 \leq L \leq M - 1)$ such that $y_l^L \leq y_l' \leq y_l^{L+1}$. Also, in step 2, compute $y_l = \sum_{i=1}^{M} f_l^i y_l^i / \sum_{i=1}^{M} f_l^i$ by initially setting $f_l^i = (\underline{f}^i + \overline{f}^i)/2$ for $i = 1, \ldots, M$, and in step 4 compute $y_l = \sum_{i=1}^{M} f_l^i y_l^i / \sum_{i=1}^{M} f_l^i$ with $f_l^i = \overline{f}^i$ for $i \leq L$ and $f_l^i = \underline{f}^i$ for $i > L$ (Mendel et al.).

The defuzzified output on an IT2 FLS is the average of y_l and y_r,

$$y(\mathbf{x}') = \frac{1}{2}(y_l(\mathbf{x}') + y_r(\mathbf{x}')).$$

$$(1.5.2)$$

In Fig. 1.7, is presented a pictorial description of the Mamdani T2FIS, for a two antecedents, one consequent system, using a minimum *t*-norm.

1.6 Artificial Neural Networks

Many advances have been achieved in developing intelligent systems, the human brain has motivated several works trying to imitate the way it works. The brain can compute efficiently many tasks that modern computer cannot do. It is a highly complex nonlinear parallel information-processing system. The human brain is especially good in solving complex perceptual problems such as recognizing a person in a crowd just looking at his face, very fast and practically effortlessly, actually, there is no computer that can do the same work at a similar speed with such performance. On the other hand, in the domain of numeric computation and related symbol manipulation, modern computers outperform humans. Researchers from many scientific disciplines have studied the differences between natural computers (biological neural system architectures) and modern digital computers based on the Von Neumann model and others, and at present time they have not developed any "intelligent architectures or programs" that can perform better in tasks where the human brain outperforms modern computers. However, they have found mathematical models that raise new computer architectures inspired in nature, in this case, the artificial neuron motivated by its counterpart the biological neuron.

Briefly, the field that studies artificial neurons, also called Artificial Neural Networks (ANN), has experienced several periods of extensive activity. The first one can be traced back to the work where the aim was to model a biological neuron. Here McCulloch and Pitts in 1943 introduced the first neural network computing model. In 1949, Hebb's book entitled *The Organization of Behavior* presented for the first time the physiological learning rule for synaptic modification. This book was an important source of inspiration that contributed to the development of new computational models of learning and adaptive systems. Many other important works were published and it is impossible to cite all of them here; however, some works that held our attention are: Ashby's book entitled *Design for a Brain: The Origin of Adaptive Behaviors* (Ashby 1960); Minsky's doctorate thesis entitled *Theory of Neural-Analog Reinforcement Systems and Its Applications to the Brain-Model Problem* published in 1954; as well as other works by Minsky, such as *Steps Toward Artificial Intelligence* and *Computation: Finite and Infinite Machines* published in 1961 and 1967, respectively. In the 1950s Taylor initialized work on associative memories, followed by Steinbuch in 1961 who introduced the learning matrix. In 1958 Rosenblatt introduced a new approach to the pattern recognition problem in his work with the perceptron which was a novel method of supervised learning. Widrow and Hoff in 1960 introduced the Least Mean-Squared algorithm (LMS) that was used to formulate the ADaptive LINnear Element (ADALINE). In 1962, Widrow and his students proposed one of the first trainable layered neural network consisting of using several (multiple) adaptive elements (ADALINEs), hence the name MADALINE. In 1969, Minski and Papert demonstrated that there are fundamental limitations about a single layer that a perceptron can compute; however, they stated that there was no reason to think that such limitation could be overcome using a multilayer version. In those times, the use of multilayer perceptrons was hampered by the problem of assigning credit to the hidden neurons in the layer, this problem was known as the "credit assignment problem" and it was first used by Minsky in 1961 textually as "Credit Assignment Problem for Reinforcement Learning". Several ideas surged to solve the aforementioned problem, but for several reasons, there was an historical lag in this field for about 20 years, although in this period, in the 1970s, self-organizing maps using competitive learning emerged. The activity in the field increased again in the 1980s, when Grossberg established a new principle of self-organization known as Adaptive Resonance Theory (ART). In 1982, Hopfield introduced several concepts based on the idea of using an energy function and recurrent networks with symmetric synaptic connections, and other important works to obtain a particular class of neural network known in our days as the Hopfield Network. In 1982, surged an important development proposed by Kohonen on self-organization maps using a one- or two-dimensional lattice structure. In 1986, the back-propagation algorithm was proposed in (Rumelhart et al. 1986). In 1989, the book *Analog VLSI and Neural Systems* provides an interesting mix of concepts drawn from VLSI technology and neurobiology. Undoubtedly, we have omitted other important works that have helped develop particular proposals in ANN such as the use of Radial Basis Functions (RBF), support vector machines, etc. (Haykin 1999); mainly, because it is of our interest to provide the reader only a synopsis of this topic.

1.6.1 A Single Neuron

The brain as a natural computer is made of a huge number of processing elements known as neurons. Similarly, an ANN is a massively parallel system with a large number of interconnected processing elements known artificial neurons, that jointly may solve a variety of challenging computational problems. Hence, to start working with complex ANN, it is important to understand that a single neuron is an information processing unit building block, that maps an input data vector into an output data vector once it was trained (learned the correct input/output behavior). There are several ways to represent schematically the architecture of a single neuron. Figure 1.8a shows the block diagram of the model of a single neuron, it provides a clear description of the elements that constitutes an artificial neuron, in this case the neuron k, where three basic elements can be identified:

1. A set of connecting links (synapses), each link has associated a weight w_{kn} and in turn has a signal input x_{kn}. Note that the subindex k refers to the current neuron, and the subindex n to the number of link.
2. The summing junction is an adder that combines the value of each input multiplied by its weight, this element is known as a linear combiner, the output of this element is the value v_k.
3. An activation function $\varphi(v_k)$ that modifies the value v_k. This modification will depend on the type of function, the main idea is to limit the amplitude value of v_k to work in the permissible limits, this function is also known as squashing function. The output of this element is the value y_k, typical amplitude ranges are the closed intervals [0,1] and [-1,1].

One of the inputs to the linear combiner of Fig. 1.8 is known as the bias input and its value is tied to a fixed value of +1; i.e., the input $x_0 = +1$ always. The corresponding weight w_{k0} can be adjusted during the learning process. This value is referred in literature as *bias* and is usually tagged as b_k where k is the current neuron. It has the effect of increasing or decreasing the value v_k which is the net input to the activation function. This is done in this way because the bias is considered as an external parameter to the neuron.

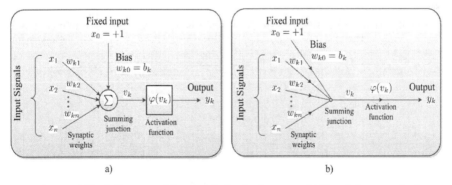

a) b)

Fig. 1.8 a) Block diagram of the model of a single neuron. b) Signal-flow graph of single neuron.

Figure 1.8b shows a signal-flow graph that simplifies the representation of an artificial neuron of Fig. 1.8a and allows us to provide the mathematical definition of a neural network (Haykin 1999):

"A neural network is a directed graph consisting of nodes with interconnecting synaptic and activation links, and is characterized by four properties:

1. Each neuron is represented by a set of linear synaptic links, an externally applied bias, and a possible nonlinear activation link. The bias is represented by a synaptic link connected to an input fixed at +1.
2. The synaptic links of a neuron weigh their respective input signals.
3. The weighted sum of the input signals defines the induced local field of the neuron in question.
4. The activation link squashes the induced local field of the neuron to produce an output."

The neuron k of Fig. 1.8, in mathematical terms can be described using Equations (1.6.1) and (1.6.2).

$$y_k = \varphi(v_k) \tag{1.6.1}$$

where

$$v_k = \sum_{j=0}^{m} w_{kj} x_j \tag{1.6.2}$$

Note, in the above equations, the induced local field (activation potential) v_k of neuron k, is given by the linear combination $u_k = \sum_{j=1}^{m} w_{kj} x_j$ and the bias b_k, so we can write $v_k = u_k + b_k$.

The activation functions $\varphi(v)$ are defined in terms of their local field; Fig. 1.9 shows the most common types that are derivations of the basic types: Threshold function, Piecewise-Linear Function, and the Sigmoid function. Note in this figure that the step and sign functions are derivations of the Threshold function; similarly, the logistic and hyperbolic tangent functions are examples of Sigmoid functions.

As we mentioned in the historical review, the single artificial neuron has been proposed in different ways, so using Fig. 1.8, we can have two different approaches—deterministic and stochastic. In a *deterministic model*, the outputs are precisely determined by the inputs, weight values and activation function. In the *stochastic model*, a probabilistic interpretation of the activation function is given. Next, the first three cases are deterministic models of neurons, whereas the fourth approach corresponds to the stochastic model.

Name	Equation	
Treshold function (step)	$\varphi(v) = \begin{cases} 1 & \text{if } v \geq 0 \\ 0 & \text{if } v < 0 \end{cases}$	
Treshold function (sign)	$\varphi(v) = \begin{cases} 1 & \text{if } v \geq 0 \\ 0 & \text{if } v = 0 \\ -1 & \text{if } v < 0 \end{cases}$	
Piecewise-Linear function	$\varphi(v) = \begin{cases} 1 & \text{if } v \geq +\frac{1}{2} \\ v & \text{if } +\frac{1}{2} > v > -\frac{1}{2} \\ 0 & \text{if } v \leq -\frac{1}{2} \end{cases}$	
Logistic function	$\varphi(v) = \frac{1}{1+\exp(-av)}$	$a > 0$
Hyperbolic tangent function	$\varphi(v_j) = a\tanh(bv_j)$	$(a,b) > 0$

For all the functions $-\infty < v < \infty$

Fig. 1.9 The most common activation functions.

1. *The Perceptron*. It is a nonlinear neuron known as the McCulloch-Pitts model of a neuron. It uses as activation function the hard limiter sign function, providing a binary output. The output y_k is used to perform the training, so it only provides information to know if the obtained output value is wrong or not. The error surface is very rough (very steep) because of the binary output that the activation function provides.

2. *The ADALINE*. It typically uses bipolar activation function for its input signals and target output. The output of the linear combiner v_k is used to achieve the training; hence the output in the training phase is linear. The output provides real numbers in the valid output range. The error surface is smooth.

3. *The modern artificial neuron*. It is in essence a perceptron that uses a *nonlinear smooth* function as the activation function, in contrast with the binary output of the McCulloch-Pitt model. This modification offers similar benefits than the ADALINE providing real numbers at the output but using a nonlinear smooth function. The training is performed in the same way than in the perceptron, i.e., the output y_k is used for training. Hence, the error surface is smooth.

4. *Stochastic model of a single neuron*. The McCulloch-Pitts model can enable or disable using a probabilistic decision; the neuron can only reside in one of the two states: +1 or –1.

The problems that a simple neuron can solve are very limited, therefore, to solve complex problems it is necessary to join several neurons to form a specific architecture known as ANN. It consists of massively parallel distributed processors that enable the ANN to store knowledge, allowing us to use it when it is required. Analogue to the brain, the ANN acquires the knowledge through a learning process, saving it in its synaptic weights. Therefore, it is also necessary to have algorithms to adjust the synaptic weights and learning paradigms that dictates how the network is going to operate in an environment.

1.6.2 ANN Architectures

There exists an intimate relationship between the manner in which the neurons of an ANN are connected and the learning algorithms are used to train the network. A network architecture refers precisely to the structure that forms the diverse interconnections of neurons. According to this, there are three different structures: Single layer Feedforward networks, multilayer feedforward networks and recurrent networks (Haykin 1999).

1. *Single layer feedforward networks.* All the neurons are organized in a single layer. Using the input layer of source nodes, the inputs patterns (examples) are presented to the network, the neurons (nodes) form the output layer of neurons and they compute the output values, this is not true viceversa, so this is the reason why it is called feedforward (or acyclic type), in this case strictly feedforward. In Fig. 1.10, an example of this network is shown; note that each node is used in abbreviated form to indicate a neuron of the type illustrated in Fig. 1.8.

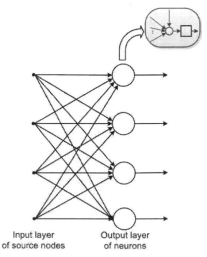

Input layer
of source nodes

Output layer
of neurons

Fig. 1.10 A single layer network.

2. *Multilayer feedforward network.* Figure 1.11 shows a network example of this category, this kind of network have one or more hidden layers, each hidden layer is composed by hidden neurons. The function of the hidden layers is to interact between external inputs to the network, and the network outputs. The more hidden layers have a network the higher statistic features can be extracted. The network shown in the figure is a fully connected feedforward network with one input layer with "m" source nodes (*m*), two hidden layers with two hidden nodes (h_1 and h_2), and one output layer with q neurons at the output. The nomenclature "$m - h_1 - h_2 - q$" is a common way to refer to them, following the nomenclature, the network of Fig. 1.11 is a 4-4-4-4 network.

3. *Recurrent networks.* Different from feedfoward ANN, a recurrent ANN has at least one feedback loop, which involves the use of unit-delay operators. Figure 1.12 illustrates a recurrent network with no hidden neurons and no-feedback loops, the last issue means that the output of any neuron is fedback into its own input. Figure 1.13 illustrates a different recurrent network with hidden neurons and self-feedback characteristics.

Besides the three different structures mentioned above, we might enumerate a fourth structure derived from feedforward networks named as *Time-lagged feed-forward network* (TLFN). The idea here is to incorporate the time into the design of the ANN implicity or explicity. For implicit

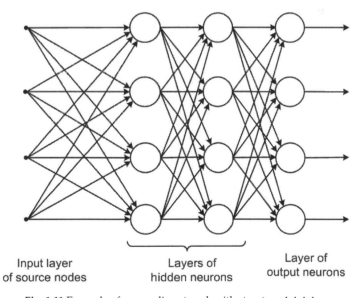

Input layer
of source nodes

Layers of
hidden neurons

Layer of
output neurons

Fig. 1.11 Example of an acyclic network with structure 4-4-4-4.

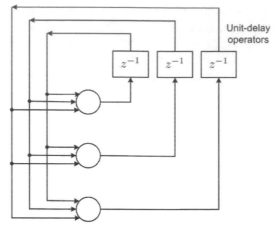

Fig. 1.12 Recurrent network with no hidden neurons and no feedback loop.

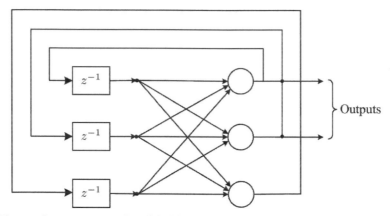

Fig. 1.13 Recurrent network with hidden neurons and self-feedback characteristics.

representation, a straightforward method is to add a short-term memory at the input of a static ANN, such configuration is called focus TLFN and its use is limited to stationary dynamical process. For nonstationary dynamical process, we can use distributed TLFNs at convenience (Haykin 1999).

1.6.3 The learning process

A primordial property of an ANN is to have the ability of learning from its environment with the aim to improve its performance, in supervised learning it is necessary to present several times the same pattern to the network several times, we say, we are training the network to adapt its behavior to perform a task of interest. An ANN is trained for different learning task, they are: Pattern association, patter recognition, function approximation, control, filtering and beamforming. Learning in the context of ANN is defined as (Haykin 1999):

> Learning is a process by which the free parameters of an artificial neural network are adapted through a process of stimulation by the environment in which the network is embedded. The type of learning is determined by the manner in which the parameter changes takes place.

The above definition implies the next three events: Stimulation of the ANN by its environment, changes in the free parameters (weights) of the ANN as a result of the stimulation, a different response of the network to the environment as a consequence of the changes in its internal structure.

To make a network learn, there are two things to consider:

1. *To use a learning algorithm.* There are several learning algorithms, but all of them have the same objective that is to modify (adjust) the synaptic weights of the neurons of the ANN. Some of the algorithms are: The error-correction, memory-base, Hebbian, Boltzmann and competitive learning algorithms.
2. *To use a learning paradigm.* It is the way in which the ANN relates with the environment to learn. Figure 1.14 shows a classification of the learning paradigms, the two main classifications are the supervised and unsupervised learning approaches. In the first one, it is necessary to have a "teacher" which is usually known as the input-output data to be used to train the ANN, this kind of networks works by correcting the network output using an error signal given by (1.6.3), where the value $d(n)$ is provided by the teacher, and the value $y(n)$ is the actual network output. To train an ANN, it is necessary to present several patterns to the network, hence n in (1.6.3) is the n_{th} training pattern presented to the network, note that this equation has been written in vectorial form because there can be p inputs and q outputs. For an unsupervised learning there is no external teacher to guide the

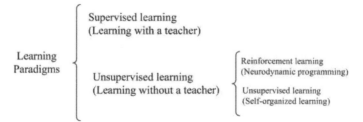

Fig. 1.14 Learning paradigms classification.

learning process, differently to the supervised case, there are no labeled examples of the function (pattern) to be learned, here the learning method allows to adjust networks themselves with new input patterns, the synaptic weights may be adjusted using Hebbian and competitive learning, or information-theoretic models. For the case of reinforcement learning, it is necessary to use a high quality heuristic reinforcement signal constructed by a critic. Self-organized learning methods use a competitive process, where a neuron can win or not (Haykin 1999).

$$\mathbf{e}(n) = \mathbf{d}(n) - \mathbf{y}(n) \tag{1.6.3}$$

For simplicity, first we are going to explain how to train the neuron of Fig. 1.8 using an error-correction learning algorithm in a supervised learning scheme. Training this neuron is very simple regardless it is a perceptron, ADALINE or a "modern artificial neuron", because we can calculate the error signal at the neuron output and use this value directly to correct the synaptic weights. The most widespread form is to use the Least-Mean-Square algorithm (LMS), this method is based on the steepest descent algorithm but it uses the error signal $\mathbf{e}(n)$, and the input vector $\mathbf{x}(n)$ to calculate an estimate of the gradient vector, so this method is also known as the stochastic gradient algorithm. In Eq. (1.6.1) the network output is $y_k = \mathbf{x}^T(n)\mathbf{w}(n)$, where $\mathbf{w}(n)$ is a vector containing the weights of the neuron, hence we have,

$$\mathbf{e}(n) = \mathbf{d}(n) - \hat{\mathbf{w}}^T(n)\mathbf{x}(n) \tag{1.6.4}$$

The LMS algorithm is formulated using the following delta rule:

$$\hat{\mathbf{w}}(n+1) = \hat{\mathbf{w}}(n) + \eta\mathbf{x}(n)\mathbf{e}(n) \tag{1.6.5}$$

where η is known as the learning rate parameter. In formulas (1.6.4) and (1.6.5) we used the nomenclature $\hat{\mathbf{w}}$ in place of $\mathbf{w}$ to indicate that the LMS produces and estimated weight vector which is different to the one that the steepest descendent method can produce. The LMS algorithm consists in performing several iterations, say N, computing formulas (1.6.4) and (1.6.5) until we fulfill a stop criterion, which usually is defined by a cost function based on the average of the squared error ξ_{av}. The calculation of ξ_{av} depends on the way we present the training patterns to the network, basically, it can be done in two ways: Sequential and batch mode.

From (1.6.3), we have that for neuron j the error signal is given by $e_j(n) = d_j(n) - y_j(n)$ and its instantaneous value of the error energy is given by $\frac{1}{2}e_j^2(n)$, therefore to calculate the instant value at the output of an ANN with q outputs we use Eq. (1.6.6)

$$\xi(n) = \frac{1}{2}\sum_{j=1}^{q} e_j^2(n) \tag{1.6.6}$$

The sequential mode is also known as pattern-by-pattern, on-line, or stochastic mode. In this mode the update of the synaptic weights is achieved after the presentation of each training example using their respective errors, therefore the arithmetic average changes of the weights of all the individuals in a complete training set, so Eq. (1.6.6) for this training method is used, and we obtain practically for each pattern, an estimate of the changes that are produced with the use of the cost function ξ_{av} that includes the whole training set.

In the batch mode, the synaptic weights update is achieved after all the training examples that constitutes an epoch have been presented to the network; here, ξ_{av} is calculated once for each epoch using all the particular $e(n)$ of each training example, the advantage resides in updating the weights only one time per epoch, so for N patterns presented to the network, the ξ_{av} is calculated by the summing of all the individual instant values of error given by (1.6.7).

$$\xi_{av} = \frac{1}{N} \sum_{n=1}^{N} \xi(n) \tag{1.6.7}$$

Equation (1.6.5) is not the only option to update the weights, there are modifications to this formulation and other algorithms such as the Levenberg-Marquardt, conjugate gradient, and others. Formula (1.6.6) shows a modification of the LMS, the term momentum has been included to increase the rate of learning.

$$\hat{w}(n+1) = \hat{w}(n) + \alpha[w(n-1)] + \eta x(n)e(n) \tag{1.6.8}$$

Training multilayer networks requires to know the errors at the neurons in the hidden layer, unfortunately it is not possible to know the desired response at each neuron (node) to calculate them, so we have to face to the credit assignment problem. The back-propagation algorithm was developed to cope with this problem in an elegant way. This algorithm is divided in two steps or passes, the forward pass and the backward pass, hence two types of signals are required. When we are in the forward pass, the information flows from the inputs to the outputs of the ANN such as they were defined for the current use, we refer to these signals as "function signals". In the backward pass, the information goes from the output layer of the ANN to the inputs in reverse form, the idea is to propagate the total error from the output to the input layer by layer, and these signals are called error signals.

1.6.4 The Back-propagation Algorithm

The back-propagation algorithm considers two cases:

Case 1: The artificial neuron j is an output node. Here the error is calculated using Eq. (1.6.3).

Case 2: The artificial neuron j is a hidden node. At this node we cannot know the desired response, so the back-propagation algorithm, using error signals, propagate the error through the nodes making possible to estimate the error named δ, at each hidden node; in short, the formula deduction of the error at the hidden nodes is given by (1.6.7), where the index k identifies neuron k, which is connected to the output of neuron j, and the term w_{kj} is the synaptic weight associated with these neurons.

$$\delta_j(n) = \varphi'_j(v_j(n)) \sum_k \delta_k(n) w_{kj}(n) \tag{1.6.9}$$

Therefore the back-propagation algorithm is summarized as follows:

Algorithm 1.2 Pseudocode of the back-propagation algorithm

---

1. Do while a stop criteria is fulfill.
2. Start with a random set of synaptic weights.
3. Calculate $\mathbf{y}(n)$ by propagating $\mathbf{x}(n)$ through the network.
4. Calculate the error $\mathbf{e}(n)$.
5. Adjust the weights using an optimization algorithm such as the LMS with momentum,

$$w_{ji}^{(l)}(n+1) = w_{ji}^{(l)}(n) + \alpha[w_{ji}^{(l)}(n-1)] + \eta\delta_j^{(l)}(n)y_i^{(l-1)}(n) \quad (1.6.10)$$

   where the local gradient depends on

$$\delta_j^{(l)} = \begin{cases} e_j^{(L)}\varphi'(v_j^L(n)) & \text{for a neuron } j \\ & \text{that is in the output layer } L \\ \varphi'(v_j^l(n))\sum_k \delta_k^{l+1}(n)w_{kj}^{(l+1)}(n) & \text{for a neuron } k \text{ in hidden layer } l \end{cases} \quad (1.6.11)$$

6. Endwhile.

To explain the algorithm, we have to suppose that the output of neuron (node) i is an input (synaptic weight) of neuron j, moreover the letter l is used to mean the number of the layer, in this way $(l-1)$ is the layer that is before of layer l, and layer $(l+1)$ is a layer that is after l; in other words, the nodes of $(l-1)$ are the inputs of the nodes in l, which in turn are the inputs of the nodes at $l+1$. In the algorithm, formula (1.6.10) was derived from (1.6.8); in (1.6.10) we are using individual node notation, where w_{ji} is the synaptic weight in the synaptic link that connects the output of neuron i to the input of neuron j. The stop criterion may be one or several, for example, we can use the number of epochs, or a specific value of ξ_{av}, or both.

1.7 Adaptive Neuro-Fuzzy Inference System

The systems that combine different soft computing techniques provide greater benefits to solve many problems in a more efficient way, because each method provides different advantages and features that solve certain types of problems, that is the case of the Adaptive Neuro-Fuzzy Inference System, better known as ANFIS, which combines artificial neuronal nets and fuzzy logic techniques.

Over the years, fuzzy logic has been used in a wide variety of applications, however there are some basic elements that are important to understand:

- There are no standard methods to transform human knowledge or experience into a base rule of a fuzzy inference system.
- There is a need for an effective method to adjust the parameters of the membership functions, to minimize the output error or maximize the performance index.

An important area of application of fuzzy logic is the fuzzy controllers, however the selection of the number and type of MFs, the rules to represent the desired behavior of the controller, the parameters of the MFs, etc., is a difficult task in most cases. ANFIS can be used to design a set of IF—THEN rules, with the appropriate MFs to generate the input-output pairs set. In the ANFIS model, the neuronal nets are used to represent the fuzzy inference systems, the same that are used as the decision making systems.

1.7.1 ANFIS

ANFIS is an adaptive network functionally equivalent to a Sugeno and Tsukamoto fuzzy models. Its working mechanism is similar to a neural network, but it adjusts the parameters of the MFs instead of the weights on the connection between neurons. ANFIS facilitates learning and adaptation since it combines fuzzy logic with its inference mechanism on the uncertainty, and the computational advantages of the neural networks such as learning, fault tolerance, parallelism, etc. Although fuzzy logic can encode the knowledge through linguistic labels, usually it takes a long time to define and adjust the parameters of the MFs. Learning techniques of neural networks can automate this process and reduce the time and development cost, improving the performance of the model.

The hybrid learning algorithm in ANFIS is based in the minimization of the error, in the forward pass it uses the least-squares optimization method to adjust the parameters of the consequent, in the backward pass, the premise parameters are optimized by the gradient descent method.

As was mentioned earlier, ANFIS has its advantages, however a drawback is that sometimes the learning process is relatively slow, because it requires a lot of training epochs.

1.7.2 ANFIS Architecture

The ANFIS model was proposed by Jang (Jang et al. 1997), to explain its operation it is assumed that the first order Sugeno fuzzy model has two inputs x and y and one output f, with the following IF—THEN rules:

Rule 1: IF x is A_1 and y is B_1, THEN $f_1 = p_1 x + q_1 y + r_1$,

Rule 2: IF x is A_2 and y is B_2, THEN $f_2 = p_2 x + q_2 y + r_2$.

ANFIS architecture has five layers, as can be seen in Fig. 1.15, where the squares represent adaptive nodes and the circles represent fixed nodes.

Layer 1. Each node i is an adaptive node with the following node function:

$$O_{1,i} = \mu_{A_i}(x), \quad \text{for } i = 1, 2 \text{ or}$$
$$O_{1,i} = \mu_{B_{i-2}}(y), \quad \text{for } i = 3, 4$$

$$(1.7.1)$$

where:

x or y is the input to node i

A_i or B_{i-2} is the linguistic label

$O_{1,i}$ is the membership grade of the fuzzy set A.

The parameters of this layer are known as the premise parameters.

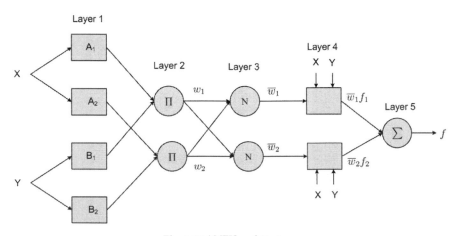

Fig. 1.15 ANFIS architecture.

Layer 2. The nodes of this layer are fixed nodes labeled Π, because the output is the product of the input signals:

$$O_{2,i} = w_i = \mu_{A_i}(x)\mu_{B_i}(y), \quad i = 1, 2 \tag{1.7.2}$$

Layer 3. In these nodes we have fixed nodes labeled N, its function is to calculate the ratio of the ith rule's firing strength to the sum of all rules' firing strength, so the output in each node is a normalized firing strength:

$$O_{3,i} = \bar{w}_i = \frac{w_i}{w_1 + w_2}, \text{ for } i = 1, 2 \tag{1.7.3}$$

Layer 4. The nodes in this layer are adaptive nodes described by:

$$O_{4,i} = \bar{w}_i f_i = \bar{w}_i (p_i x + q_i y + r_i), \tag{1.7.4}$$

in this case $\bar{w}_i$ represents the normalized firing strength from layer 3, where p_i, q_i, r_i are the parameters of this node, which are known as *consequent parameters*.

Layer 5. This a single node, which is a fixed node labeled $\sum$, which calculates the overall output as following:

$$O_{5,1} = \sum_i \bar{w}_i f_i = \frac{\sum_i w_i f_i}{\sum_i w_i} \tag{1.7.5}$$

As it was explained ANFIS is an adaptive network that is functionally equivalent to a Sugeno fuzzy model.

1.7.3 Hybrid Learning Algorithm

The ANFIS learning algorithm is based in the minimization of the error function. It is recommended the minimization using an hybrid algorithm, where the output of the net is linear in the consequent parameters p_i, q_i, r_i, so they can be identified by the least squares method, meanwhile the premise parameters are updated by gradient descent. The reason for the use of hybrid learning algorithm is that the convergence is faster than using only gradient descent methods.

Therefore, the algorithm consists of two steps, one forward and one backward, as shown in the Table 1.1.

Table 1.1 Steps of the hybrid learning algorithm for ANFIS.

	Forward pass	Backward pass
Premise parameters	Fixed	Gradient descent
Consequent parameters	Least-squares estimator (LSE)	Fixed
Signals	Node outputs	Error signals

There are two types of parameters that are updated in an ANFIS:

- The premise parameters, which describe the MFs of the system.
- The consequent parameters, which represent the parameters of the output f, in the rules.

From the ANFIS architecture in Fig. 1.15, if the premise parameters are fixed, the output can be represented as:

$$f = \frac{w_1}{w_1 + w_2} f_1 + \frac{w_2}{w_1 + w_2} f_2 = \bar{w}_1 f_1 + \bar{w}_2 f_2$$
$$= \bar{w}_1(p_1 x + q_1 y + r_1) + \bar{w}_2(p_2 x + q_2 y + r_2) \tag{1.7.6}$$
$$= (\bar{w}_1 x)p_1 + (\bar{w}_1 y)q_1 + (\bar{w}_1)r_1 + (\bar{w}_2 x)p_2 + (\bar{w}_2 y)q_2 + (\bar{w}_2)r_2$$

and f is linear in the consequent parameters p_1, q_1, p_2, q_2, r_1 and r_2.

So, the learning hybrid algorithm described earlier can be applied directly as follows:

- In the forward pass, the node outputs go forward until layer 4, and the consequent parameters are updated by the least squares method, and the premise parameters are fixed.
- In the backward pass, the error signals propagate backward and the premise parameters are updated with the gradient descent method.

This algorithm is much faster because it reduces the search space dimensions of the original pure back-propagation method (Jang et al. 1997).

Acknowledgments

The authors would like to thank the "Instituto Politécnico Nacional (IPN)", "Comisión de Operación y Fomento de Actividades Académicas del IPN (COFAA-IPN)" and the Mexico's entity "Consejo Nacional de Ciencia y Tecnología (CONACYT)" for supporting their research activities.

References

Ashby, R.W. (1960). Design for a Brain: The Origin of Adaptive Behaviour. New York: John Wiley & Son, Inc., 2nd. Revised Edition.

Cordón, O., F. Herrera, F. Hoffmann and L. Magdalena (2001). Genetic Fuzzy Systems. Evolutionary Tuning and Learning of Fuzzy Knowledge Bases. Advances in Fuzzy Systems—Applications and Theory Vol. 19, Singapore, World Scientific Publishing Co.

Hagras, H. and C. Wagner (2012). Towards the Widespread Use of Type-2 Fuzzy Logic Systems in Real World Applications. IEEE Computational Intelligence Magazine.

Haykin, S. (1999). Neural Networks. A Comprehensive Fundation. Second Edition. Upper Saddle River, New Jersey: Prentice Hall.

Haykin, S. (1999). Neural Networks: A Guided Tour. In: N.K. Sinha and M.M. Gupta, Soft Computing and Intelligent Systems. Theory and Applications. San Diego, Ca.: Academic Press Series in Engineering.

Jang, J.-S.R., C.-T. Sun and E. Mizutani (1997). Neuro-Fuzzy and Soft Computing, A Computational Approach to Learning and Machine Intelligence. Upper Saddle River, NJ: Prentice-Hall Inc.

Jech, T. (2006). Set Theory. The Third Millenium Edition, Revised and Expanded. New York, Springer-Verlag Berlin.

Karnik, N. N. and J. M. Mendel (1998). Introduction to Type-2 Fuzzy Logic Systems. Proceedings of 1998 IEEE FUZZ Conf., pp. 915–920, Anchorage, AK.

Karnik, N.N. and J.M. Mendel (1999). Applications of Type-2 Fuzzy Logic Systems to Forecasting of Time-series. Information Sciences, 120(1-4): 89–111.

Karnik, N.N. and J.M. Mendel (2001). Operations on type-2 fuzzy sets. International Journal on Fuzzy Sets and Systems, 122(2): 327–348.

Karnik, N.N., J.M. Mendel and Q. Liang (1999). Type-2 Fuzzy Logic Systems. IEEE Transactions on Fuzzy Systems, 7(6).

Karnik, N.N. and J. Mendel (2001). Centroid of a Type-2 Fuzzy Set. Information Sciences, 195–220.

Klir, G.J. and B. Yuan (1995). Fuzzy Sets and Fuzzy Logic: Theory and Applications. In: G.J. Klir. Upper Saddle River, New Jersey: Prentice-Hall.

Liang, Q., and J.M. Mendel (2000). Interval Type-2 Fuzzy Logic Systems: Theory and Design. IEEE Trans on Fuzzy Systems, 8(5). 535–550.

Mamdani, E. (1993). Twenty years of Fuzzy Control: Experiences Gained and Lessons Learned. Proceedings of 2nd. IEEE International Conference on Fuzzy Systems, San Francisco, CA. pp. 339–344.

Melin, P., L. Estudillo, O. Castillo, F. Valdez and M. Garcia (2013). Optimal design of type-2 and type-1 fuzzy tracking controllers for autonomous mobile robots under perturbed torques using a new chemical optimization paradigm. Expert Systems with Applications, 3185–3195.

Mendel, J. (2007). Type-2 Fuzzy Sets and Systems: an Overview. IEEE Computational Intelligence Magazine, 2: 20–29. US.

Mendel, J.M. (1998). Type-2 Fuzzy Logic Systems: Type-reduction. IEEE System, Man, Cybernetic Conference. San Diego.

Mendel, J.M. (1999). Computing with words, when words can mean different things to different people. ICSC Congress on Computational Intelligence Methods and Applications. Rochester, N. Y.

Mendel, J.M. (2000). Uncertainty, Fuzzy Logic, and Signal Processing. Signal Processing Journal, 913–933.

Mendel, J.M. (2001). Uncertain Rule-Based Fuzzy Logic Systems: Introduction and New directions. USA: Prentice Hall.

Mendel, J.M. (2003). Fuzzy Sets for Words: a New Beginning. Proc. of IEEE International Conference on Fuzzy Systems, 10: 37–42. St. Louis.

Mendel, J.M. and R. Bob John (2002). Type-2 Fuzzy Sets Made Simple. IEEE Transactions on Fuzzy Systems, 10(2): 117–127.

Mendel, J.M. and G.C. Mounzouris (1999). Type-2 Fuzzy Logic Systems. IEEE Transactions on Fuzzy Systems, 643–658.

Mendel, J., H. Hagras and R. John (n.d.). Standard Background Material About Interval Type-2 Fuzzy Logic Systems That can be Used By All Authors.

Montiel, O., J. Quiñones and R. Sepúlveda (2012). Designing High-Performance Fuzzy Controllers Combining IP Cores and Soft Processors. Advances in Fuzzy Systems, 2012 (Article ID 475894), 1–11.

Ngo, L.T., D.D. Nguyen, L.T. Pham and C.M. Luong (2012). Speedup of Interval Type-2 Fuzzy Logic Systems Based on GPU for Robot Navigation. Advances in Fuzzy Systems, 2012 (Article ID 698062).

Rumelhart, D.E., G.E. Hinton and R.J. Williams (1986). Learning Representations of back-propagation errors. Nature, 533–536.

Sahab, N. and H. Hagras (2011). Adaptive Non-singleton Type-2 Fuzzy Logic Systems: A Way Forward for Handling Numerical Uncertainties in Real World Applications. International Journal of Computers, Communications and Control, 503–529.

Sepúlveda, R., O. Montiel, O. Castillo and P. Melin (2012). Embedding a high speed interval type-2 fuzzy controller for a real plant into an FPGA. Applied Soft Computing, 988–998.

Wagenknech, M. and K. Harmann (1988). Application of Fuzzy Sets of Type-2 to the Solution of Fuzzy Equations Systems. Fuzzy Sets and Systems, 25: 183–190.

Zadeh, L. (1975a). The Concept of a Linguistic variable and its application to Approximate Reasoning, Part II. Information Sciences, Vol. 8, No. 4, pp. 301–357.

Zadeh, L.A. (1975b). The Concept of a Linguistic Variable and its Application to Approximate Reasoning III. Information Sciences, Vol. 8, No. 1, pp. 43–80.

Zadeh, L.A. (1975c). The concept of a linguistic variable and its application to approximate reasoning, Part 1. Information Sciences, 8: 199–249.

Zadeh, L.A. (1971). Similarity Relations and Fuzzy Ordering. Information Science, 3: 177–206.

Zadeh, L.A. (1965). Fuzzy Sets. Information and Control, 338–353.

Zadeh, L.A. (1973). Outline of a New Approach to the Analysis of Complex Systems and Decision Processes. IEEE Transactions on Systems, Man, and Cybernetics, 3: 28–44.

2

Introduction to Novel Microprocessor Architectures

Oscar Montiel,* Roberto Sepúlveda, Jorge Quiñones
and Ulises Orozco-Rosas

ABSTRACT

In this chapter, the novel microprocessor architectures for general-purpose computers are explored. The aim is to emphasize their main characteristics that allow achieving high-performance computations. Present research has common goals that threat to exploit parallelism from different points of view, sometimes hidden within programs because the processor can perform a long sequence of instructions, given the illusion to the programmer, that it is executing only one instruction, or explicit, where the programmer can improve the performance of a program using threads taking advantages of novel multicore microprocessor's architectures. Hence, this chapter will introduce the reader with the most popular microprocessor architectures that are using multicore technology, as well as important concepts about performance measures.

2.1 Introduction

Computer technology has made incredible tremendous strides in the last sixty years since the first commercial general purpose computer was introduced. Nowadays, one can purchase a personal computer for a few hundred dollars that offers more performance, memory, and storage

Instituto Politécnico Nacional, CITEDI, Tijuana, México.
  Email: oross@ipn.mx
* Corresponding author

capability than a million-dollar computer of the 80s. The huge increase in today's computational capabilities has made it possible to have simulation as a third discipline of Science, since it allows us to achieve experiments and perform the evaluation of a wide range of systems, which otherwise would not be accessible; for example, in quantum computation using simulation is possible to evaluate proposals and new algorithms even though at present time there is still not a physical quantum computer for testing them. Therefore, knowing the fundamentals of computation is an indispensable issue for scientists because they provide them with the mathematical and technical tools to improve the performance of existing theoretical models and simulations, or to create new ones.

Nowadays, the term "high-performance computing" (HPC) is gaining popularity in the scientific community with those who are working with computers with very high processing capacities for super-fast calculation speeds. This term is also rapidly gaining popularity because the diversity and complexity of mathematical models that need simulation is increasing; as well, there are more technical problems that require to process data at high speed. HPC is an invaluable tool for analysts, engineers, and scientist because it offers the resources that they need to make vital decisions, to speed up research and developments, to promote product innovations, and to reduce time to market.

The development of HPC has become in the last years, a growing industry that continuously offers to science and engineering new technological innovations that embrace a huge diversity of applications. There are several trends to increase a computers' performance, the use of parallelism being one of the favorites; therefore, many efforts are being made to add more cores and memory to them, as well as the use of graphics processing unit (GPU).

Today the use of GPU is an integral part of mainstream computing systems. A modern GPU is not only a powerful graphics engine, but also a highly parallel programmable processor featuring peak arithmetic and bandwidth that substantially outpaces its CPU counterpart. The GPUs rapid increase in both programmability and capability has spawned a research community that has successfully mapped a broad range of computationally demanding, complex problems to the GPU.

Understanding the performance of programs is very complex today because writing a program with the aim of improving its performance involves a good deal of knowledge about the underlying principles and the organization of a given computer. The performance of a program is a combination of excellent algorithms, their good codification using a computer language, the compiler who serves to translate programs to machine instructions, and the efficacy of the computer in executing those instructions, which may include input/output (I/O) operations.

Among the diversity of software existing for a computer system, there are two software components that have a great impact in the application performance:

1. Every computer system today has an operating system such as Windows, MacOS, Linux, etc., that provides the interface between the user's program and the computer hardware, offering a variety of services and supervisory functions, such as, handling basic-input-output operations, allocating storage and memory, handling the appropriated operations to allow the use of multiple applications running simultaneously, etc.
2. Software developers use compilers to translate a program that was written in a high-level language such as C/C++, Java, Matlab, etc., into instructions that the hardware can execute.

Hardware and software designers use hierarchical layers where each layer hides details from the above layer, to cope with today's complex computer systems. In general, there exists the goal of finding a language that facilitates building the hardware and the compiler who maximizes performance minimizing cost.

This chapter addresses an overview of the computer processors starting with a brief history of processor's architectures that embraces from early processors to the modern ones that are installed in everyday personal computers of today. In Section 2.2, we start with the history of computer processors, talking about outstanding processors and its main features. Section 2.3 presents the basic components of a computer in terms of its architecture. Section 2.4 talks about the main characteristics of modern processors. Section 2.5 details the classification of computer architectures, where we anticipate that basically the classification is in four architectures that are based on instructions and data handling. Section 2.6 talks about the concepts concerning memory architectures that are handled in parallel computing. Section 2.7 explains the measuring performance on processors to understand why a piece of software can be implemented to work better than another, or how some hardware features affect the software performance; the computer performance is a significant attribute when we choose among different computers, so, it is important to have methods to measure accurately and compare dissimilar computers to know their limitations. Finally, Section 2.8 deals with summing up the chapter.

2.2 History of Computer Processors

Moore's law observes that the number of transistors on a chip doubles every two years. Based on this law, the development of the computer processors started in the year 1971. In the following two blocks, we are going to take a brief tour talking about the most outstanding processors and its main features.

2.2.1 Beginning of the Computer Processors

In 1971, the company Intel made the presentation of the first commercial microprocessor on a single chip. This was the i4004 model, a 4-bit CPU, shown in Fig. 2.1, it was intended to be used in calculators, and was built with 2250 transistors (Floyd 2000).

In April of the following year, Intel introduced the 8008 processor, with an improved architecture extending its word size to 8 bits, 48 instructions, addressing 16KB, PMOS technology and a maximum operating clock of 740Khz, which could perform 300,000 operations per second (Intel, 40 years of Microprocessors).

Consecutively in the mid-70s was launched the Models 8080 and 8085 of 8 bits. The 8080 processor was first used in a computer system to the public, the Altair computer.

In 1975, the Motorola Company launched an 8-bit processor with a clock rate of two MHz; it had the ability to route 16 KB of memory.

Based on the design of Intel 8080 and 8085, in 1978 and 1979, the processor 8086 completely compatible in assembler level with the 8080 made its appearance. Its architecture forms the base design of all Intel processors. The 8086 had a 16-bit architecture and could address up to 1MB of memory and represented a significant improvement over its predecessor, turn all versions operated at a frequency of 5MHz, 8MHz and 10MHz. The 8088 with minor modifications passed to be the one used by the IBM PC.

The 68000 of Motorola is a 16-bit processor and incorporated 32-bit registers and could address a 24-bit bus, 16MB of memory and operating at 16MHz.

Fig. 2.1 i4004 processor.
Color image of this figure appears in the color plate section at the end of the book.

Intel 80186 integrates new capabilities that makes it more versatile, including the increase in the frequency clock at 12.5MHz, interrupt controllers and direct memory access (DMA) as the main features.

The 80286 has an external data bus of 16 bits; the address bus could also be increased to 24 bits and could address up to 16 MB. It implements the protected mode for operating systems MS-Windows, UNIX or OS/2 from IBM.

Since 1982, the company AMD (Advanced Micro Device) has created compatible processors with Intel architecture, including models 8086, 8088, 286, 386 and 486.

On 16 October 1985, Intel made a considerable improvement of the 80386 architecture, increasing the word size to 32 bits for the data bus and address, this capability makes possible to address up to 4 GB of memory. This processor was widely used in the mid-80s, its maximum operating frequency was 40 MHZ, and it was built with 275,000 transistors. The 386 remained compatible with previous models; it was the first model where the **pipeline** feature was implemented. In the same year Motorola released in the market the 68020 processor, its first 32-bit architecture and at the same time with the Intel 80386 could address 4GB of memory at a maximum frequency of 33MHz. The internal memory cache is part of the 80486 processor with 8KB capacity reducing latency when accessing memory banks, also incorporated an inner math coprocessor reaching speeds up to 100 MHz (Domeika 2008).

The first Pentium (80586) was launched in 1993 with many features from the 486 including a 32-bit architecture. It developed the execution of super-scalar instructions by implementing a double pipeline getting to execute two instructions at a time in one clock cycle; this processor has been shown in Fig. 2.2. It reduced the supply voltage at 3.3V reaching a frequency of

Fig. 2.2 Pentium processor.

Color image of this figure appears in the color plate section at the end of the book.

200MHz. Regarding its cache passes from 8KB to 16KB for instruction and data memory. The Pentium Pro is an improved version planned for use in high gamma computers but eventually centered in network application's servers and workstations.

The PowerPC 601 processor shown in Fig. 2.3, is a processor developed in conjunction by Motorola Semiconductor (now Freescale) and IBM, in order to compete with Intel processors. The PowerPC 601 processor has a 64-bit architecture, it is based on a reduced instruction set computing (RISC), it supports floating-point data types of 32 and 64 bits and it is optimized to run three instructions per clock cycle. It is available with 32 KB of cache and built with 3.6V CMOS technology maintaining compatibility with TTL devices (IBM-Freescale 1995).

Intel Pentium II and Pentium III were launched in 1997 and 1999 respectively for domestic use. They introduced MMX technology which helps significantly to improve the video processing on compression and decompression stage for image manipulation. Pentium III incorporates 70 new instructions to boost multimedia performance, and is built with 9.5 million transistors and a separation between components of 0.25 microns. This processor also incorporates a second cache of 512KB called L2.

On November 2000 and 2004, Intel launched Pentium 4 and Pentium Prescott, shown in Fig. 2.4, with the manufacturing technology of 90nm and 65 nm. They are based on X86 architecture with the intention of offering an enhanced experience in PC and workstations for the Internet and multimedia applications, through its NetBurst microarchitecture that offers great performance at high frequencies. The Pentium 4 was finally released in 2008 and then replaced by Intel Core duo. Among the main features the 400MHz bus system for communication with main memory is highlighted. It executes instructions in a pipeline of 20 levels compared to

Fig. 2.3 PowerPC 601 processor.

Color image of this figure appears in the color plate section at the end of the book.

Fig. 2.4 Pentium 4 processor.

Color image of this figure appears in the color plate section at the end of the book.

10 levels in the Pentium III; moreover, 144 new instructions could operate from 1.5 to 3.8 GHz. Prescott processors can handle 64-bit instructions with an improved system of instructions (SSE3) (Intel, Desktop Performance and Optimization for Intel® Pentium® 4 Processor 2001).

2.2.2 Modern Commercial Multicore Processors

2.2.2.1 Cell processor

The Cell processor (Cell Broadband Engine Architecture) for high-performance computing was created in 2005 jointly by Sony, IBM and Toshiba (Arevalo and Peri 2008). It contains 234 million transistors with a separation of 90 nm providing a multicore architecture based on the 64-bit PowerPC processor. The cell processor offers a high-speed data-transfer which is ideal for games, multimedia applications, and servers; it executes single instructions with multiple data (SIMD) and it incorporates a power processing element (PPE) as a master processor and 8 coprocessor slaves called synergistic processing elements (SPEs) with a process speed of 4 GHz, connected together, through the element interconnect bus (EIB) in which it can exploit parallel computing and even support several operating systems simultaneously, as is shown in Fig. 2.5.

Intel launched in the year of 2005 Pentium D shown in Fig. 2.6, which adopted the multicore architecture, including two Pentium 4 (Prescott) embedded in a single package, built with technology of 90 and 65 nm, operating at frequencies of 2.8, 3 and 3.2GHz, with two cache levels L1 and L2. With this, dual-core architecture provides to users the ability of simultaneously computing obtained from their two native processors unlike HT (HyperThreading) Intel technology, which includes only the necessary hardware to emulate a single core and two logical processors implementing programs with multiple threads. In the same year, Intel released the Core 2

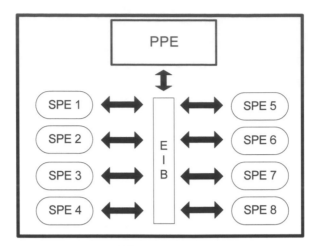

Fig. 2.5 Cell processor architecture.

Fig. 2.6 Pentium D processor.
Color image of this figure appears in the color plate section at the end of the book.

series, with versions of 1 (solo), 2 (duo) and 4 (quad) cores, 64-bit architecture and x86-64 instruction set (Intel, 40 years of Microprocessor).

The Intel Atom processor appeared in the year of 2008 designed for use in low-power devices and small size (60% smaller) products of the manufacturing process of 45 nanometers. Its area of application includes mobile internet, mobile smart, Netbooks, etc. The Atom family has models with one and two cores with architecture x86-64; each core has two threads (Intel, Intel Atom Processor 2010).

From 2010 to 2012 emerged the first, second and third generation of multicore processors Core i3, i5, i7 and the Extreme edition of Intel line with a set of 64-bit instructions with a manufacturing technology of 22 nanometers, designed for desktop computers. These processors include Intel Turbo Boost technology, which can increase the clock frequency for computing demand; this characteristic has great advantages in thermal and power consumption. The Core i3 and i7 have Hyper-Threading support (Table 2.1) for each core doubling the number of kernel threads. It also includes an integrated memory controller (IMC) for DDR3 memory technology (Intel, 3rd Generation Intel® Core™ Desktop Processor Family 2012).

Table 2.1 Overview of the third generation of Intel processors.

	CORE i7 Extreme edition	**CORE i7**	**CORE i5**	**CORE i3**
Number of cores	6	4	4	2
Threads	12	8	4	4
Cache L3	15	8MB	6MB	3MB
Hyper-Threading Technology	Yes	Yes	No	Yes
Intel® Turbo Boost Technology	Yes	Yes	Yes	No

2.2.2.2 SPARC64 VI/VII/VII+ processors

The SPARC64 VI/VII/VII+ are examples of SPARC architecture and were designed by Sun Microsystems and distributed by the Fujitsu Company (Fujitsu 2012). The SPARC64 VI processor has two cores, with double thread per core with the aim of reducing waiting time for an efficient execution. The SPARC64 VII processor is shown in Fig. 2.7. Both this processor and the processors SPARC64 VII+ include 4 cores technology.

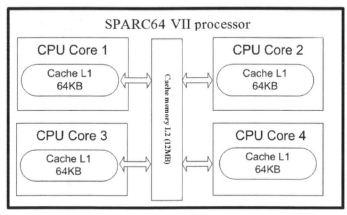

Fig. 2.7 SPARC64 VII.

2.2.2.3 AMD processors

The first commercial multicore processor of the company AMD is the Athlon X2, with a 64-bit architecture, technology of 90 and 65 nm, and operating at frequencies 2GHz to 3.2GHz. It integrates 2 cores with their own caches L1 (64KB) and L2 (1MB). The Phenom is another series of processors listed in the year 2006 and recently updated in 2012 to 5 models named as FX processors, APU Series A, Phenom II, Athlon II and Semprom. The FX series built on 32nm can be found with a native 8 cores architecture and Turbo CORE technology that delivers more performance when needed, including a last level cache L3 with 8MB and a bandwidth for inputs and outputs with HyperTransport at 16GB/s. APU series models contain architecture called APU (Accelerated Processing Unit) that combines a CPU and a GPU (Graphics Processing Unit) in the same package, giving maximum performance and acceleration on graphics processing. The model Apu A10 series contains 4 native cores with a clock frequency of 4.2GHz and 384 cores AMD Radeon (AMD 2012).

2.3 Basic Components of a Computer

Computer Architecture is the science and art of selecting and interconnecting hardware components to create computers that meet functional, performance and cost goals. It may also refer to the practical art of defining the structure and relationship of the subcomponents of a computer. As in the architecture of buildings, computer architecture can comprise many levels of information.

The Central Processing Unit (CPU) is the main hardware device of a computer system (Floyd 2000). It is composed of a control unit, an instruction-decoding unit, and an arithmetic-decoding-unit. Figure 2.8 shows a basic blueprint of the stored-program digital computer conceptual architecture; it is the basis for all present mainstream computer systems. This elementary scheme comes from the Von Neumann paradigm, but it has two inherent problems. (1) The control and arithmetic units of the

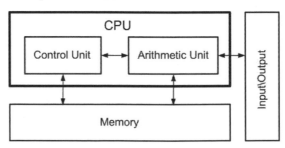

Fig. 2.8 The stored-program computer model.

CPU must be continuously fed with instruction and data; this problem is known as the von Neumann bottleneck. (2) The architecture is inherently sequential; this concept is known as Single Instruction Single Data (SISD). Despite the above drawbacks, no other architectural computer concept has found a similar widespread use in nearly 70 years of digital electronics (Hager and Wellein 2011).

In the past, generic computer systems were big, energy demanding and expensive. Such drawbacks were bigger for scientific computers because this kind of equipment required special characteristics to perform calculations faster. Nowadays, the HPC world offers solutions at low cost using multicore processors such as Intel Core i7. The mother board shown in Fig. 2.9a has only one processor (monoprocessor system), whereas a multiprocessor system is illustrated in Fig. 2.9b, the mother board has several processors in the same mother board. Similarly, Fig. 2.10a shows that a monoprocessor system can have more than one CPU (Core) or many processors with some cores, as is illustrated in Fig. 2.10b (Domeika 2008).

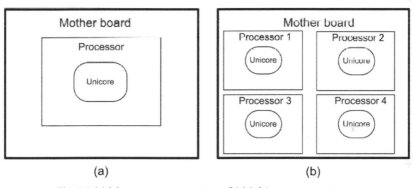

Fig. 2.9 (a) Mono processor system, (b) Multiprocessor system.

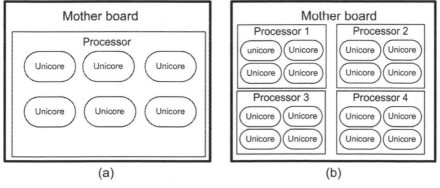

Fig. 2.10 Multicore systems.

Typically, a processor is characterized by its word length, operation frequency and memory capacity. The first characteristic indicates the number of bits that a computer system can handle to perform low level operations, in particular; modern personal computers use 32 and 64 bits. However, there are other kinds of computer systems based on Microcontroller's units (MCUs), or in Digital Signal Processors (DSPs) that use 4, 8, 24 and 32 bits of word length.

The clock frequency measured in Hertz is another significant parameter of a processor. It is important since it is a measure that indicates the speed of executing basic operations of a particular computer system. Although high clock speeds are associated with faster code execution, it does not necessarily mean a performance increase because the processor temperature rises, reducing the computational performance. Nowadays, multicore technology is used to obtain a good performance in the frequency ranges of 1 GHz to 4 GHz.

The two basic processor's architectures are the von Neumann model and the Harvard model. They are based on how the memory blocks are connected to the CPU to store data and programs. The von Neumann model also known as the Princeton Architecture is shown in Fig. 2.11. In this architecture, there are two important components, the CPU and the memory to store both data and programs; the CPU consists of an arithmetic logic unit, processor registers, a control unit, an instruction register, and a program counter. Other significant components are the external mass storage, and input and output mechanism. The meaning of the term has evolved over several years to mean a stored-program computer in which data operations and instruction fetch cannot occur at the same time because they share a common bus. This characteristic limits the performance of the system, and it is referred as the von Neumann bottleneck.

In the Harvard model shown in Fig. 2.12, the program memory and data memory are separated and accessed using different buses. This characteristic improves the bandwidth over traditional von Neumann architecture (Floyd 2000).

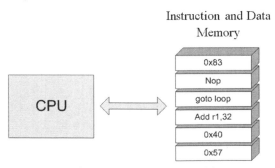

Fig. 2.11 Von Neumann model.

The functional unit also known as execution unit or fetch/decode unit is a part of the CPU responsible for retrieving individual instructions in a serial way from their location in memory, then translating them into commands that the CPU can understand. These commands are usually referred to as machine language instructions, but also they are called micro-operations. When the translation is complete, the control unit sends the micro-operations to the executing unit for processing, for example, to be transferred to the peripherals through input/output ports to interact with the user (see Fig. 2.13).

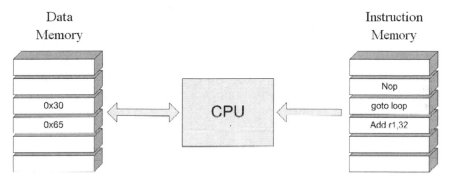

Fig. 2.12 Harvard model.

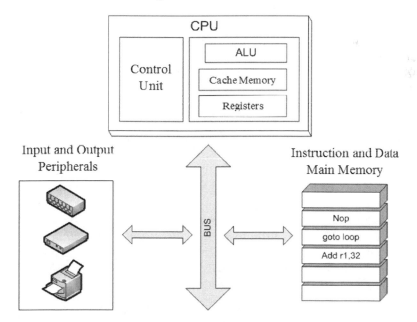

Fig. 2.13 Basic architecture of a computer system.

Memory is the part of the computer that serves to store programs and data while the CPU is processing them, and is known as random-access memory (RAM). It is typically an integrated circuit or a package of integrated circuits that are used to store binary data. However, the correct use of the word memory includes any kind of storing device of binary data, such as transistors, capacitors, and other electronic components. It can be external to the CPU such as the RAM, or within the circuitry of the CPU itself such as registers, and L1 cache. Historically, the terms "primary storage" and "secondary storage" have been used to denote memory and mass storage devices such as hard drives. The memory together with the CPU, is regarded as one of the most important factors in the overall computer performance.

Memory which can be classified as volatile and non-volatile, refers to the ability to store data with or without the presence of electrical power. In other words, the content of the volatile memory is not permanent; the stored data is lost when power is removed.

In the **volatile memory** category, we have the Synchronous Dynamic Random-Access Memory (**SDRAM**), the Dynamic RAM (**DRAM**), and the Static Random-Access Memory (**SRAM**). Conventional DRAMs are asynchronous; they are the type that has been used in PCs since IBM PC days, and do not need to be synchronized to the system clock, which controls how often the processor can receive data; they work fine in low-speed memory bus systems but is not nearly as suitable for use in high-speed (>66 MHz) memory systems. A newer type of DRAM is the SDRAM which is synchronized to the system clock; all signals are tied to the clock system; therefore, timing is much tighter and better controlled. SDRAMs are much faster than asynchronous DRAM and are used to improve the performance of the system. SDRAMs are inexpensive to produce since the circuits inside them use one transistor and one capacitor to create a memory cell to represent a single bit of data. The capacitors of the SDRAM need to be recharged (refreshed) thousands of times per second for maintaining the content, the refreshment cycles produce slow access time. The **SRAM** consists of flip-flops that holds each one a bit of memory, and each flip-flop consists of four to six transistors in a special arrangement of wires; an advantage is that SRAM does not need to be refreshed, which results in much faster access times; one disadvantage is that it needs more space on chip than SDRAM since each memory cell requires more components. As a consequence, the user gets less memory per chip, making SRAM chips larger and more expensive than SDRAM chips. A typical use of SRAM is the computer's L2 cache. The SRAM can be purchased in integrated circuits (IC) in the form of DIP, or module packages similar to DIMMs. The market for SRAM is mainly reserved for electronic hobbyists and IC manufacturers, because this kind of memory is of no practical use to the average personal computer (PC) consumer as the L2 cache in a PC cannot be upgraded.

The content of a **non-volatile** memory is permanent; the stored data remains even after power is removed. Two examples of this category are Electrically Erasable Programmable Read-Only Memory (**EEPROM**) and the Flash Memory (**Flash**). The EEPROM IC chips are built in such a way that it can trap electrons inside the circuit, which allows holding onto its data without an external power supply. The electronic device known as flash memory is a particular type of EEPROM. The difference between flash memory and EEPROM is that flash memory is logically structured like a storage device, with data divided in sectors of 512-byte blocks; therefore, data is read/write 512 bytes at a time; while in EEPROM data can be accessed (read/write) one byte at a time. Flash memory is often used in portable electronics, such as smartphones, digital cameras and digital music devices. This technology is useful also for computer basic input/output systems (BIOS) and personal computer memory cards.

In computer systems, the memory is organized hierarchically; it consists of multiple levels of memory with different speed and sizes. The faster memories are more expensive per bit than the slower memory and thus smaller. Up to the present time, there are three primary technologies used in building memory hierarchies. The main memory is implemented from DRAM, while levels closer to the processor (caches) use SRAM. The third technology used to implement the larger and slowest level in the hierarchy is the **magnetic disk**. A memory hierarchy can have multiple levels, but data is copied between only two adjacent levels at a time. The processor's closer level is smaller and faster (because it uses more expensive technology) than the lower level. Undoubtedly, performance is the major reason for having a memory hierarchy, the time to service **hits** and **misses** is very important. The time to access the upper level of the memory hierarchy is known as the **hit time**, it includes the time needed to determine if the access is a hit or a miss. The time to replace a block in the upper level with the corresponding block from the lower level, plus the time to deliver this block to the processor is known as the miss penalty. All modern computers make use of cache memory, in most cases it is implemented on the same die as the microprocessor that forms the processor. To close further the gap between high clock speeds of modern processors and the relatively long time required to access DRAM, many processors have incorporated an additional level of catching. This second level can be located on the same chip or in a separate set of SRAMs out of the chip processor; it is accessed if a miss occurs in the primary cache. In the two-level cache structure, the primary cache allows to concentrate on minimizing hit time to yield a shorter clock cycle while the secondary cache allows concentrating on miss rate to reduce the penalty of long memory access time.

In Fig. 2.14a memory is one word wide, and all the access is achieved sequentially. In Fig. 2.14b the bandwidth is increased by widening the system memory and buses to connect memory to the processor with the aim to allow parallel access to all the words in the block; a multiplexor is used on reads as well as in control logic to update the appropriated words of the cache on writes. The bandwidth is proportionally enlarged by increasing the width of the memory and the bus, which decrease both the access time and transfer time portions of the miss penalty. One option is to organize in banks to read or write multiple words in one access time rather than reading or writing a single word at a time, this instead of making the entire path between the memory and the cache wider. In Fig. 2.14c the bandwidth is increased by widening the memory but not the interconnection bus; each bank can be one word wide with the aim of not changing the width of the bus and the cache (Patterson 2012).

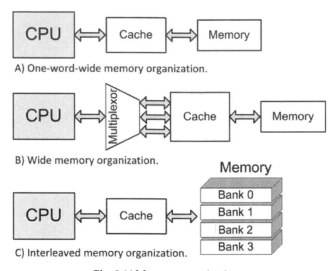

Fig. 2.14 Memory organization.

2.4 Characteristics of Modern Processors

The basic scheme of a computer system shown in Fig. 2.15 is a single-processor (uniprocessor) system. This system based on the Von Neumann model is not optimal for many applications of present-day, since it is inherently sequential; the entire task to solve a problem is executed by the CPU one instruction at a time, as is shown in Fig. 2.15 (Herlihy and Shavit 2008). Moreover, instructions and data must be continually provided by the CPU, etc. To overcome the main drawbacks of this kind of computers,

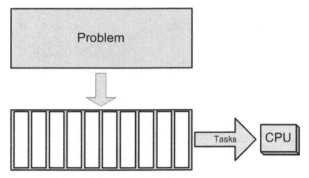

Fig. 2.15 Computing with monoprocessors.

many efforts have been made; some of them are related to improve the CPU machinery, others are related to integration, clock rate increase, pipelining, superscalarity, multicore architectures, multithreading, etc.

Regarding integration, Moore's law continues scaling technology by improving transistor performance, increasing transistor integration capacity to implement more complex architectures, and reducing power consumption per logic operation for keeping heat dissipation within limits. Performance increase because of microarchitecture alone is governed by Pollack's rule, which states that the increase is roughly proportional to the square root increase in complexity; hence, for efforts that devote r Base Core Equivalent (BCE) resources will result in sequential performance of $\sqrt{r}$. In other words, if the logic of a processor core is doubled, then it delivers only 40% more performance; on the other hand, a multi-core processor has the potential to provide near linear performance improvement with complexity and power; for example, the large monolithic processor for which we obtained a 40% improvement can be easily defeated using two smaller processor cores that can potentially provide 70–80% more performance (Borkar 2007). Note that the microarchitecture of multicore processors follows Moore's law since the processor chip contains the cores. Figure 2.16 illustrates how to solve a problem using a parallel computer such as a multiprocessor or a multicore system; the problem can be divided into several tasks that can be solved in parallel.

The clock rate increase is another strategy that designers have followed to increase performance of a computer system since they are reducing the time of every clock cycle, hence the execution time. At the beginning, this increase grew up exponentially, but this growth in 2006 collapsed because the clock rate and power consumption have a cubic relation, so higher the clock rate, larger the power consumption.

Pipelining in an implementation technique that consist of making distinct functional parts of the processor (CPU) perform different stages of more complex tasks. For example, multiple instructions can be overlapped

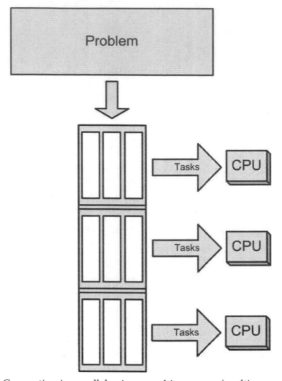

Fig. 2.16 Computing in parallel using a multiprocessor/multicore system.

in execution. Designers have included special instruction sets to take advantage of the CPU pipelines; they are known as **Microprocessor without Interlocked Pipeline Stages (MIPS)** instructions. A MIPS instruction classically has five steps (Patterson 2012):

1. Fetch instructions from memory.
2. Read registers while decoding instructions; i.e., the MIPS instruction allows to read and decode simultaneously.
3. Calculate an address or execute the operation.
4. Look for (access) an operand in data memory.
5. The result is written into a register.

However, there are three situations called hazards when the following instruction cannot be executed following the schedule in the next clock cycle. They are (García-Risueño and Ibáñez 2012):

1. *Structural hazards.* The hardware cannot support the combination of instructions that are going to be executed in the same clock cycle. The MIPS instruction set was designed to avoid this situation.

2. *Data hazards.* They occur when the pipeline must be stalled because one step is obligated to wait for another to complete. To overcome this situation, designers have included a method called forwarding or bypassing, which basically consists of adding extra hardware to retrieve the missing item from the internal resources. This problem can also be solved at the compiler level.
3. *Control hazards.* They occur when a decision must wait because it is based on the results of one instruction while other are executing. CPUs include branch prediction methods to overcome such situations.

Superscalarity is a capacity of the CPU to provide more than one result per clock cycle; it is mainly based on hardware replication, and it is a special form of parallel execution and a variant of parallelism (Hager and Wellein 2011). Computer architects have the following next goals in mind for designing new superscalar CPUs:

- To fetch and decode multiple instructions concurrently.
- Address and other integer calculations are achieved in multiple integers (mult, shift, add, etc.).
- To run in parallel several floating-point pipelines.
- Cache memories fast enough to support more than one load or store operation per cycle.

The design of Multicore architectures for processors is a powerful solution that computer architects had to develop to overcome the limitations given by the sequential nature of computers based on the Von Neumann model and power limitations. This paradigm proposes to include more than one CPU (cores) in the same socket; the easier way to manufacture a multicore processor is to place separate CPU dies in a shared package. Conventional PCs have a single socket, while computers that are used as servers have more than one socket, all sharing the same memory. There are different ways in how the cores on a socket (chip) can be arranged, for example:

- The cores can have separate caches or share certain levels.
- The socket of a multicore processor can include an integrated memory controller to reduce memory latency and allows the addition of fast intersocket networks like the QuickPath or HyperTransport.
- The socket can have fast data paths between caches, enabling the processor to read and write in a fast way.

To take advantage of multicore technology it is crucial to use the hardware resources by means of an efficient parallel programming. There are certain challenges that multicore technology imposes, most of them are regarding the reduction in main memory bandwidth, and cache memory size per core.

Many modern processors have threading capabilities, they are also known as Hyper-threading or Simultaneous Multithreading (SMT). Multithreaded processors have multiple architectural states that embrace all data, status and control registers, including stacks and instruction pointers. However, execution resources such as arithmetic-logic units (ALUs), cache memories, queues, etc., are not duplicated. The CPU, thanks to the multiple architectural states, can execute several instructions streams (threads) in parallel, no matter if they belong to the same program or not, it appears to be composed of several cores, such cores are known as logic processors. Here, the hardware is responsible for keeping track about which instructions belong to which architectural state; therefore, all threads share the same execution resources, so it is possible to fill bubbles that might exist in a pipeline due to stalls in threads. The use of SMT can enhance the throughput, and there are different ways to implement it (Hager and Wellein 2011).

The use of vector processor machines to improve performance is a strategy that has been used since the 70s. The vector processor also known as an array processor is a CPU with an instruction set which contains an instruction set that operates on one-dimensional arrays of data called vectors. This is in contrast to scalar processors whose instructions operate on single data items. Vector processors can provide a big improve in performance on certain workloads and numerical simulations.

2.5 Classification of Computer Architectures

In 1966 Michael Flynn proposed a characterization of computer systems that was slightly expanded in 1972. Today it is known as Flynn's Taxonomy (ABD-El-Barr and El-Rewini 2005). This proposal is a methodology to classify general forms of parallel operation that are available within a processor, the aim is to explain the types of parallelism either supported in the hardware by a processing system or available in an application; therefore, the classification can be achieved from the point of view of the machine or the application by the machine language programmer. Considering the point of view of the assembly language programmer, parallel computers are classified by type concurrency in processing sequence known as streams, data, and instructions. The classification is as follows (Flynn 2011):

1. SISD (Single Instruction, Single Data)
2. SIMD (Single Instruction, Multiple Data)
3. MISD (Multiple Instruction, Single Data)
4. MIMD (Multiple Instruction, Multiple Data).

2.5.1 SISD

These are the conventional systems that contain one CPU (uniprocessor) and hence can accommodate one instruction stream that is executed serially; Von Neumann computers are classified as SISD systems. Therefore, they are sequential computers, which exploit no parallelism in either the instruction or data streams. A SISD computer can be seen as a Finite State Machine where moving to the next instruction is a transition between states, early CPU designs of the family 8086 with a single execution unit belong to this category; other examples are the IBM 704, VAX 11/780, CRAY-1. Figure 2.17 shows several instructions that are going to be processed sequentially by the CPU.

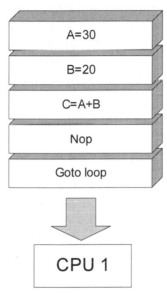

Fig. 2.17 Processing instructions using a computer based on SISD architecture.

2.5.2 SIMD

This architecture is essential in the parallel computer world. It can manipulate large vectors and matrices; so it offers greater flexibility than SISD and opportunities for better performance in video, audio and communications. Examples of common area of application are 3-D graphics, video processing and theater quality audio and high performance scientific computing. SIMD units are present on all G4, G5, the XBOX CPU processors and Intel Core i7 processors. To take advantage of SIMD, typically an application must be reprogrammed or at least recompiled; most of the time is unnecessary to rewrite the entire application. The power of this architecture can be appreciated when the number of processor units is equivalent to the size of the vector. In such situations, component-wise addition and multiplication of vector elements can be done simultaneously. The power of this architecture compared to a sequential one is huge, even when the size of the vector is larger than the number of processors; Fig. 2.18 shows a block diagram of this architecture. The SIMD architecture is twofold: True SIMD and Pipelined SIMD, each one has its own advantages and disadvantages.

There are two types of True SIMD architecture: True SIMD with distributed memory and True SIMD with shared memory. In the distributed memory case, the SIMD architecture is composed by a single control unit (CU) with multiple processor elements (PE); each PE works as an arithmetic unit (AU), so the PEs are slaves of the control unit. In this situation, the only processor which can fetch and interpret instruction codes is the CU, the only capability of the AUs is to add, subtract, multiply and divide; each AU has access only to its own memory. If one AU requires information from a different AU, the AU needs to request the information to the CU which needs to manage the transferring process. A disadvantage is that the CU is responsible for handling the communication transfers between the AUs memory. For the shared memory case, the True SIMD architecture is designed with a convenient configurable association between the PEs and the memory modules. Here, the local memories that were attached to each AU are replaced by memory modules, which are shared by all the PEs with the aim of sharing their memory without accessing the control unit. It is evident that the shared memory True SIMD architecture is superior to the distributed case.

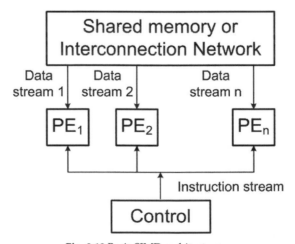

Fig. 2.18 Basic SIMD architecture.

The Pipelined SIMD architecture (vector machine) consists of a pipeline of AUs with shared memory. The pipeline is a first in first out (FIFO) kind. To take advantage of the pipeline, the data to be evaluated must be stored in different memory, pipelining of arithmetic operations divides one operation into many smaller functions to execute them in parallel on different data, the data to the pipeline must be fed as fast as it can be possible. These kinds of computers can be also distinguished according to the number and types of pipelines, processor/memory interaction and implementation arithmetic.

2.5.3 MISD

In the MISD architecture, there are *n* processor units where a single data stream is fed into multiple processing units. In this architecture, a single data stream is fed into several processing elements (PE); each PE operates on the data individually using independent instructions. Figure 2.19 shows the architecture of this kind of computers.

MISD did not exist when Flynn was categorizing the machines. It might have been added for symmetry in his chart. Its applications are very limited and expensive and currently there seems to be no commercial implementation. However, it is a research interest topic. One example is a systolic array with matrix multiplication like computation, and with rows of data processing units (cells) sharing the information with their neighbors immediately after processing.

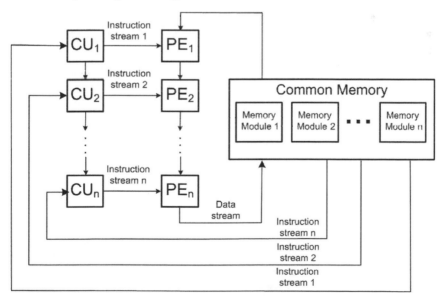

Fig. 2.19 MISD architecture.

2.5.4 MIMD

It is a kind of parallel computer where there are independent and asynchronous processors running different instructions and data flow. Figure 2.20 shows its architecture. These processors are common in modern supercomputers, cluster, and SMP multiprocessor and multicore processors.

An extension of Flynn's taxonomy was introduced by D. J. Kuck in 1978. In his classification, Kuck extended the instruction stream further to single (scalar and array) and multiple (scalar and array) streams. The data stream in Kuck's classification is called the execution stream and is also extended to include single (scalar and array) and multiple (scalar and array) streams. The combination of these streams results in a total of 16 categories of architectures.

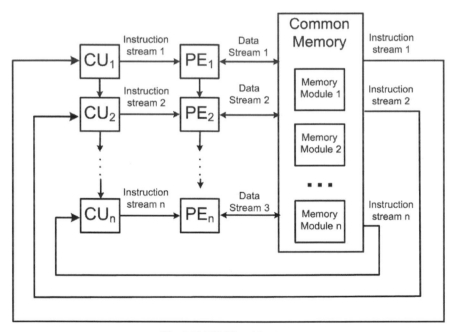

Fig. 2.20 MIMD architecture.

2.6 Parallel Computer Memory Architectures

The definition between supercomputers and high-performance computers is rather arbitrary and because of technology advances, today's supercomputers are common PCs of tomorrow. At the moment, most of the modern computer systems have many processors which can be single core or multicore. Therefore, we can have three different layouts for multiprocessing: a multicore system, multiprocessor system, and a multiprocessor/multicore system.

The processors of shared memory use a single address space; this resource is managed by software through multiple processors in the system by global variables. Because all processors can access memory, it is necessary to coordinate this access through a synchronized mechanism known as **lock**, where only one processor can have access at that point in time and the other processors have to wait until the resource is released. Shared memory systems are divided into two classes: UMA (Uniform Memory Access) and NUMA (Non-Uniform Memory Access) (Patterson 2012).

The UMA memory system also known as symmetric processors (SMP) and shown in Fig. 2.21. This system allows identical processors with the same access time, to share the same memory resources to perform different tasks regardless of the memory location that it is requested to manage, with only one condition that only one processor can access at a time. This is accomplished by a coordinating mechanism called synchronization.

Shared memory system NUMA, shown in Fig. 2.22, consists commonly of several physical union SMP processors, with the main feature that a processor can access directly the memory of another processor, but without having the same access time.

The distributed memory shown in Fig. 2.23 is obtained by sharing the local memory bank of each independent processor via a communications network. This is achieved by sending and receiving routines called "message

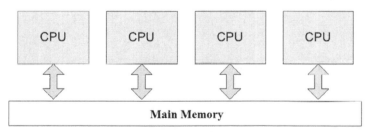

Fig. 2.21 UMA shared memory.

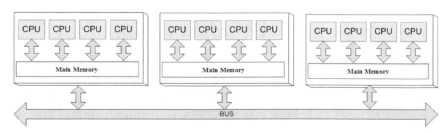

Fig. 2.22 NUMA shared memory.

passing". One of the greatest advantage of this type of memory is that it is scalable to the number of computers on the network.

The hybrid distributed memory shown in Fig. 2.24 consists of multiple processors UMA, NUMA and GPU (Graphics Processing Unit) connected by a data network and is known for mixing different processing technologies, including graphics coprocessors (Patterson 2012).

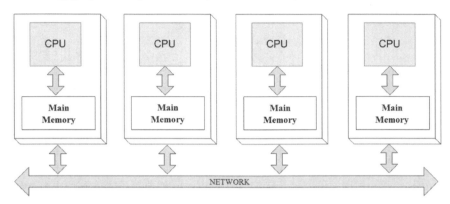

Fig. 2.23 Distributed memory.

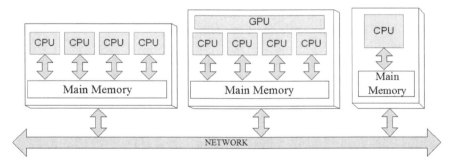

Fig. 2.24 Hybrid distributed memory.

2.7 Measuring Performance

Modern computers are very complex because there is a wide range of performance improving techniques at the hardware level, in combination with the big degree of sophistication of some software, which makes a challenge to assess the performance of such systems. In fact, it is necessary to consider several aspects of a computer system to determine which ones are the most significant in order to improve and evaluate the performance of applications. Therefore, diverse types of applications need different performance metrics.

2.7.1 Computer Performance

Computer performance can be specified in several ways because the measurement and evaluation of the performance are aimed to ensure that a minimum amount of effort, expense and waste are incurred in the production of a task. For example:

- Performance in embedded systems is often characterized by real-time constraints (deadlines) because it is common that they work with critical missions, that is, certain application-specific events must occur within a limited amount of time. Regarding real-time, there are two different kinds of real-time scheduling, hard and soft real-time. With hard-real time, you must not miss the deadline ever. Soft-real time systems can tolerate delays.
- Using the processing speed is also known as the response time, we can measure the processing speed. Here, we need to measure the time that a complete task last.
- Measuring the throughput which is the amount of work that can be performed or the amount of output that be produced by a system or component in a given period of time. A work may include more than one task.

It is common to use the term "faster" when we are trying to compare computer qualitatively, and it is a fact that performance and execution time are reciprocal quantities. Increasing the performance means reducing the execution time, so expression (2.7.1) is valid to measure the performance of Computer A.

$$\text{Performance}_A = \frac{1}{\text{Execution time}_A} \tag{2.7.1}$$

We can say that computer A has better performance than computer B if the performance of A is greater than the performance of B using (2.7.2), or we can say that A is n times faster than B using (2.7.3).

$$\text{Perfomance}_A > \text{Perfomance}_B \tag{2.7.2}$$

$$\frac{\text{Performance}_A}{\text{Performance}_B} = n \tag{2.7.3}$$

Computer performance is measured in time; therefore, the faster computer is the one that achieves the same amount of work in less time. Program execution time is measured in seconds per program. The time to complete a task disk and memory accesses, I/O activities, operating system, etc., can be evaluated in different ways depending on what the evaluator is interested to measure. The most common definition is called **wall-clock-time**; other synonyms are **elapsed time** and **response time**.

Other important definitions regarding performance and time are the **CPU execution time (CPU time)**, which is divided in **user CPU time** and **system CPU time**. These concepts are significant since usually computers execute several tasks at the same time sharing the processor. In such cases, the system may try to optimize the throughput instead of minimizing the elapsed time of a program. The **user CPU time** is the time that the processor spends computing our task without including any other program or event. The **system CPU time** is the time that the CPU spends achieving the operating system task on behalf of the program. Although there is the conceptual difference it is easy to view, to have an accurate measurement of each one separately is difficult since it is very hard to assign all the responsibilities to all the activities. Therefore, for consistency's sake, the term **elapsed time** and **system performance** will be handled as synonymous, and the term **CPU performance** to indicate the **user CPU time**.

The CPU execution time for a program can be calculated using (2.7.4)

$$\begin{array}{l} \text{CPU execution time} \\ \text{for a program} \end{array} = \begin{array}{l} \text{CPU clock cycles} \\ \text{for a program} \end{array} \times \text{Clock cycle time} \qquad (2.7.4)$$

The formula (2.7.5) states that the hardware designer can improve performance reducing the clock period (increasing the clock rate), or reducing the number of clock cycles required by a program.

$$\begin{array}{l} \text{CPU execution time} \\ \text{for a program} \end{array} = \frac{\text{CPU clock cycles for a program}}{\text{Clock rate}} \qquad (2.7.5)$$

The CPU clock cycles that a program requires can be calculated using (2.7.6) where the term CPI means **clock cycles per instruction** which is the average time that an instruction requires to be executed; each instruction performance is dependent on the number of clock cycle. In all these formulations, the clock cycle time = 1/(clock rate), and clock rate is given in Hertz, and it is equal to cycles/sec.

$$\text{CPU clock cycles} = \begin{array}{l} \text{Instructions for} \\ \text{a program} \end{array} \times \underbrace{\begin{array}{l} \text{Average clock cycles} \\ \text{per instruction} \end{array}}_{\text{CPI}} \qquad (2.7.6)$$

Therefore, the basic CPU time performance equation can be written as (2.7.7) or (2.7.8),

$$\text{CPU time} = \frac{\text{Instruction count} \times \text{CPI}}{\text{Clock rate}} \qquad (2.7.7)$$

$$\text{CPU time} = \text{Instruction count} \times \text{CPI} \times \text{Clock cycle time} \qquad (2.7.8)$$

In (2.7.7) there are some problems that must be addressed to calculate the time. Although there is no specific order to obtain the parameters' values, the easier one is **the clock cycle time** which is usually a known parameter that is given in the computer's documentation. It mainly depends on the hardware technology which includes the CPU microarchitecture. The **CPU execution time** can be measured by running the program several times to take an average value. **The Instruction count** may be obtained using benchmark software (Mellor-Crummery and LeBlanc 1989), although the greatest benchmark programs are real applications, and the best way to evaluate them is by a user who has experience running the same applications during a long period of time. The **CPI** depends on several details such as the **Instruction Set Architecture** (ISA), the microarchitecture, and the algorithm implementations of the same ISA; this parameter is obtained by performing a detailed simulation of an implementation.

It is possible to obtain the CPU clock cycles required by a program using the formula (2.7.9). Here the index i indicates the class; C_i is the amount (count) of the number of instructions of class i, and CPI is the average number of cycles per instruction of class i.

$$\text{CPU clock cycles} = \sum_{i=1}^{n} (\text{CPI}_i \times C_i) \tag{2.7.9}$$

In general, formula (2.7.10) provides an overview of the different relations (Patterson 2012),

$$\text{CPU Time} = \underbrace{\frac{\text{time}}{\text{Program}}}_{} = \underbrace{\frac{\text{time}}{\text{Clock cycles}}}_{\text{Clock cycle time}} \times \underbrace{\frac{\text{Clock cycles}}{\text{instruction}}}_{\text{CPI}} \times \underbrace{\frac{\text{instructions}}{\text{program}}}_{\text{Instruction count}} \tag{2.7.10}$$

Sometimes, the result of running the application just one time is not enough, in such cases it is necessary to perform statistical tests to obtain the conclusion. One accepted test to obtain the execution time is the arithmetic mean (AM) of execution times given by (2.7.11)

$$\text{AM} = \frac{1}{n} \sum_{i=1}^{n} (\text{CPU time})_i \tag{2.7.11}$$

where $(\text{CPU time})_i$ is the execution time for the ith program of a total of n in the workload. Since it is the mean of the execution time, a smaller mean indicates a smaller average execution time and thus improved performance. The arithmetic mean is proportional to execution time, assuming that the programs in the workload are each running an equal number of times (Patterson 2012).

An alternative metric to time is the **native MIPS** (Million of Instruction per Second), note that the acronym MIPS is preceded by the word *native* to distinguish this concept to the previous used in section 1.4. Here, native MIPS is an instruction execution rate given by (2.7.12)

$$\text{MIPS} = \frac{\text{Intruction count}}{\text{Execution time} \times 10^6} \tag{2.7.12}$$

The use of native MIPS has following three problems:

1. Native MIPS does not consider the instruction capabilities but the instruction execution rate. It is not useful to compare computers with different instruction sets because the instruction count is different.
2. Native MIPS might vary between programs on the same computer which means that a computer can have different MIPS for different programs.
3. Native MIPS can vary inversely with performance.

2.7.1.1 Example 1. Calculating the CPI of a program

In this example, we show how to calculate the average CPI of a program. For this purpose we use the data shown in Table 2.2. The instruction of the program was classified in five classes as is illustrated in column one of the table. Column two shows the cycles per instruction, and column three the percentage of the whole program as of each class of instructions. For example, for the ALU class each instruction lasts five cycles. Forty percent of the instruction in the program are of this class; so, using the definition of CPI yields

$$\text{CPI} = 5 \times 0.20 + 5 \times 0.20 + 5 \times 0.35 + 3 \times 0.10 + 6 \times .15 = 4.95 \text{ cycles}$$

Therefore, the average clock cycles per instruction of the program is 4.95 cycles.

Table 2.2 Percentage of instruction count for a program.

Instruction	# cycles	% Instr. Count
Load	5	20
Store	5	20
ALU	5	35
Branch/Jump	3	10
Others	6	15

2.7.1.2 Example 2. Which computer is faster?

Consider two different computers, A and B, with the following characteristics (Table 2.3)

Table 2.3 Characteristics of two different computer systems.

Characteristics	A	B
Clock cycle time (ns)	1.0	2.0
CPI	4.95	4.0

We will evaluate them using two implements of the same ISA. Applying (2.7.8) we have:

For computer A:

CPU time (A) = Instruction count $\times$ 1 $\times$ 4.95 = 4.95 $\times$ Instruction count

For computer B:

CPU time (B) = Instruction count $\times$ 2 $\times$ 4.0 = 8.0 $\times$ Instruction count

Now, we need to calculate the performance ratio using (2.7.3),

$$\text{Performance ratio} = \frac{\text{CPU time (B)}}{\text{CPU time (A)}} = 1.61$$

Therefore, computer A is 1.61 times faster than computer B.

2.7.2 Analyzing the Performance in Multicore/Multiprocessor Systems

The Amdahl's law was proposed four decades ago by Gene Amdahl. It was a mathematical formulation focused to know the expected improvement of a system, considering a given improvement. Originally, this law was defined for the special case of using n processors (cores) in parallel. He used a limit argument to assume that a fraction f of a program's execution time was infinitely parallelizable with no scheduling overhead, while the remaining fraction $(1 - f)$ was totally sequential. Without presenting any equation, Amdahl noted that the speedup on n processors is dictated by (2.7.13). He argued that typical values of $(1 - f)$ were large enough to favor single processors. Today, this law still valid for enhanced machines where the computation problem size does not change, this is because the fraction of the program that is parallelizable remains fixed.

$$\text{Speedup}_{\text{parallel}}(f, n) = \frac{1}{(1 - f) + \frac{f}{n}} \tag{2.7.13}$$

The modern version of the Amdahl's law states that if you enhance a fraction f of a computation by a speedup s, the overall speedup is given by (2.7.14) (Hill and Marty 2008).

$$\text{Speedup}_{\text{enhanced}}(f, s) = \frac{1}{(1 - f) + \frac{f}{s}} \tag{2.7.14}$$

Gray (2003) indicated that the scalability problem is priority in more than a dozen long-term information technology research goals, since systems with chip multiprocessors (multicores) are emerging as the dominant technology for computing platforms. In (Hill and Marty 2008) Amdahl's law for multicore hardware was argued by constructing a cost model for the number and performance of cores that the chip can support. They started with a generic unit of cost called **Base Core Equivalent** (BCE) which is a basic unit of "cost" that depending on the context, could be area, power, design effort, dollars, or some combination of multiple factors. Using one unit of BCE it is possible to build a single processor that delivers a single unit of baseline performance (performance =1). For a budget of n BCEs, it is possible to build a single n-BCE core, or n single-BCE cores, or in general n/r cores where each consume r BCEs. They concluded that for obtaining optimal multicore performance, it is necessary to achieve further research in extracting more parallelism, as well as in making sequential cores faster.

Figure 2.25a and b show two common ways founded in literature of drawing a BCE. In (a), a core architecture is shown (a single core processor with only a single primary calculations core), in (b) only the acronym BCE is written to indicate that the block contains a base core equivalent, in Fig. 2.25c the core has more features, therefore new core area $r > 1$ and it is faster with performance $perf(r)$, but the question is: what is the value of $perf(r)$, and for which performance $perf(r)$ function is this large core better than multiple small ones? The first question can be answered with the Pollack's rule which states that the performance increase is roughly proportional to square root increase in complexity; hence, for efforts that devote r BCE resources will result in sequential performance of $\sqrt{r}$. In other words, if the logic of a processor core is doubled, then it delivers only 40% more performance; on the other hand, a multi-core processor has the potential to provide near linear performance improvement with complexity and power; for example, the large monolithic processor for which we obtained a 40% of improvement can be easily defeated using two smaller processor cores that can potentially provide a 70–80% more performance (Borkar 2007). To answer the second question, it is necessary to know more about the large core, for example, we need to know the micro-architecture of the chip: it could be a single core processor with a single primary calculation core with added features that are not considered cores, such as pipeline and cache memories.

In Fig. 2.26 different architectures of multicore chips are shown. In Fig. 2.26(a) the system is a symmetrical multicore model with 16 one BCE cores, in (b) the system is a symmetrical multicore model with four BCE cores, and in (c) the system is an asymmetric multicore model with one four-BCE and 12 one-BCE cores. In a symmetrical multicore chip, all the cores have the same cost. According to Amdahl's law, the speedup of a symmetric

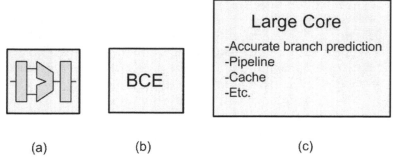

(a)   (b)   (c)

Fig. 2.25 (a) and (b) are two ways of illustrating a BCE. In both, the performance is equal to 1. In (c) the core (large) has more features, so, the *r* values are *r* > 1; this model might correspond to an *r*-BCE core model.

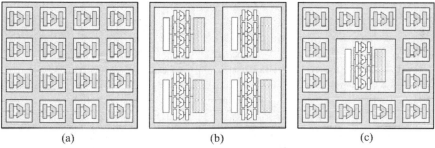

(a)   (b)   (c)

Fig. 2.26 Three hypothetical multicore chips. (a) Symmetrical architecture, sixteen cores with one-BCE each. (b) Symmetrical architecture, four cores, each core has 4 one-BCE (4/4-BCE). (c) Asymmetrical multicore architecture, there are 12/1-BCE and 1/4-BCE.

multicore chip considering that each core is a single BCE, depends on three things: (1) the fraction of software that is parallelized (f), (2) the total of available resources in BCEs, i.e., the n value of the relation n/r, and (3) the BCE resources, i.e., the r value of resources that are devoted to increase the performance of each chip.

For the symmetric case, we have a performance value $perf(r)$ when the chip uses one core to execute the software sequentially; in (Hill and Marty 2008) they assumed the value $perf(r) = \sqrt{(r)}$. Since there is a total of n/r cores, the parallelizable fraction of the program are further speedup obtaining $perf(r) \times n/r$, the overall speedup can be calculated using (2.7.15). The optimal number of cores depends on the value of f, and it can vary significantly for different values. For small f values, the sequential code of cores dominates,

$$\text{Speedup}_{\text{symmetric}}(f, n, r) = \cfrac{1}{\frac{1-f}{perf(r)} + \frac{f \cdot r}{perf(r) \cdot n}} \qquad (2.7.15)$$

The symmetrical (heterogeneous) multicore chip is an alternative to the symmetrical multicore chip. In the asymmetric architecture, one or more cores are more powerful than others; this is shown in Fig. 2.26 (c). For example, considering that we have a budget of $n = 16$ BCEs, an asymmetric multicore chip can have one-nine BCE and seven one-BCE core, or one-four BCE and 12 one-BCE cores. As a general rule, the chip can have $1 + n - r$ cores since the larger core uses r resources and leave $n - r$ resources for the one-BCE cores. Therefore, the Amdhal's law shows different effects on an asymmetric multicore chip. This chip uses the one core with bigger resources to execute the sequential part at performance $perf(r)$. In the parallel fraction, it gets performance $perf(r)$ for the large core and performance 1 for each of the $n - r$ base cores. The speed up of the asymmetric multicore chip under Amdahl's law is given by (2.7.16).

$$\text{Speedup}_{\text{asymmetric}}(f, n, r) = \cfrac{1}{\frac{1-f}{perf(r)} + \frac{f \cdot r}{perf(r)+n-r}} \tag{2.7.16}$$

A third possibility is to have dynamic multicore chips. The idea here is to combine dynamically up to r cores to boost performance of only the sequential components, in sequential mode, the dynamic multicore chip can execute the code with a performance $perf(r)$ considering that the dynamic techniques can use r BCEs. In parallel mode, the same dynamic multicore chip can get performance n using all its base cores in parallel. The overall speedup of this mode under Amdahl's law can be calculated using (2.7.17).

$$\text{Speedup}_{\text{dynamic}}(f, n, r) = \cfrac{1}{\frac{1-f}{perf(r)} + \frac{f}{n}} \tag{2.7.17}$$

The dynamic multicore chips can provide better but never worse speedups than asymmetric chips; however, with the Amdahl's sequential parallel assumption to achieve better speedups than asymmetric multicore chips will require dynamic techniques that exploit more cores for sequential mode.

There are other works related to investigate Amdahl's law, such as (Woo and Lee 2008) where the Hill's work (Hill and Marty 2008) was developed by taking power and energy into account. The revised models provided developers with a better understanding of multicore scalability; however, it is still required to continue with more theoretical analysis, because existing works are all achieved using programs and experiments. In (Kepner 2009) interesting plots and result about the use of the Amdahl's law are shown; moreover, the topic of problem size is introduced to analyze scaling.

A different point of view is presented by (Gustafson 1998); here the concept of scalable computing and the **fixed-time speedup model** is introduced. In the Amdahl's law, the tacit assumption is that the problem size or the workload is fixed to that which runs on the unenhanced system. The speedup emphasizes time reduction of a given problem, so it **is a fixed-size speedup model**; in (2.7.13) when n increases to infinity, the speedup bound is given by (2.7.18),

$$\lim_{n \to \infty} \left[\text{Speedup}_{\text{parallel}}(f, n) = \frac{1}{(1 - f) + \frac{f}{n}} \right] = \frac{1}{1 - f} \qquad (2.7.18)$$

Since most applications have sequential portions of code that cannot be parallelized, under the Amdahl's law, parallel processing is not scalable. For example, for an application where 90% of code can be parallelized, and the rest (10%) cannot, for 8 to 16 processors; the 10% of sequential code will contribute about 50%–80% of the total execution time, and if we decide to add more processors we will have a diminish effect (saturation effect). In (Sun and Chen 2010) Amdahl's law and Gustafason's law are analyzed; in this paper, both point of views are discussed; moreover, the scalability of multicore architectures were studied, and they provided three performance models: **fixed-time, memory-bounded,** and **fixed-size** for symmetric architectures.

The **fixed-time speedup model:** Speedup$_{FT}$, disputes that using of powerful computers to solve large problems, the problem size must scale up with the increase of computing capabilities. There are many practical workloads, for example, real time applications, where the problem size scale-up is bounded by the time execution. Therefore, the fixed-time speedup is defined by (2.7.19).

$$\text{Speedup}_{FT} = \frac{\text{Sequential Time of Solving Scaled Workload}}{\text{Parallel Time of Solving Scaled Workload}} \qquad (2.7.19)$$

If we consider that the original workload w, and the scaled workload w', both finish in the same amount of time using sequential processing and parallel processing with n processors, and we assume that the scale of the workload is only in the parallel processing part, we have $(1 - f)w + fnw$, hence, using (2.7.19), we can obtain the equation (2.7.20) known as the Gustafason's law which states that the fixed-time speedup is a linear function of n whether the workload is scaled up to maintain a fixed execution time. This law suggests that it is good to build a large-scale parallel system as the speedup can grow linearly with the system size.

$$
\begin{aligned}
\text{Speedup}_{\text{FT}} &= \frac{\text{Sequential Time of Solving } w'}{\text{Parallel Time of Solving } w'} \\
&= \frac{\text{Sequential Time of Solving } w'}{\text{Sequential Time of Solving } w} \\
&= \frac{w'}{w} = \frac{(1-f)w + fnw}{w} = (1-f) + nf
\end{aligned}
\tag{2.7.20}
$$

There are many applications that due to some physical constrains cannot scale up to the time bound constrain. Memory limitation is one such constraints; so, in (Sun and Ni 1990, Sun and Ni 1993) the memory-bounded speedup model $\text{Speedup}_{\text{MB}}$, was proposed. In this proposal the speedup is defined as

$$
\text{Speedup}_{\text{MB}} = \frac{\text{Sequential Time of Solving } w^*}{\text{Parallel Time of Solving } w^*}
\tag{2.7.21}
$$

where, in (2.7.21) the scaled workload under memory space constraint is w^*. Here, the authors assumed that each computing node is a processor-memory. Therefore, increasing the number of processors, the memory capacity is also increased.

Considering that $y = f(x)$ is a function that reflects the parallel workload increase factor, as the memory capacity increases n times. Let N be the memory capacity of one node, and

$$
w = g(N)
\tag{2.7.22}
$$

and

$$
w^* = g(m \cdot N)
\tag{2.7.23}
$$

by combining (2.7.22) and (2.7.23) we obtain

$$
\text{Speedup}_{\text{MB}} = \frac{(1-f)w + f \cdot g(n \cdot g^{-1}(w)}{(1-f)w + \frac{f \cdot g(n \cdot g^{-1}(w))}{n}}
\tag{2.7.24}
$$

Many algorithms have polynomial computational complexity and always the highest degree term to represent such complexity is used; in this case, it is used to represent the complexity of memory requirements. If $g(x)$ represents such a complexity, it could take the form of $g(x) = ax^b$ for any rational numbers a and b, using the formula deduction explained in (Sun and Chen 2010), we have,

$$
g(nx) = a(nx)^b = n^b \cdot ax^b = n^b g(x) = \bar{g}(n)g(x)
\tag{2.7.25}
$$

In (2.7.25) $\overline{g}(m)$ is the power function, its coefficient value is 1, therefore (2.7.24) can be simplified as follows,

$$\text{Speedup}_{\text{MB}} = \frac{(1 - f) + f \cdot \overline{g}(n)}{(1 - f) + \frac{f \cdot \overline{g}(n))}{n}} \tag{2.7.26}$$

The **fixed-size** speedup model of a multicore system is based on the speedup definition

$$\text{Speedup} = \frac{\text{Enhanced performance}}{\text{Original Perfomance}} = \frac{T_{\text{Original}}}{T_{\text{Enhanced}}} \tag{2.7.27}$$

Considering that for a single BCE core, $r = 1$, the performance value is 1, such as it was stated in (Hill and Marty 2008); for a problem size w, the original execution time is given by (2.7.28)

$$T_{Original} = w/perf(1) = w \tag{2.7.28}$$

To calculate the enhanced execution time for a n-BCE multicore system (2.7.29) is used ,

$$T_{\text{Enhanced}} = \frac{(1 - f)w}{perf(r)} + \frac{fw}{\frac{n}{r} \cdot perf(r)} \tag{2.7.29}$$

Assuming that in an n/r-core there are n-BCE resources with performance $perf(r)$ for each core, the formula (2.7.30) is valid.

$$\text{Speedup} = \frac{\text{Enhanced performance}}{\text{Original performance}} = \frac{\frac{w}{perf(1)}}{\frac{(1-f)w}{perf(r)} + \frac{fwr}{n \cdot perf(r)}} \tag{2.7.30}$$

$$= \frac{1}{\frac{(1-f)}{perf(r)} + \frac{fr}{n \cdot perf(r)}}$$

In (Sun and Chen 2010) only the symmetric multicore structures were studied because the asymmetric systems are more complex and they were left for a future study.

2.8 Conclusions

This chapter was structured to introduce the reader to the field of architecture of high-performance computers; however, for people interested in getting a deeper knowledge, further reading of specialized books as well as computer architecture journals is needed. Up to this point we have presented an overview of the main characteristics that modern processors have to provide better performance with the aim to offer to the reader the basic foundations to read the rest of this book. Complementary information about this topic is given through the chapter, for example; Chapter 4 is devoted to show the advantages of using a Graphic Processing Unit (GPU) in scientific intensive computing using the Computer Unified Device Architecture (CUDA) which is one of the most utilized GPU hardware and software architecture.

Acknowledgments

The authors would like to thank the "Instituto Politécnico Nacional (IPN)", "Comisión de Operación y Fomento de Actividades Académicas del IPN (COFAA-IPN)" and the Mexico's entity "Consejo Nacional de Ciencia y Tecnología (CONACYT)" for supporting their research activities.

References

ABD-El-Barr, M. and H. El-Rewini (2005). Fundamentals of Computer Organization and Architecture. Hoboken, New Jersey: John Wiley & Sons.

AMD(2012). AMD. Retrieved September 3, 2012, from www.amd.com

Arevalo, A. and E. Peri (2008). Programming the Cell Broadband Engine. USA: ibm.com/ redbooks.

Borkar, S. (2007). Thousand Core Chips—A Technology Perspective. Proceedings of the 44th annual Design Automation Conference. ACM, pp. 746–749.

Domeika, M. (2008). Software development for embedded multi-core systems. Burlington, MA: Newnes-Elsevier.

Floyd, T.L. (2000). Digital fundamentals. Prentice Hall.

Flynn, M. (2011). Flynn's Taxonomy. In: D. Padua, Encyclopedia of Parallel Computing. New York/ Heidelberg, USA: Springer, pp. 689–697.

Fujitsu (2012). Multi-core multi-thread processor SPARC6 Series. Retrieved Noviembre 3, 2012, from http://www.fujitsu.com/global/services/computing/server/sparcenterprise/ technology/performance/processor.html.

García-Risueño, P. and P.E. Ibáñez (2012). A review of High Performance Computing foundations for scientists. arXiv:1205:5177v1[physics.comp-ph]. Cornell University Library, pp. 1–33.

Gray, J. (2003). A dozen information-technology research goals. Journal of the ACM, 50(1): 41–57.

Gustafson, J.L. (1998). Reevaluating Amdahl's Law. Communications of the ACM.31, New York, NY, USA: ACM, pp. 532–533.

Hager, G. and G. Wellein (2011). Introduction to High Performance Computing for Scientists and Engineers. Boca Raton, FL: CRC Press.

Herlihy, M. and N. Shavit (2008). The Art of Multiprocessor Programming. Burlington, MA: Morgan Kaufmann-Elsevier.

Hill, M.D. and M.R. Marty (2008). Amdahl's Law in the Multicore Era. Computer, 41(7): 33–38.

IBM-Freescale. (1995). PowerPC™ 601 RISC Microprocessor User's Manual. Retrieved Octubre 12, 2012, from http://www.freescale.com/files/32bit/doc/user_guide/MPC601UM. pdf

Intel. (n.d.). Retrieved Noviembre 2012, 4, from 3rd Generation Intel® Core™ Desktop Processor Family: http://www.intel.la/content/dam/www/public/us/en/documents/ product-briefs/3rd-gen-core-desktops-brief.pdf

Intel. (2001). Desktop Performance and Optimization for Intel® Pentium® 4 Processor. Retrieved Octubre 5, 2012, from http://download.intel.com/design/Pentium4/papers/24943801.pdf

Intel. (n.d.). 40 years of Microprocessor. Retrieved November 5, 2012, from http://www.intel.la/content/dam/www/public/lar/xl/es/documents/40_aniversario_del_procesador.pdf

Intel. (n.d.). Intel Atom Processor. Retrieved Octubre 26, 2012, from http://www.intel.la/content/dam/www/public/us/en/documents/product-briefs/atom-netbooks-brief.pdf

Kepner, J. (2009). Parallel Matlab for Multicore and Multinode Computers. Philadelphia, USA: SIAM.

Mellor-Crummery, J.M. and T.J. LeBlanc (1989). A Software Instruction Counter. ASPLOS-III Proceedings—Third International Conference on Architectural Support for Programming Languages and Operating Systems. ACM, pp. 78–86.

Patterson, D. (2012). Computer Organization and Design. The Hardware/Software Interface. Waltham, MA: MK-Elsevier.

Sun, X. and L. Ni (1993). Scalable problems and memory-bounded speedup. Journal of Parallel and Distribuited Computing, 19: 27–37.

Sun, X.-H. and Y. Chen (2010). Reevaluating Amdahl's law in the multicore era. Journal of Parallel and Distribuited Computing, 183–188.

Sun, X.-H. L.M. and Ni (1990). Another view on parallel speedup. Proceedings of the 1990 ACM/IEEE conference on Super computing. New York, USA: Super computing '90. pp. 324–333.

Woo, D.H. and H-.H. Lee (2008). Extending Amdahl's Law for Energy-Efficient Computing in the Many-Core Era. Computer, 41(12): 24–31.

3

High-Performance Optimization Methods

Roberto Sepúlveda, * *Oscar Montiel, Daniel Gutiérrez, Jorge Nuñez and Andrés Cuevas*

ABSTRACT

In this chapter, we will provide the reader with the basic mathematic knowledge about optimization with the aim to illustrate how classical iterative optimization algorithms such as those based in gradient descendent methods and Newton's method, as well as global optimization algorithms, in this case a Memetic Algorithm, can be programmed in parallel to take advantage of the technology to improve the performance of the mentioned algorithms. To illustrate the benefits of parallelizing, we chose for Iterative Algorithms the multiplication operation, and for Memetic Algorithm we parallelized the population evaluation and the crossover operation. For all the algorithms, the parallelized part was programmed in C/C++ and converted to a MATLAB mex file with the aim to use it in the algorithms implemented in MATLAB. For the iterative algorithms we used threads for a multicore CPU; for the Memetic Algorithm, the parallel implementation was achieved for a multicore processor and for the CUDA, they were also converted to MATLAB mex files.

Instituto Politécnico Nacional, CITEDI, Tijuana, México.
  Email: rsepulvedac@ipn.mx
* Corresponding author

3.1 Introduction

Optimization is a fundamental discipline in science, engineering, economics, or in any other field where there is a problem involving decision-making. Optimization Problems (OP) can be classified based on the nature of design variables (deterministic or stochastic), existence of constraints, nature of the equation involved (linear, nonlinear, etc.), and other criteria. A valid classification is illustrated in Fig. 3.1. The decision variables of an OP can take continuous or discrete values, hence the objective function can be continuous or discrete and they can be linear or nonlinear. The constrained OP are subject to one or more constraints, whereas for the unconstrained OP there are no constraints. Constrained discrete OP can be also classified as polynomial and non-polynomial hard (Arora and Barak 2010).

In general, optimization methods search for the best solution or solutions for a given problem, it can have one or several objective functions to be optimized as well as a set of constrictions to deal with.

Recently, the interest in finding novel methods and original algorithms that take advantage of innovative technologies has increased. The advent of new high-performance multicore processors and GPU technology; as well as software that helps to exploit modern processors and computer architectures, has allowed the rapid development of several areas, for example, computer simulation, graphics, high-speed digital signal processing and video, artificial intelligence, etc. In all of these areas, there are complex optimization problems that require very fast algorithms to offer results in real time applications, or at least in a reasonable short time to be useful considering all the constrictions.

Similar to the classification of OP, the Optimization methods can be also classified in several ways; they can be organized as single-objective or multiple-objective methods; derivative-based methods or derivative-free optimization methods, local search or global search, etc. Global search methods have mechanisms to escape from local optima, while a local search does not (Fig. 3.2).

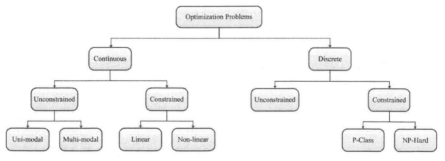

Fig. 3.1 Classification of optimization problems.

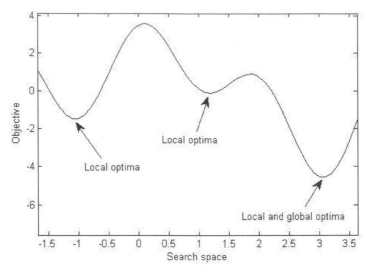

Fig. 3.2 Local and global solutions in a search space.

A general classification and some of its representative techniques are:

- *Mathematical programming or optimization techniques*: Calculus methods, linear programming, non-linear programming, geometric programming, quadratic programming, dynamic programming, multi-objective programming, etc.
- *Stochastic process techniques*: Statistical decision theory, Markov processes, etc.
- *Statistical methods*: Regression analysis, cluster analysis, etc.
- *Non-traditional optimization techniques*: Genetic Algorithms (GA), Memetic Algorithms (MA), Ant Colony Optimization, particle swarm optimization, etc.

In this chapter, we are going to explore single objective optimization problems using iterative algorithms, as well as MA. So it is important to distinguish between direct search and iterative methods, in (Baldick 2008) the next classification is given:

1. *Direct search methods*. They are known since the 1950's, the field flourished with some algorithms like the Hooke and Jeeves algorithm where the term "direct search" was established. They were considered very good methods since their ability to search and find good solutions in non-smooth surfaces using easy to understand heuristics; these methods avoid the calculation of gradients of the objective function. For several years, there were a lack of interest in the field by the mathematical optimization community for several reasons, however,

they remained used by the scientific and engineering communities (G. Kolda et al. 2003).

2. *Iterative methods.* Here, the methods begin with an initial value, and they keep looking to improve the actual solution through a series of iterations, until we reach the optimal value or we fulfill with a stop criterion. Iterative methods typically involve a line search at each iteration. In general, these kind of algorithms try to minimize the objective function $f: \mathbb{R}^n \to \mathbb{R}$ using equation (3.1.1), under proper conditions it must be fulfilled equation (3.1.2),

$$x_{(i+1)} = x_i + \alpha_i d_i \tag{3.1.1}$$

$$f(x_{(i+1)}) < f(x_i) \tag{3.1.2}$$

where $x \in \mathbb{R}^n$, x_i, is the actual point, $x_{(i+1)}$ is the next point to be calculated, d_i is a search direction vector, and d_i is an step value that might be fixed or variable that it is used to control the approximation rate to the optimal point. In the iterative methods, the search direction is obtained using gradient information of the objective function, hence any method based on derivatives can be considered in this group, some examples are, steepest descent family methods and Newton's method, conjugate direction method, etc.

To find the global optima for a specific problem using an iterative optimization method, we need to move from a solution at the actual point to a better solution in the search space. The problem is that in multimodal function, we can get stuck in local optima, therefore, local search optimization methods are not enough to find the global solution for such problems. Figure 3.2 illustrates the concepts of local and global optima.

For this chapter, since there is not a precise distinction between direct search methods (G. Kolda et al. 2003), then they can perform iterations in the quest of the optimal solution without using gradients. We will use the term iterative methods for all those methods based on derivatives and Taylor series.

3.2 Unconstrained Iterative Optimization Methods

In this section, we will describe the most common iterative algorithms for solving unconstrained optimization problems. The pseudo-code of the gradient descent, Newton, and conjugate gradient methods is given. Iterative methods are widely used in intelligent systems applications, for example, in the optimization of the weights of neural networks; to optimize parameters of membership functions in the Adaptive Neuro-Fuzzy Inference System (ANFIS), etc. These kinds of applications usually may have hundreds of decision variables to be optimized; in those cases, sequential programming becomes inefficient. There are cases that just for a few variables, valid solutions are generated after several minutes or in some cases hours or days. For these reasons, in Section 3.3 we have shown time comparisons of sequential programming vs. parallel programming of the iterative methods. We chose to parallelize the matrix multiplication operation to illustrate the process; however, a similar procedure can be applied to parallelize a different operation that the algorithms involve.

3.2.1 Gradient Descent Method

This is a simple method based on the information provided by the descent gradient of the objective function f. In other words, the search direction is given by $-\nabla f(x)$. The Pseudocode 3.1 explains the computation steps to minimize a function f using this method. First, in line 3 the iteration counter i is defined. On line 4, the x variables are initialized with random values, preferably. On lines 5 and 6, the constant parameters ε and α_i are initialized conveniently for the problem, a typical value ε is 10^{-6}; in the case of α_i a value of 10^{-6} might be chosen for small steps, and a value of 10^{-3} for bigger steps, emphasizing that small steps will take more time to reach the optimal value providing a smooth path, while using bigger values the algorithm is faster, with the risk to get a zig-zag trajectory, see Fig. 3.3. On line 7, the next value is computed using the derivative information $(-\nabla f(\mathbf{x}))$ of the objective function, when $\nabla f(\mathbf{x}) \approx 0$, the x values will not change, this information will be used for the stop criteria, such as it is indicated on line 8, so, lines 9 and 10 will iterate until we fulfill the stop condition.

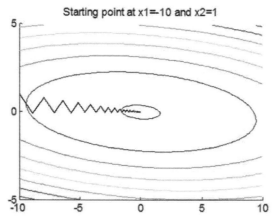

Fig. 3.3 Optimization of the function $f(x_1, x_2) = (x_1 + 10x_2)2 + 5(x_1 - x_2)^2 - 15$. Here the zig-zag trajectory is shown.

Color image of this figure appears in the color plate section at the end of the book.

On line 8, it was necessary to verify for a specific threshold value since in practice, it is difficult to achieve one of the stop condition because of the numeric precision. There are other criterion that might be used in iterative algorithms to know if there are significant value changes in two consecutive iterations, for example, we can check for the absolute value of the objective functions of two iterations using the condition given in (3.2.1), or in (3.2.2) or in (3.2.3).

$$\|f(x_{i+1}) - f(x_i)\| < \varepsilon$$

(3.2.1)

$$\frac{|f(x_{i+1}) - f(x_i)|}{|f(x_i)|} < \varepsilon$$

(3.2.2)

$$\frac{\|x_{i+1} - x_i\|}{\|x_i\|} < \varepsilon$$

(3.2.3)

Pseudocode 3.1 Gradient Descent Method.

---

1.   Minimize $f(x)$
2.   initialize variables
3. $i := 0$
4. $x_i := K_1$
5. $\varepsilon := K_2$
6. $\alpha_i := K_3$
7. $x_{i+1} = x_i - \alpha_i \nabla f(x_i)$
8.   While $\|x_{i+1} - x_i\| > \varepsilon$
9. $x_{i+1} = x_i - \alpha_i \nabla f(x_i)$
10. $i := i + 1$
11. end while
12. Return x_{i+1}

3.2.2 Newton's Method

Gradient descent methods use first-order derivatives to determine a suitable direction d to find the next point using equation (3.1.1), the Newton's methods use second order derivatives of the objective function $f(x)$ for the same purpose. The Newton's methods to solve a problem is reduced to solving a sequence of quadratic problems, they are iterative methods that start in an initial point x_0, sufficiently close to a local minimum, a quadratic approximation to the objective function based on the Taylor series is constructed, that matches the first and second order derivatives values at the point of interest. The idea is to minimize the quadratic function $q(x)$, given by the second-order Taylor approximation, instead of the original objective function $f(x)$, see Eq. (3.2.4).

$$f(x) \approx f(x_i) + (x - x_i)^T \mathbf{g}_i + \tfrac{1}{2}(x - x_i)^T \mathbf{F}(x_i)(x - x_i) \triangleq q(x) \quad (3.2.4)$$

where $\mathbf{g}_i = \nabla f_i(\mathbf{x})$, and $\mathbf{F} = [\nabla^T \frac{\partial f}{\partial x_1} \cdots \nabla^T \frac{\partial f}{\partial x_n}]^T$ is the Hessian matrix of $f(x)$.

Then, to minimize $q(x)$ it is necessary to fulfill the optimality condition $\nabla q(\mathbf{x}^*) = 0$, therefore

$$\nabla q(\mathbf{x}) = \mathbf{g}_i + \mathbf{F}(\mathbf{x}_i)(\mathbf{x} - \mathbf{x}_i) = 0 \tag{3.2.5}$$

Let $\mathbf{x} = \mathbf{x}_{i+1}$, and using (3.2.5), for $\mathbf{F}(\mathbf{x}_i) > 0$, the quadratic approximation $q(x)$ can achieve its minimum by applying the iterative formula (3.2.6),

$$\mathbf{x}_{i+1} = \mathbf{x}_i - \mathbf{F}(\mathbf{x}_i)^{-1}\mathbf{g}_i \tag{3.2.6}$$

In **Pseudocode 3.2** we illustrate the main computational steps to implement this method.

Pseudocode 3.2 Newton's Method.

---

1. Minimize $f(x)$
2. initialize variables
3. $i := 0$
4. Solve $x_{i+1} = x_i - \mathbf{F}(x_i)^{-1}g_i$
5. While $\|x_{i+1} - x_i\| > \varepsilon$
6. $x_{i+1} = x_i - \mathbf{F}(x_i)^{-1}g_i$
7. $i := i + 1$
8. end while
9. Returnx_{i+1}

3.2.3 Conjugate Gradient Method

The conjugate gradient method is used to minimize a function of the type

$$f(x) = \tfrac{1}{2}x^T Q x - x^T b, x \in R^n \tag{3.2.7}$$

where $Q = Q^T > 0$ and Q is a symmetric, positive-definite matrix. The steepest-descent method use the next equation to calculate the first search direction d_0 for x_0.

$$d_0 = -g_0$$

then x_1 is calculated with

$$x_1 = x_0 + \alpha_0 d_0$$

where α_0 is obtained with the next equation,

$$\alpha_0 = \arg\min_{\alpha \geq 0} f(x_0 + \alpha d_0) = -\frac{g_0^T d_0}{d_0^T Q d_0} \tag{3.2.8}$$

The next step specifies that d_1 is Q-conjugated to d_0. Then d_{i+1} is a linear combination of g_{i+1} and d_i. In that case d_{i+1} is obtained with the next equation,

$$d_{i+1} = -g_{i+1} + \beta_i d_i, \quad i = 0,1,2,\ldots$$

As d_{i+1} is Q-conjugated to $d_0, d_1,\ldots,d_i$, the coefficient β_i, $i = 1,2\ldots$ is obtained with the next equation.

$$\beta_i = \frac{g_{i+1}^T Q d_i}{d_i^T Q d_i} \tag{3.2.9}$$

The conjugated gradient algorithm is given in Pseudocode 3.3.

Pseudocode 3.3 Conjugate Gradient Method (Chong and Zak 2001).

---

1. Set $i := 0$; select the initial point x_i
2. $g_i = \nabla f(x_0)$ If $g_0 = 0$, stop, else set $d_0 = -g_0$
3. $\alpha_i = -\frac{g_i^T d_i}{d_i^T Q d_i}$
4. $x_{i+1} = x_i + \alpha_i d_i$
5. $g_{i+1} = \nabla f(x_{i+1})$ If $g_{i+1} = 0$, stop
6. $\beta_i = \frac{g_{i+1}^T Q d_i}{d_i^T Q d_i}$
7. $\quad d_{i+1} = -g_{i+1} + \beta_i d_i$
8. Set $i := i + 1$; go to step 3

3.3 Unconstrained Optimization Using Parallel Computing

Parallel computing, in essence, is the process of calculating multiple tasks simultaneously (Gottlieb 1989). Before multi-core systems, better performance of software depended mostly on CPU clock speed, but in recent years increasing CPU speed has become harder to accomplish. Furthermore, high clock rates have serious drawbacks such as increased heat and power consumption. One way of improving performance is to use multiple CPUs and make them work together to accomplish a specific task.

To take full advantage of multi-core processors, programmers have to think and design codes taking into account multiple threads in execution and forget about the traditional sequential way of programming. The way of doing this is to identify the parts of the code that can run simultaneously without interfering with one another and make them run in parallel. Although writing parallel programs is considerably more difficult than writing the sequential ones, depending on the application, the speed and efficiency gained by parallel processing can be worth the effort (Hennessy and Patterson 1999).

The flow charts in Fig. 3.4 show the way a task is processed using sequential and parallel programming, which is not always faster than sequential programming. Its benefits are limited by the number of processors in the system, as well as the application itself. Also, it is worth mentioning that there is no speed gained by using parallel code on a single processor system because there are resources consumption to synchronize the worker threads.

Another way in which parallel programming may become slower is by using nested loops, it is best to parallelize only the outer loop, unless the inner loop is known to be very long lasting, in which case, it will probably

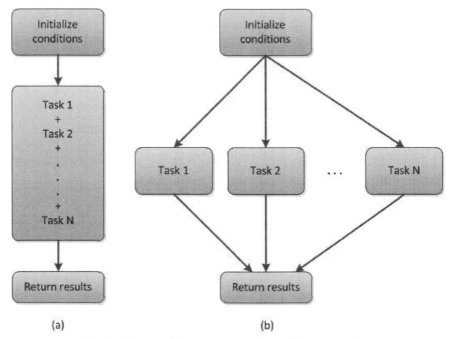

(a)   (b)

Fig. 3.4 a) Sequential programming. b) Parallel programming.

result in faster result. The best way to determine the optimum performance is to test and measure.

3.3.1 MATLAB mex-file Implementation

For implementing the three different optimization methods, the Windows "process.h" header file was used, so we can work with threads and processes, mainly for the creation of threads using the Windows built in function **_beginthreadex**, the code was written in C/C++ with the purpose to convert it to MATLAB mex-file. The implemented C/C++ code for a simple two array multiplication, has been included in the Appendix (Program 3.2) of this chapter; this code is called from MATLAB every time that an array multiplication is required by the optimization method (Quarteroni, Sacco and Saleri 2000). Figure 3.5 illustrates the global procedure to integrate a C/C++ program to the MATLAB through a MEX-file; considering that the C function is MatrixOperation.c, it is converted to mex-file using the MATLAB command "mex MatrixOperation.c" the output file is "MatrixOperation.mexw64" which is illustrated in Fig. 3.5.

Let us consider the next generic C/C++ prototype function:

void mexFunction (int nlhs, mxArray *plhs[], int nrhs, const mxArray *prhs[])

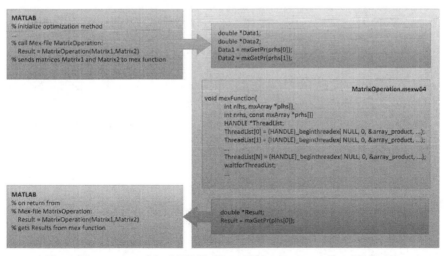

Fig. 3.5 Integration of the MEX-file MatrixOperation.mexw64 to MATLAB.

where the meanings of the arguments of the mex function are described in Table 3.1:

Table 3.1 Mex function arguments.

C/MEX	Meaning	M-code equivalent
Nhls	Number of output variables	nargout
Plhs	Array of mxArray pointers to the output variables	varagout
Nrhs	Number of input variables	nargin
Prhs	Array of mxArray pointers to the input variables	varagin

For our example, the mex function is called from MATLAB as follows:

`Result=MatrixOperation(Matrix1,Matrix2)`

where we have, `nhls` = 1 because `Result` is only one variable calculated in the mex function, and `plhs[0]` is a pointer (`type mxArray*`) pointing to `Result`. Also the number of input variables are given by `nrhs` = 2, because we have two input variables (`Matrix1` and `Matrix2`), where `prhs[0]` and `prhs[1]` are pointing respectively to `Matrix1` and `Matrix2`. Initially, the output variable is unassigned; the mex function will create it. In the case of nlhs=0, the mex function will return the value `plhs[0]` that represents the `ans` variable of MATLAB.

Figure 3.6 illustrates the use of two threads to multiply two matrices, A and B. The result is written in the matrix C by the corresponding thread. One thread calculates the even rows of the matrix A multiplied by the columns of matrix B. similarly, the odd rows of the matrix A are multiplied by the columns of B by the other thread. The results are written in the matrix C by the corresponding thread. In Program 3.1 a code section to illustrate the matrix multiplication of Fig. 3.6, in Program 3.2 the complete code is shown.

Program 3.1 Code section to illustrate how the threads are used to perform a parallel 2x2 matrix multiplication.

```
1. if (Thread = 1)
2.  for( i = 0; i < 1;  i = i + +)
3.      for( j = 0; j < 1;  i = i + +)
4.          for( k: = 0; i < 1;  i = i + +)
5.              C_ij += A_ik * CB_ki
6. if (Thread = 2)
7.  for( i = 1; i < 2;  i = i + +)
8.      for( j = 1; j < 2;  i = i + +)
9.          for( k: = 1; i < 2;  i = i + +)
11.             C_ij += A_ik * CB_ki
```

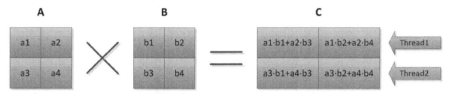

Fig. 3.6 Matrix multiplication done by two threads.

3.3.2 Testing algorithm performance

When testing a parallel algorithm, the speedup is one of the measurements to determine its performance. It is defined as $Sp = Ts/Tp$, where Ts is the algorithm execution time when the algorithm is completed sequentially, and Tp is the algorithm execution when the algorithm is executed in parallel. For this academic example, for the iterative methods gradient descent, Newton and conjugate gradient the sequential and parallel execution times were recorded (Absil, Mahony and Sepulchre 2007). The parallel versions of the algorithms use parallel matrix operation; the results are shown in Table 3.2.

Table 3.2 Parallel programming of the iterative algorithms clearly outperforms the sequential implementation.

Algorithm name	Sequential Time (seconds)	Parallel 2-threads Time (seconds)	Speedup
Gradient descent	1.184	0.478	2.477
Newton	0.375	0.072	5.20
Conjugate Gradient	0.029	0.004	7.25

3.4 Global Optimization Using Memetic Algorithms

A memetic algorithm (Krasnogor and Smith 2005) is a stochastic global search heuristic optimization method that combines an evolutionary algorithm to perform the global search, with a local search technique such as an iterative algorithm (Talbi 2009) (Youngjun, Jiseong, Lee and Kim 2010). The objective of combining techniques is to accelerate the exploitation of the search space in local search for which evolution alone can take a lot of time (Moscato 1989), see Fig. 3.7. The pseudocode and code listing to implement the MA can be found in (Brownlee 2011). Here we show only the basic pseudocode of this algorithm in Pseudocode 3.4.

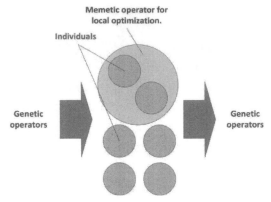

Fig. 3.7 The memetic operator is performing local search on some individuals of the population.

Pseudocode 3.4 Memetic Algorithm (Brownlee 2011).

1. **Input**: ProblemSize, Pop_{size}, MPop_{size}
2. **Output**: S_{best}
3. Population ← InitializePopulation(ProblemSize, Pop_{size})
4. **while** ¬StopCondition() **do**
5. **foreach** S_i ∈ Population **do**
7. Si_{cost} ← Cost(S_i)
8.   **end**
9. S_{best} ← GetBestSolution(Population)
10.   Population ← StochasticGlobalSearch(Population)
11.   MemeticPopulation ← SelectMemeticPopulation(Population, MPop_{size})
12.   **foreach** S_i ∈ MemeticPopulation **do**
13. S_i ← LocalSearch(S_i)
14.   **end**
15. **end**
16. **return** S_{best}

3.4.1 Set up of the Implemented MA

In this work, we implemented an MA with the following characteristics.

For the global search part:

1. **Codification:** Binary, 16 bits to represent a real number.
2. **Population size:** Variable. It depends on the test.
3. **Selection method:** 25% of total population is selected by elitism, the rest is selected using the roulette method.
4. **Crossover:** random, one-point.
5. **Mutation rate:** 0.001

For the local search part:

At each iteration, the gradient descent is applied only to the best individual. The MA was tested using the De Jong's functions (De Jong 1975) known as Sphere function (F1), the Rosenbrock function (F2), and the Fox holes (F5), which are illustrated in Fig. 3.8, Fig. 3.9 and Fig. 3.10 respectively.

F1: Sphere function

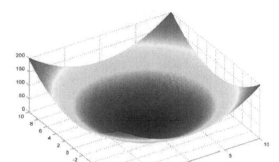

$$F1(X) = \sum_{i=1}^{N} x_i^2$$

Fig. 3.8 Sphere Function (F1).

Color image of this figure appears in the color plate section at the end of the book.

F2: Rosenbrock function

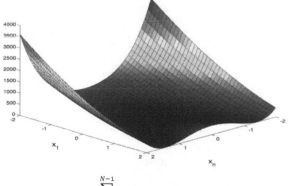

$$F2(X) = \sum_{i=1}^{N-1} [(1 - x_i)^2 + 100(x_{i+1} - x_i^2)^2]$$

Fig. 3.9 Rosenbrock Function (F2).

Color image of this figure appears in the color plate section at the end of the book.

F5: Foxholes function

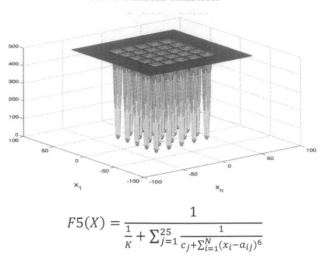

$$F5(X) = \cfrac{1}{\frac{1}{K} + \sum_{j=1}^{25} \cfrac{1}{c_j + \sum_{i=1}^{N}(x_i - a_{ij})^6}}$$

Fig. 3.10 Fox holes Function (F5).

Color image of this figure appears in the color plate section at the end of the book.

3.4.2 Experiments and Results

Two different sets of experiments were achieved:

1. The objective of the first set is to compare the performance of the sequential implementation of the MA vs. the GA. Since the MA is a combination of global search and local search (GA + local search), to test the GA part, we simply removed the local search section. For the first set, the GA was tested for a different number of generations (10 to 200); the time and objective value for each test function and test were recorded, in Table 3.3 the results are shown. The same experiments were done for the MA; the results are presented in Table 3.4. A population size of 2000 individuals was used.
2. The objective of the second set is to compare two different models of parallelization, parallel for multicore and parallel using the CUDA.

In the second set of experiments, for each test function (F1, F2 and F5) the algorithm ran 2000 generations. For F1 and F2, the algorithm was tested for 64, 128 and 256 variables; we vary the number of variables for different population sizes. For function F5, we tested the algorithm for 2, 4 and 8 variables. For all the experiments, the used population sizes were

400, 800, 1600, 3200 and 6400. The results for F1, F2 and F5 are shown in Tables 3.5, 3.6 and 3.7, respectively. Figure 3.5 illustrates for the test function F5, shows population size vs. the execution time for the three different implementations: sequential, parallel using multicore technology, and for the CUDA.

Table 3.3 Results of the GA programmed in sequential mode. Four variables were used for all the test functions.

Number of generations	F1		F2		F5	
	Time (ms)	Objective value	Time (ms)	Objective value	Time (ms)	Objective value
10	47.00	972.55	47.00	2433.11	78.00	2921.77
20	78.00	965.18	94.00	2367.45	140.00	614.38
30	125.00	807.19	125.00	1663.99	187.00	354.19
40	156.00	730.12	172.00	1586.94	249.00	169.49
50	203.00	716.21	219.00	1575.36	312.00	164.55
60	234.00	716.21	250.00	1570.46	374.00	161.34
70	265.00	699.25	281.00	1352.53	421.00	160.29
80	296.00	236.35	328.00	1316.78	483.00	160.29
90	343.00	236.35	359.00	368.15	530.00	140.22
100	374.00	202.93	390.00	368.15	577.00	138.89
200	780.00	28.01	780.00	22.18	1123.00	3.10

Table 3.4 Results of the MA programmed in sequential mode. There were used four variables for all the test functions..

Number of generations	F1		F2		F5	
	Time (ms)	Objective value	Time (ms)	Objective value	Time (ms)	Objective value
10	47.00	0.00	47.00	7.08	62.00	22.00
20	78.00	0.00	109.00	6.72	125.00	22.00
30	109.00	0.00	172.00	6.34	172.00	22.00
40	156.00	0.00	219.00	5.92	234.00	22.00
50	187.00	0.00	265.00	0.92	296.00	21.00
60	218.00	0.00	328.00	0.67	343.00	21.00
70	265.00	0.00	375.00	0.44	406.00	21.00
80	296.00	0.00	437.00	0.25	452.00	21.00
90	327.00	0.00	484.00	0.11	515.00	21.00
100	359.00	0.00	531.00	0.04	562.00	21.00
200	702.00	0.00	936.00	0.00	1092.00	1.00

Table 3.5 Comparative results of the MA. There were used N variables for the F1 test function.

Population	N	F1	Sequential time (ms)	Parallel time (ms)	GPU time (ms)
400	64		7.28	4.16	4.07
	128	$F1(X) = \displaystyle\sum_{i=1}^{N} x_i^2$	14.19	6.97	4.86
	256		27.76	12.24	6.82
800	64		15.16	8.15	4.83
	128		27.99	13.17	6.13
	256		54.90	23.69	9.20
1600	64		30.20	15.10	6.53
	128		55.69	25.62	8.79
	256		107.56	46.36	15.11
3200	64		60.73	29.65	11.25
	128		111.49	50.36	14.70
	256		217.09	91.79	25.63
6400	64		121.83	58.64	22.26
	128		224.78	100.10	26.82
	256		430.38	182.88	47.74

The workload in all the experiments for the multicore parallel implementation was achieved dividing the number of individuals in the population by the number of used threads, four threads for our experiments. For the CUDA, a block for each individual was used, and we further divide the workload in each block as follows: For each individual, the objective function is split in terms, then each term is evaluated in a cell of the block, finally all the terms are added using a thread.

Table 3.6 Comparative results of the MA. There were used N variables for the F2 test function.

Population	N	F2	Sequential time (ms)	Parallel time (ms)	GPU time (ms)
400	64		22.80	17.96	5.35
	128	$F2(X) = \displaystyle\sum_{i=1}^{N-1} [(1 - x_i)^2 + 100(x_{i+1} - x_i^2)^2]$	77.75	67.69	7.62
	256		283.55	261.63	216.99
800	64		31.62	20.81	18.81
	128		92.91	71.93	12.76
	256		308.79	268.94	20.37

Table 3.6 contd....

Table 3.6 contd.

Population	N	F2	Sequential time (ms)	Parallel time (ms)	GPU time (ms)
1600	64		46.45	25.34	20.79
	128		119.67	81.58	13.26
	256		360.97	286.21	62.00
3200	64		78.00	36.44	25.05
	128		177.50	99.54	19.57
	256		472.91	324.51	50.51
6400	64		147.92	61.69	32.99
	128		300.99	138.65	86.54
	256		692.40	399.74	319.91

Table 3.7 Comparative results of the MA. There were used N variables for the F5 test function.

Population	N	F5	Sequential time (ms)	Parallel time (ms)	GPU time (ms)
400	2	$$\frac{1}{F5(X)} = \frac{1}{K} + \sum_{j=1}^{25} \frac{1}{f_j(X)}$$	1.26	2.37	3.16
	4		2.15	2.57	3.23
	8	*where*	25.21	9.71	5.98
800	2	$$f_j(X) = c_j + \sum_{i=1}^{N}(x_i - a_{ij})^6$$	2.56	3.41	3.44
	4		4.27	3.82	3.50
	8	*and*	53.67	18.30	7.71
1600	2	$a_{ij} = \begin{bmatrix} -32, & -16, & 0, & 16, & 32, & -32, & -16, & \cdots, & 0, & 16, & 32 \\ -32, & -32, & -32, & -32, & -32, & -16, & -16, & \cdots, & 32, & 32, & 32 \end{bmatrix}$	5.27	5.59	5.14
	4	with $c_j = j$ and $K = 500$	8.51	6.59	5.28
	8		98.65	32.82	11.15
3200	2		10.77	10.13	9.50
	4		17.17	12.17	9.74
	8		202.97	66.79	17.79
6400	2		21.95	19.48	18.32
	4		36.28	23.80	25.02
	8		401.50	132.15	31.22

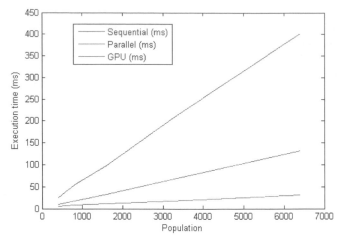

Fig. 3.11 Population size vs. Execution times for the function F5 (Fox holes).
Color image of this figure appears in the color plate section at the end of the book.

3.5 Concluding Remarks

For the iterative optimization methods, the parallel implementations of the algorithms reduce the execution time for all the methods, the best speed up was obtained with the conjugate gradient algorithm.

For the global optimization methods, for the first set of experiments, comparing the MA vs. the GA, the MA could reach better values in less time, practically, the MA always found the optimal values whereas the GA demonstrated slow convergence rate when it was near the optimal value. For the second set of experiments, it was observed that for all the cases, sequential, parallel for multicore and for the CUDA, the execution time grew linearly with the population size, the smallest rate of growing was obtained using the CUDA, and the biggest rate using sequential programming. The faster MA implementation was obtained with the CUDA. The execution time of the parallel multicore implementation is between the other two.

The results illustrate how the parallel implementation outperformed the sequential version, and that parallelizing the algorithms outperforms sequential implementation; we also showed that CUDA implementation always provided solutions in less time, however, the CUDA implementation only worthwhile for huge population sizes and high complexity functions.

Acknowledgments

The authors would like to thank the "Instituto Politécnico Nacional (IPN)", "Comisión de Operación y Fomento de Actividades Académicas del IPN (COFAA-IPN)" and the Mexico's entity "Consejo Nacional de Ciencia y Tecnología (CONACYT)" for supporting our research activities.

3.6 Apendix

Program 3.2 This code performs the multiplication of two arrays using parallel threads. The code is prepared to be called from MATLAB.

```c
#include "mex.h"
#include "stdio.h"
#include "math.h"
#undef EXTERN_C
#include <windows.h>
#include <process.h>

// Variables used to detect if the threads are finished
static volatile int WaitForThread1;
static volatile int WaitForThread2;

unsigned __stdcall array_product(double **Args)
{

    int i=0;
    double *Data1, *Data2, *Result, *ThreadID, *DataSize;
    Data1 = Args[0];
    Data2 = Args[1];
    Result = Args[2];
    ThreadID = Args[3];
    DataSize = Args[4];

    // array_product all even numbers in the first thread
    if(ThreadID[0]==1)
    {
        for (i=0; i<(int)DataSize[0]; i=i+2)
        {
            Result[0] += (Data1[i] * Data2[i]);
        }
    }

    // array_product all odd numbers in the second thread
    if(ThreadID[0]==2)
    {
        for (i=1; i<(int)DataSize[0]; i=i+2)
        {
```

Program 3.2 contd....

Program 3.2 contd.

```
            Result[1]+=(Data1[i]* Data2[i]);
    }
}

// Set the thread finished variables
if(ThreadID[0]==1){ WaitForThread1 =0;}
if(ThreadID[0]==2){ WaitForThread2 =0;}

// explicit end thread, helps to ensure proper recovery of resources allocated for the thread
    _endthreadex(0);
return0;
}

// The matlab mex function
void mexFunction(int nlhs, mxArray *plhs[],
int nrhs,const mxArray *prhs[])
{
double*Data1;// The input data
double*Data2;// The input data
double*Result;// The output data
    mwSize Dcols, Drows;// Dimensions of input data
double**ThreadArgs1,**ThreadArgs2;// double pointer array to store all needed function
                                    //variables
double ThreadID1[1]={1};// ID of first Thread
double ThreadID2[1]={2};// ID of second Thread
double DataSize[1]={0};// Total size of the input data
    HANDLE *ThreadList;// Handles to the worker threads


// Reserve room for handles of threads in ThreadList
    ThreadList =(HANDLE*)malloc(2*sizeof( HANDLE ));

// Reserve room for 5 function variables(arrays)
    ThreadArgs1 =(double**)malloc(5*sizeof(double*));
    ThreadArgs2 =(double**)malloc(5*sizeof(double*));
// Get the dimensions of the input data
    Dcols =(mwSize) mxGetN(prhs[0]);
    Drows =(mwSize) mxGetM(prhs[0]);
// Get total data size
    DataSize[0]=(double)Drows*Dcols;
// Link the array data to the first input array
    Data1 = mxGetPr(prhs[0]);
    Data2 = mxGetPr(prhs[1]);
// Initialize and link the output matrix
    plhs[0]= mxCreateDoubleMatrix(1,1, mxREAL);
    Result = mxGetPr(plhs[0]);

// create first thread to call the array_product() function.
    WaitForThread1 =1;
    ThreadArgs1[0]=Data1;
    ThreadArgs1[1]=Data2;
    ThreadArgs1[2]=Result;
    ThreadArgs1[3]=ThreadID1;
    ThreadArgs1[4]=DataSize;
    ThreadList[0]=(HANDLE)_beginthreadex(NULL,0,&array_product, ThreadArgs1 ,0,NULL);

// create first thread to call the array_product() function.
    WaitForThread2 =1;
    ThreadArgs2[0]=Data1;
    ThreadArgs2[1]=Data2;
    ThreadArgs2[2]=Result;
    ThreadArgs2[3]=ThreadID2;
    ThreadArgs2[4]=DataSize;
    ThreadList[1]=(HANDLE)_beginthreadex(NULL,0,&array_product, ThreadArgs2 ,0,NULL);
```

Program 3.2 contd....

Program 3.2 contd.

```
// waiting on Threads to finish
while( WaitForThread1||WaitForThread2)
{
        Sleep(1);
}
// Add the two thread results to get the final product
    Result[0]+= Result[1];
// Destroy the thread objects
    CloseHandle( ThreadList[0]);
    CloseHandle( ThreadList[1]);
}
```

References

Absil, P.A., R. Mahony and R. Sepulchre, R. (2007). Optimization algorithms of matrix manifolds. Princeton: Princeton University Press.

Arora, S. and B. Barak (2010). Computational Complexity. A Modern Approach. USA: Cambridge University Press.

Baldick, R. (2008). Applied Optimization. Formulation and Algorithms for Engineering Systems. New York: Cambridge University Press.

Barzilai, J. and J. Borwein (1988). Two point step size gradient methods. IMA Journal of Numerical Analysis, 141–148.

Brownlee, J. (2011). Clever Algorithms: Nature-Inspider Programing Recipes.

Chong, E. and S. Zak (2001). An introduction to optimization. New York: Wiley.

De Jong, K.A. (1975). Analysis of the Behaviour of a Class of Genetic Adaptive Systems. Hampton, Virginia: Langly Research Center. Technical Report No. 185.

G. Kolda, T., R. Michael Lewis and V. Torczon (2003). Optimization by Direct Search: New Perspectives on Some Classical and Modern Methods. Society for Industrial and Applied Mathematics, 385–482.

Gottlieb, G.S. (1989). Highly Parallel Computing. Redwood City, CA, USA.: Benjamin-Cummings Publ. Co. Inc.

Hennessy, J.L. and D.A. Patterson (1999). Computer organization and design : the hardware/ software interface (2. ed., 3rd print. ed.). San Francisco: Kaufmann. ISBN 1-55860-428-6.

Krasnogor, N. and J. Smith (2005). A tutorial for Competent Memetic Algorithms: Model Taxonomy and Design Issues. IEEE.

Moscato, P. (1989). On Evolution, Search Optimization, Genetic Algorithms and Martial Arts Towards Memetic Algorithms. Caltech Concurrent Computartion Program.

A. Quarteroni, R. Sacco and R.R. Saleri (2000). Numerical Mathematics. Secaucus: Springer

Talbi, E.-G. (2009). Metaheuristics: From Design to Implementation. Lille: University of Lille.

Youngjun, A., P. Jiseong, C.-G. Lee and J.-W.S.-Y. Kim (2010). Novel Memetic Algorithm implemented With GA (Genetic Algorithm) and MADS (Mesh Adaptive Direct Search) for Optimal Design of Electromagnetic System.

4

Graphics Processing Unit Programming and Applications

Oscar Montiel, * *Juan J. Tapia, Francisco Javier Díaz-Delgadillo* and *Nataly Medina Rodríguez*

ABSTRACT

In this chapter, we shall introduce the reader to the field of graphic processing unit programming (GPU) and applications. The main goal is to show the advantages of using GPUs for problem-solving of scientific applications that require intensive computation. Specifically, we have focused this chapter in giving an introduction to the Compute Unified Device Architecture—CUDA™—which is a parallel computing platform and programming model of the NVIDIA Company. CUDA is one of the most well-known and used GPU programming framework. As a case study, we present the computational intensive task of generating the Mandelbrot fractal programmed sequentially, and the application of the CUDA™ for breaking down this computational problem with the aim of showing the benefits of using GPUs.

4.1 Introduction

The use of Graphics Processing Units for general parallel computation is an ongoing research field. Parallel algorithms running on GPUs can often achieve a speedup of hundreds of times over their respective sequential CPU

Instituto Politécnico Nacional, CITEDI, Tijuana, México.
  Email: oross@ipn.mx
* Corresponding author

algorithms. The GPUs can be used in applications for physics simulations, signal processing, financial modeling, neural networks, image processing and many other fields. The characteristic that allows the use of the GPUs in such high performance scientific computations is their very large scale of parallelization; some GPUs have hundreds and even thousands of embedded cores.

The key success of the use of GPUs in general computing applications lies on two main facts. The first one is their relation cost/performance when compared with other processor types, i.e., multicore architectures, and the second one is the wide distribution of programming frameworks for general computing—such as NVIDIA CUDA—from the main GPU developers around the world. Hence, the GPUs constitute a relatively cheap high performance computing tool that offers a theoretical peak number of GFLOPS that is nearly six times greater than CPUs and a memory bandwidth that is practically five-times higher than CPUs (García-Risueño and Ibáñez 2012) but at a fraction of the price of these others architectures. The reason behind this gap in the computational capacity between a GPU and a CPU comes from their design; a GPU devotes more transistors and resources to data processing, rather than data caching and flow control, as in a CPU. All these factors, in addition to the available programming frameworks, are the reasons behind the wide use of GPU computing by a diverse scientific community of many different research areas. In the textbook (Sanders and Kandrot 2010) the authors give a detailed explanation of the concepts of CUDA programming and the main characteristics of the GPUs with very intuitive examples. In (Kirk and Hwu 2013) is presented the history of GPU computing as well the main concepts of parallel programming in GPU, some applications show the development process in the use of GPUs.

In the field of mathematical optimization, many authors are dedicating their efforts on bringing some of the most widely used algorithms to the CUDA platform. In (Robilliard et al. 2009) was presented one of the earlier works on Genetic Programming in GPUs. There the authors provided a faster evaluation process and thus improved the general performance of the algorithm. In (Weiss 2011) the methodology needed in order to parallelize Ant Colony Optimization Algorithm is shown.

Pallipuram et al. (2012) compares the two most popular GPU programming models (CUDA and OpenCL) and two GPU architectures from different vendors (NVIDIAS's Fermi and AMD/ATI's Radeon 5870), using a two-level character recognition network developed employing four spiking neural network models.

Another field of application is the real time simulation and graphical representation of physical and chemical processes. Venetillo and Celes (2007) isone of the earliest works that simulates particles in confined environments, including support for inter-particle collisions, constraints, and particle–obstacle collisions, and also shows data sets of thousands of particles whereas maintaining a great number of frames per second.

In (Faber 2011) the author provides a compendium of algorithm implementations for machine learning, evolutionary computation and computer vision methods, maximizing the benefit of GPU architecture.

As can be seen, CUDA has been used in many different fields of research, and every day additional fields of study and application are emerging. This reflects the great success of the GPU programming. Therefore, in this chapter we will give a general overview of CUDA, its hardware and programming model. Also, a detailed explanation of the CUDA implementation for the Mandelbrot fractal generation is given, as an example of application.

4.2 CUDA: Compute Unified Device Architecture

CUDA™, introduced on November 2006 by NVIDIA, is a parallel computing architecture for general purpose as well as a parallel programming model that enables the use of NVIDIA GPUs in the solution of many complex problems in a more efficient way than using a CPU (NVIDIA 2012b). These applications are known as General-Purpose computing on Graphics Processing Units (GPGPU). CUDA allows developers to use C as a high-level programming language for GPUs.

4.2.1 Programming model

The heterogeneous CUDA programming model enables the use of a GPU as a co-processor of the CPU. In this context, the GPU is called the device and the CPU is called the host. A CUDA program is composed of sequential code sections for the host and parallel code sections for the device. In the parallel code sections, thousands of CUDA threads are executed concurrently on the GPU in order to reduce the overall computation time (NVIDIA 2012b).

The base of the CUDA programming model is the transparent scalability of applications through GPUs with different number of cores. Such scalability is based on the following three abstractions:

- Hierarchy of thread groups
- Shared memory administration and
- Barrier synchronization.

The above three key points provide the developer a methodology, in which the task at hand can be separated into sub-problems that can be solved independently in parallel by thread blocks, and each one of these thread blocks contains a specific number of CUDA threads. The thread blocks can be computed independently in the available GPU cores, in any order and in parallel or sequential manner. This independence of the thread block execution is the characteristic that allows the scalability of the CUDA applications.

The threads within a block are able to cooperate and communicate when solving each sub-problem while conserving the trait of automatic scalability. A compiled CUDA program can be executed in any GPU with any given number of multiprocessors, being only the host system responsible to know the physical amount of multiprocessors (NVIDIA 2012b).

4.2.2 Kernel functions

Kernels are CUDA C subroutines that, when they are called, are executed in parallel N number of times by N different CUDA threads. A kernel is defined with the specific keyword __global__. This keyword denotes that the function is a kernel and that once it is called from the host, it will generate a user-specified number of threads. The threads generated by a kernel are grouped in thread blocks, and the total of thread blocks within a kernel is called a grid. The threads within a block and also the blocks within a grid can be arranged in a 1-D, 2-D or 3-D manner, based upon the type of application.

Independently of the described threads / blocks hierarchy, at a hardware level, the threads created by a kernel are executed in groups of 32 threads, called warps, in a similar manner as the Single-Program Multiple-Data (SPMD) parallel programming style. Therefore, the code of a kernel runs simultaneously on many warps with different input data for each thread in order to obtain faster results (NVIDIA 2012a).

The kernel calls are asynchronous, which implies that after the invocation of a kernel, the host can execute the rest of the sequential code or just wait for the termination of the kernel execution.

4.2.3 Thread hierarchy

Because all threads in a grid execute the same kernel code, they require a set of unique coordinates (thread ID) to identify the appropriate portion of the data they need to process. The reserved keywords **threadIdx.x**, **threadIdx.y** and **threadIdx.z** refer to the indices of the thread for 1-, 2- and 3-dimensions respectively, allowing a multi-dimensional thread organization within a block. Likewise, the reserved keywords **blockIdx.x**, **blockIdx.y** and **blockIdx.z** are the indices of a block for a multi-dimensional block organization within a grid. All these keywords are built-in variables that are initialized by the runtime system (NVIDIA 2012b).

Their values can be accessed within the kernel functions, returning them to form the coordinates of the thread.

The maximum number of threads per block is determined by the compute capability, which is the characteristics set, both hardware and software features, of the GPU. The compute capability also determines the maximum grid size, the maximum x-, y- or z-dimension of a block and many other characteristics that must be taken into account by the programmer, because they affect the data mapping and the parameters of the kernel. In this chapter we consider the compute capability 3.5, which is the latest version available (NVIDIA 2012a).

The total size of a block is limited to 1024 threads, being the dimensional distribution of these threads customizable as long as the total number of them does not exceed the limit. Valid examples are (1024, 1, 1) or (32, 32, 1); a non-valid definition would be (32, 32, 2) since the total number of threads would be 2048. The same reasoning applies to blocks distribution within a grid. Figure 4.1 illustrates an example of how a thread is identified. The blocks in the grid are organized in a 2-dimensional array. Inside each block, the threads are arranged in a 3-dimensional array. In this example, the block is organized into 5x3x2 arrays of threads, giving a total of 30 threads per block. For the particular case of thread (4, 2, 0), its coordinates are given by threadIdx.x = 4, threadIdx.y = 2, and threadIdx.z = 0. This example has six blocks within the grid, and 30 threads per block giving a grand total of 180 threads in the grid.

CUDA program

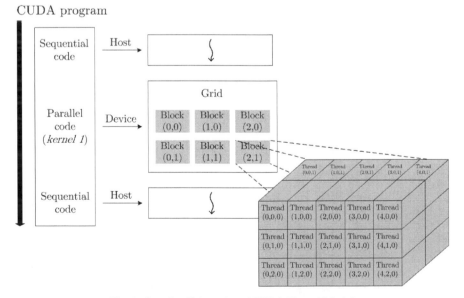

Fig. 4.1 Levels of hierarchy of CUDA Thread Model.

The number of threads per block, the quantity of blocks within a grid and their respective array are defined by parameters given at the kernel launch. These parameters are `dimBlock` and `dimGrid`. Figure 4.2 shows the corresponding kernel definition for the array showed in Fig. 4.1. Note that the z-value for `dimGrid` is ignored as grids are only 2-D arrays.

```
dim3 dimBlock(5, 3, 2);
dim3 dimGrid(3, 2);
Kernel1<<<dimGrid, dimBlock>>>(...);
```

Fig. 4.2 Defining dimBlock and dimGrid parameters.

4.2.4 Memory spaces and hierarchy

CUDA threads work with data that must be transferred from the host memory to the device memory, called global memory. From the device memory, each thread can access the appropriate section of data that it requires by using its thread ID. However, there are other memory spaces inside the GPU that have a specific purpose.

As stated, global memory is a general purpose memory. It is the main and biggest memory space in the device, but is also the slower one. It can be accessed by any thread and by the host.

The registers are the fastest memory in the GPU. They are used to store frequently used variables by the threads. They are located on-chip, i.e., physically near the CUDA cores.

The local memory is located inside the global memory, so it is also very slow. It is used to store local variables for each thread.

The constant memory space is a read-only memory that stores variables that remain constant through all the application lifetime. It is located on the global memory but has a cache memory space on-chip.

Shared memory is an on-chip memory space visible only to the threads within its block and it has the same lifetime as the corresponding block. Because it is significantly faster than global memory, shared memory is used to reduce the bottleneck that the global memory access represents.

Textures are read-only memories located in the global memory. They are represented by 2-dimensional arrays which are used in visualization operations and in applications where specific memory accesses are needed. As constant memory, it is cached on-chip.

Figure 4.3 shows the described memory hierarchy for the CUDA programming model. An important rule to optimize a GPGPU application is to minimize the data transfer operations between the CPU and the GPU, because these memory operations have a lower memory bandwidth than

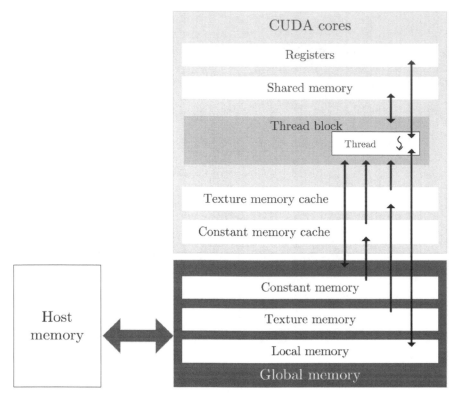

Fig. 4.3 CUDA memory model and hierarchy.

the internal transfers within the GPU (NVIDIA 2012a). It is also important to minimize the kernel access to the global memory and maximize the use of the shared memory and the other specialized memory spaces.

4.3 CUDA Programming: Fractal Generation and Display

In the CUDA toolkit version 5.0, several excellent examples are provided, which helps to new CUDA programmers comprehend the many virtues of GPU programming. It is highly recommended heading over to the official website and downloads this development package (https://developer. NVidia.com/cuda-toolkit). As a case of study in this chapter, we will use the Mandelbrot example that was chosen due to its clear distinction between utilizing a GPU and a CPU for computational intensive tasks such as an iterative algorithm. Additionally, this example also implements various parallel computing concepts, such as the utilization of a global memory from which all threads work from.

The task at hand consists of the calculation of the Mandelbrot set depicted in Fig. 4.4. The program does this by iterating values, which are evaluated by a Mandelbrot function. From here, the CUDA implementation consists of launching a fixed number of CUDA blocks corresponding to the maximum number of blocks allowed by CUDA, and having each one of those blocks iterate through the available Mandelbrot's image blocks (the image is partitioned into blocks or sections) until the task is finished.

The iterative process is done through a counter stored in the global memory; the counter is incremented by each worker (CUDA) block as its job is done.

The programming code is divided in two sections: the code that runs on the CPU or Host and the code that runs on the GPU or Device. The initialization code, parameter definition and starting values are always defined in the host code, while the task itself is executed on the device (GPU).

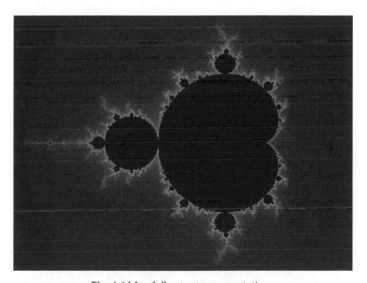

Fig. 4.4 Mandelbrot set representation.

Color image of this figure appears in the color plate section at the end of the book.

The code follows the next main steps:

- **Initialization and parameter definition:** In this step, we take the time to perform sanity checks over our hardware, including but not limited to the existence of a CUDA device and what version of the hardware is installed. This step is performed in the host. The used Key functions are:

—findCudaDevice: It allows to detect, if any, CUDA devices attached to the computer. Multiple devices can be found and they can be used at a given time.

—checkCudaCapabilities: It allows to detect what major and minor version of CUDA the GPU is capable of running.

—chooseCudaDevice: It allows to choose which CUDA device attached to the CPU will be used.

- **Block and thread distribution:** In this step, we require to planify how we will divide our workload between the available hardware resources, mainly the number of blocks and threads per block. Both parameters are defined by our hardware; additionally, the thread allocation must comply with the CUDA thread model, described in a previous section. Because the generation of this particular fractal is in 2D space, we are able to evenly divide the work in areas of the image as grids, as it is shown in the Fig. 4.5. Once the workload is divided, we specify CUDA code to handle each of this grid blocks with the use of the build in variables threadIdx and blockIdx.

- **Allocate and transfer any necessary data from host to device:** In order for the device to work with specific data, it must be declared and transferred from the host to the device. This is done mainly with

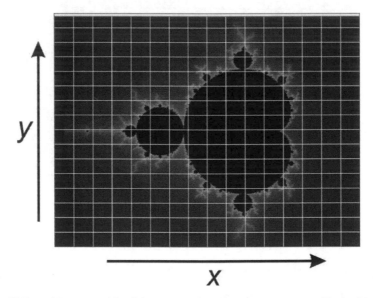

Fig. 4.5 2D Fractal Image workload division. Each grid section represents a block of data that will be processed by each worker thread. If the image size is 800x600 pixels, this will generate a grid of blocks of 50x37.5 pixels, rounded up to 38 for a total of 11900 blocks. The excess pixels are denoted with yellow and are not calculated as we are out of bounds.

Color image of this figure appears in the color plate section at the end of the book.

the following function, in our particular case; we only require using a variable to store our current global block position.

—cudaMemcpyToSymbol: It allows to either copy from the host to the device (use of cudaMemcpyHostToDevice parameter), or form the device to the host (use of cudaMemcpyDeviceToHost parameter) an allocated memory space.

- **Invoking and running the device code:** Calling the CUDA function is straightforward, once we know the number of blocks and the thread model we are using. The function must contain the reserved word __global__ before its definition and must be called using the format functionName<<number of blocks, threads per block>>(parameters).

4.3.1 Code analysis

The function RunMandelbrot0 is executed on the host, and it is responsible to plan the workload distribution described in the previous section. It does so, by creating two variables of type dim3 named threads and grid as it is shown in Fig. 4.6, which hold x, y and z values for the thread and grid distribution, respectively. In this example, each CUDA block has a size of 16x16x1 threads, which are defined in the variables blockdim_x_dynamic and blockdim_y_dynamic. Ideally the workload of rendering the whole image should be divided equally; however, depending on the width and height of the image this may not give a rounded number. In this case, we need another block to work with and for this reason, a helper function called iDivUp is used to determine if the division between the width, height and the block dimensions require to be rounded up, and creating an additional block layer which increases the grid size by 1 as shown in Fig. 4.6, displaying the fractal image workload division. This can be done using a three dimensional variable of type dim3 named as grid using the x and y values of the fractal image.

```
dim3 threads(blockdim_x_dynamic, blockdim_y_dynamic);
dim3 grid(iDivUp(imageW, blockdim_x_dynamic),
iDivUp(imageH, blockdim_y_dynamic));
```

Fig. 4.6 Variables of type dim3 which hold the thread and grid distribution.

Next, a variable named as hBlockCounter, that will help us to keep track of the image grid block position while calculating the fractal, is defined. We must pass a reference to the CUDA device with the function cudaMemcpyToSymbol with a reference to the variable hBlockCounter, by using the function cudaMemcpyHostToDevice, where a CUDA block should be working on; the code section is shown in Fig. 4.7.

```
unsigned int hBlockCounter = 0;
checkCudaErrors(cudaMemcpyToSymbol(blockCount
er, &hBlockCounter, sizeof(unsigned int), 0,
cudaMemcpyHostToDevice));
int numWorkerBlocks = numSMs;
```

Fig. 4.7 Definition of variables that keep track of the grid block position while calculating the fractal.

Once all the parameters were declared and specified, we run the main CUDA thread called Mandelbrot0, see Fig. 4.8, by utilizing the correct syntaxes containing the number of CUDA blocks, in which the function will run and the number of threads per block in the style of functionName<return value><<<Number of Blocks, Number of threads>>>(parameters).

```
Mandelbrot0<float><<<numWorkerBlocks, threads>>>(dst,
    imageW, imageH, crunch,
    (float)xOff, (float)yOff,(float)xjp, (float)yjp, (float)
    scale, colors,
    frame, animationFrame,
    grid.x, grid.x *grid.y, isJ);
```

Fig. 4.8 Mandelbrot main CUDA thread function.

Figure 4.9 shows variables names, we start by declaring the variables called the blockIndex, blockX and blockY, which will hold the block grid positioning information.

```
__shared__ unsigned int blockIndex;
__shared__ unsigned int blockX, blockY;
```

Fig. 4.9 Mandelbrot grid variables that hold block grid positioning information.

The next step is to start a "while structure instruction block", which will cycle until all the grid blocks of the image are calculated. Additionally, we must determine the grid block index which we will use. To do this, we need to implement a condition to check if we are working with the first thread of the CUDA block, and in such case, we proceed to calculate the blockIndex by using our global block counter variable blockCounter and an atomic[1] add function. Hence, to determine the x and y coordinates of the block is as simple as applying the modulus operator to the blockIndex and the grid width to get the x coordinate and blockIndex divided by the grid width as well, see Fig. 4.10.

---

[1] The term "atomic" in this context defines a type of operations which are performed without interference from any other threads. Atomic operations are often used to prevent race conditions which are common points of conflict in multithreaded applications.

```
while (1)
    {
        if ((threadIdx.x==0) && (threadIdx.y==0))
        {
            blockIndex = atomicAdd(&blockCounter, 1);
            blockX = blockIndex % gridWidth;
            blockY = blockIndex / gridWidth;
        }
```

Fig. 4.10 Main Mandelbrot loop determining which block of the fractal is to be calculated by the thread.

Figure 4.11 shows the procedure where the threads are synchronized using the build in function __syncthreads(), after that, we check for the global stop criteria of the blockIndex being equal or higher than the maximum amount of blocks of the grid.

```
__syncthreads();

if (blockIndex >= numBlocks)
{
    break;     // finish
}
```

Fig. 4.11 Synchronization and stop condition.

If the stop criteria are not met, we continue with determining the pixel of the image that we will calculate next. This is done by using the build in variables blockDim, threadIdx and the previously calculated blockX and blocky, as it is shown in Fig. 4.12.

```
const int ix = blockDim.x * blockX + threadIdx.x;
const int iy = blockDim.y * blockY + threadIdx.y;
```

Fig. 4.12 Calculating the pixel of the image to be computed.

Figure 4.13 depicts an additional checkpoint to see if we have finished with the grid block work, and if not, we continue to determine the real position of the image based on scaling and offset variables determined by the GUI.

```
if ((ix < imageW) && (iy < imageH))
{
    // Calculate the location
    const T xPos = (T)ix * scale + xOff;
    const T yPos = (T)iy * scale + yOff;
```

Fig. 4.13 Additional check point to see if we have finished with the grid block work.

With this information we can finally calculate the Mandelbrot index using the function presented in Fig. 4.14.

```
int m = CalcMandelbrot<T>(xPos, yPos, xJP, yJP, crunch,
isJ);
```

Fig. 4.14 Calling the function that calculates the Mandelbrot index.

The main Mandelbrot function in charge of generating the fractal values is shown in Fig. 4.15, this function is utilized by the GPU and CPU implementations, it does not implement a sequential or parallel strategies when computing the new fractal values, this can be used in a sequential or parallel form by using GPU and CPU devices.

```
// The core Mandelbrot calculation function
inline int CalcMandelbrot(const float xPos, const float yPos,
const float xJParam, const float yJParam, const int crunch)
{
    float x, y, xx, yy, xC, yC ;
    xC = xPos ;
    yC = yPos ;
    y = 0 ;
    x = 0 ;
    yy = 0 ;
    xx = 0 ;
    int i = crunch;

    while (--i && (xx + yy < 4.0f))
    {
        y = x * y * 2.0f + yC ;
        x = xx - yy + xC ;
        yy = y * y;
        xx = x * x;
    }
    return i;
} // CalcMandelbrot
```

Fig. 4.15 Main Mandelbrot function in charge of generating the fractal values.

The Mandelbrot set is defined by the iterated complex Equation (4.3.1)

$$P_c = z \rightarrow z^2 + c \qquad (4.3.1)$$

where:

c: pixel coordinates (complex)
z: begins at 0

The Mandelbrot set were created by changing the scale to render for the Eq. (4.3.1) which uses only two adjustable complex parameters z and c, that are plotted on the horizontal and vertical axes of the page to draw the shape. Each time we apply the equation, the previous value for z is taken (z is initialized with 0 at the beginning of the program), square it, and add a fixed constant c, i.e., when $c=1$, this gives the resulting sequence c, $c^2 + c$, $(c^2 + c)^2 + c$, ...; the sequence tends to infinity and we can see that this iteration is unbounded, so 1 is not an element of the Mandelbort set. When $c = i$, the sequence is now bounded because it is part of a complex plane, so i is part of the Mandelbrot set and this is the main reason why the `CalcMandelbrot` function return i as shown in Fig. 4.15. In this implementation, each pixel represents a point in the 2D complex plane, giving the familiar Mandelbrot shape.

Figure 4.16 illustrates the final step to convert the Mandelbrot index into a color for its graphical representation on the display.

```
if (m)
{
    m += animationFrame;
    color.x = m * colors.x;
    color.y = m * colors.y;
    color.z = m * colors.z;
}
else
{
    color.x = 0;
    color.y = 0;
    color.z = 0;
}
// Output the pixel
int pixel = imageW * iy + ix;
if (frame == 0)
{
    color.w = 0;
    dst[pixel] = color;
}
else
{
    int frame1 = frame + 1;int frame2 = frame1 / 2;
    dst[pixel].x = (dst[pixel].x * frame + color.x + frame2)/frame1;
    dst[pixel].y = (dst[pixel].y * frame + color.y + frame2)/frame1;
    dst[pixel].z = (dst[pixel].z * frame + color.z + frame2)/frame1;
}
```

Fig. 4.16 Code that assigns a color to the Mandelbrot index.

4.3.2 Benchmarking GPU vs. CPU implementation

The Mandelbrot fractal code was tested on an NVIDIA GeForce 9500M GS GPU with an Intel Core 2 Duo 2.5 GHz T9300 CPU. CUDA configuration is four blocks with 16x16x1 threads per block. Given the graphical representation of the Mandelbrot fractal, measuring the Frames Per Second (FPS) is a valid reference point to determine the performance. The testing process consists of running the code using different resolutions and recording the behavior of the software considering two conditions: static rendering and rendering while zooming/panning. Table 4.1 shows a summary of test results.

Table 4.1 Concentrated benchmarking values for the Mandelbrot algorithm, reported in Frames Per Second (FPS).

Image size (w x h pixels)	Static FPS		Zoom/Panning FPS	
	GPU	CPU	GPU	CPU
640x480	19	3	17	2
800x600	14	2	12	1
1024x768	9	1	8	.5
1280x720	6	.5	4	.3
1440x900	4	.3	2	.2
1920x1080	3	.1	2	.1

Clearly the GPU implementation of the Mandelbrot algorithm has much better performance than a CPU implementation in all the tested resolutions.

4.4 Conclusions

In this chapter, we saw an overview of the CUDA programming model, and we presented a concise example that shows the power of CUDA for computational intensive tasks such computing the Mandelbrot fractal in real time.

The experimental results of comparing the GPU and CPU implementation of the Mandelbrot algorithm are very straightforward, showing great performance increase when utilizing the GPU for the primary calculations. This result gives us validation of the benefits of GPU for solving other computing intensive tasks, such as Ray Tracing, NP combinatorial problems, real-time simulations, and a great number of intensive algorithms needed to be computed for medical applications, industry and educational applications, where scenes are often complex, and require complex spatial data structures.

Acknowledgments

The authors would like to thank the "Instituto Politécnico Nacional (IPN)", "Comisión de Operación y Fomento de Actividades Académicas del IPN (COFAA-IPN)" and the Mexico's entity "Consejo Nacional de Ciencia y Tecnología (CONACYT)" for supporting our research activities.

References

Faber, R. (2011). CUDA Aplication Design and Development. Morgan Kaufmann-Elsevier, United States.

García-Risueño, P. and P.E. Ibáñez (2012). A review of High Performance Computing foundations for scientists. Cornell University Library, pp. 1–33.

Kirk, D.B. and W.-m. W. Hwu (2013). Programming Massively Parallel Processors A Hands-on Approach (Second ed.). Elsevier. Waltham, MA, USA.

NVIDIA (2013). Retrieved 2013, from NVIDIA: http://docs.nvidia.com/cuda/pdf/CUDA_C_Best_Practices_Guide.pdf

NVIDIA (2013). Retrieved 2013, from NVIDIA: http://docs.nvidia.com/cuda/pdf/CUDA_C_Programming_Guide.pdf

Pallipuram, V., M. Bhuiyan and M. Smith (2012). A comparative study of GPU programming models and architectures using neural networks. The Journal of Supercomputing, Springer-Verlag, 61: 673–718.

Robilliard, D., V. Marion-Poty and C. Fonlup (2009). Genetic programming on graphics processing units. Genetic Programming and Evolvable Machines. Kluwer Academic Publishers Hingham, MA, USA ,447–471.

Sanders, J. and E. Kandrot (2010). CUDA by Example: An introduction to General-Purpose GPU Programminig. Addison-Wesley. Boston, MA. USA.

Venetillo, J.S. and W. Celes (2007). GPU-based particle simulation with inter-collisions. The Visual Computer, Springer-Verlag, 851–860.

Weiss, R.M. (2011). GPU-Accelerated Ant Colony Optimization. In: W.-M.W. Hwu, GPU Computing Gems Esmerald Edition. MK-Elsevier, Burlington, MA. USA, pp. 325–342.

5

GPU-based Conjugated Gradient Solution for Phase Field Models

*Juan J. Tapia** and *Rigoberto Alvarado*

ABSTRACT

This chapter presents a GPU-based solution for the Allen-Cahn differential equation which describes a phase field model. The proposed algorithm is based in the conjugate gradient method. CUDA and its libraries CUSPARSE and CUBLAS for sparse matrix operations and linear algebra operations respectively, are used to implement the solution. The proposed implementation uses the Fletcher-Reeves version of the nonlinear conjugate gradient method.

5.1 Introduction

The nonlinear problems are of especial interest in the scientific field because most physical systems and phenomena, e.g., chemical reactions, ecology, biomechanics, population growth, and multi-phase or multi-component fluids phenomena, are nonlinear in nature. The impact of a droplet on a solid surface, nuclear safety, petroleum engineering, and combustion or reaction flows are examples of multi-component and multi-phase fluids. From the point of view of traditional fluid dynamics, such fluids are treated as sharp

Instituto Politécnico Nacional, CITEDI, Tijuana, México.
  Email: jtapiaa@ipn.mx
* Corresponding author

interfaces on which a set of interfacial balance conditions must be imposed. However, this kind of mathematical modeling and its numerical solutions are difficult to develop due to the inherent nonlinearities, topological changes, and the complexity of dealing with unknown moving interfaces present in the fluids. Phase field models are an alternative approach for solving the interface problems and are used to describe and simulate this kind of fluids. In a phase field model, the interfaces are described by a mixing energy, so they are implicitly captured in a functional.

A nonlinear Partial Differential Equation (PDE) can be solved using an explicit method or an implicit one. The explicit method involves the iterative solution of an equation; the implicit method involves the iterative solution of a nonlinear system. Generally, the implicit methods offer better numerical stability than the explicit methods and allow the use of bigger time steps in the numerical solutions. For these reasons, an ever increasing proportion of modern scientific research in the nonlinear field is devoted to the design of implicit solutions for nonlinear PDE.

The conjugate gradient (CG) algorithm (M.R. Hestenes 1952) is an efficient method for solving linear systems, and it can be extended to nonlinear systems. Unlike other methods, such as Gauss-Seidel, it can be parallelized and it is highly efficient with large sparse matrices. These characteristics allow the development of CG parallel algorithms that solve large sparse linear and nonlinear systems, in particular, GPU-based algorithms that allow a faster and reliable solution.

In (Garcia 2010), a GPU-based algorithm based on the biconjugate gradient method for the solution of a power flow formulation is presented. Michels et al. derived a GPU sparse matrix solver for the heat equation based on the CG algorithm (Michels 2011). In (Galiano 2012), a GPU-based algorithm that uses a nonlinear preconditioned conjugate gradient method is derived.

In this work, we focus on a CG parallel solution for the Allen-Cahn (AC) equation via a GPU-based algorithm. Our approach is based on the sparse matrix library CUSPARSE and the basic linear algebra library CUBLAS, both supported by CUDA.

In Section 2, an introduction to the CG algorithm is presented. The theory of the phase field models, the discretization of the Allen-Cahn equation, and the GPU tools used are discussed in Section 3. The results are presented in Section 4 and finally, these results are discussed in Section 5, where some ideas for future work are highlighted.

5.2 Conjugate Gradient Method

The Conjugate Gradient method is the most widely used iterative method for solving sparse systems of linear equations that arise from the discretization of PDEs. The CG algorithm is based on the Steepest Descent (SD) method and in the generic method of Conjugate Directions (CD).

5.2.1 Solution of a linear system as an optimization problem

Suppose a system of the form

$$Ax = b \tag{5.2.1}$$

where x is an unknown $N \times 1$ vector, b is a known $N \times 1$ vector and A is a known square, symmetric positive definite (SPD) $N \times N$ matrix. The discretization of a PDE with the finite difference or finite element methods produces a system like Eq. (5.2.1). In these systems, A is the stiffness matrix, x is the unknown solution, and b is the information of the problem.

A quadratic form is a quadratic function with the form

$$f(x) = \frac{1}{2}x^T A x - b^T x + c \tag{5.2.2}$$

where A is an $N \times N$ symmetric positive definite matrix, x and b are $N \times 1$ vectors and c is a scalar. As matrix A is PD, the surface defined by $f(x)$ is shaped like a paraboloid bowl. The paraboloid form of $f(x)$ for a system with $N = 2$ is shown in Fig. 5.1. The gradient of (5.2.2) is a vector field that for a given point x, points in the direction of greatest increase of $f(x)$. At

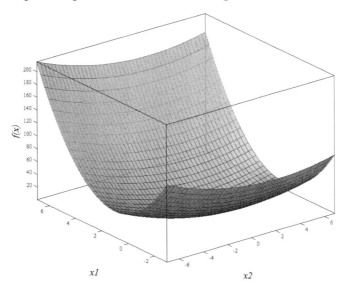

Fig. 5.1 Graph of the quadratic form $f(x)$ with a positive definite matrix.
Color image of this figure appears in the color plate section at the end of the book.

the bottom of the paraboloid bowl, the gradient is zero; at this point, $f(x)$ is minimized (Edwin and Chong 2001).

The derivative of $f(x)$ is

$$f'(x) = \frac{1}{2}A^T x + \frac{1}{2}Ax - b \qquad (5.2.3)$$

and if A is symmetric, Eq. (5.2.3) reduces to

$$f'(x) = Ax - b. \qquad (5.2.4)$$

By calculus, the minimal for $f(x)$ is found by setting $f'(x) = 0$ and this produces the linear system of eq. (5.2.1) from Eq. (5.2.4). Therefore, the solution of the system $Ax = b$ is a critical point of $f(x)$. If A is SPD, this critical point is a minimum of $f(x)$. So Eq. (5.2.1) can be solved by finding an x that minimizes $f(x)$ (Andreas Antoniou 2007).

5.2.2 Steepest Descent (SD) method

The SD method is an algorithm to find the nearest local minimum of a function which presupposes that its gradient can be computed. Considering the quadratic function $f(x)$ shown in Fig. 5.1, the SD algorithm starts at an arbitrary point x_0 over the surface and takes as many steps x_{i+1} as needed to slide down to the bottom of the paraboloid.

At each step, the direction in which $f(x)$ decreases most rapidly is chosen. This direction is opposite to $f'(x_i)$. From Eq. (5.2.4), this direction is

$$-f'(x_i) = b - Ax_i. \qquad (5.2.5)$$

The error $e_i = x_i - x$ is a vector that indicates how far the current point is from the real answer. The residual vector $r_i = b - Ax_i$ establishes the relation between the vector b and the vector Ax. When $r_i = 0$, the solution of the system (5.2.1) is reached. The residual could be written as

$$r_i = -Ae_i$$
$$r_i = -f'(x_i)' \qquad (5.2.6)$$

therefore, the residual r_i is the direction of the steepest descent, i.e., the direction that each step x_i takes to reach the bottom of the paraboloid.

Based on x_0, the next point x_1 is given by

$$x_1 = x_0 + \alpha r_0, \qquad (5.2.7)$$

where α is the step size, and r_0 is the last steepest descent direction. The value α is chosen by line search, which is a procedure that selects the α value that minimizes $f(x)$ along the steepest direction (Edwin and Chong 2001).

Figure 5.2 shows an arbitrary x_0 point and its steepest direction represented as a plane. The bottommost point of the intersection between the vertical plane and the paraboloid surface is the target, so α should be chosen to reach that point.

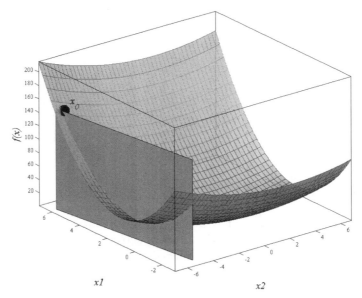

Fig. 5.2 The α value finds the point of the intersection of the surfaces that minimizes *f(x)*.

Color image of this figure appears in the color plate section at the end of the book.

By calculus, α minimizes $f(x)$ when the directional derivative equals to zero, i.e.,

$$\frac{df(x_1)}{d\alpha} = 0. \tag{5.2.8}$$

Applying the chain rule

$$\frac{df(x_1)}{d\alpha} = f'(x_1)^T \frac{d(x_1)}{d\alpha} \tag{5.2.9}$$

and considering $x_1 = x_0 + \alpha r_0$,

$$\frac{df(x_1)}{d\alpha} = f'(x_1)^T r_0. \tag{5.2.10}$$

Setting

$$f'(x_1)^T r_0 = 0 \tag{5.2.11}$$

it is clear that α should be chosen so that $f'(x_1)^T$ and r_0 are orthogonal. From Eqs. (5.2.5) and (5.2.6)

$$f'(x_1) = -r_1 \tag{5.2.12}$$

and replacing Eq. (5.2.12) in Eq. (5.2.11)

$$-r_1^T r_0 = 0. \tag{5.2.13}$$

Considering $-r_1 = Ax_1 - b$

$$(Ax_1 - b)^T r_0 = 0, \tag{5.2.14}$$

and substituting Eq. (5.2.7) in Eq. (5.2.14)

$$(A(x_0 + \alpha r_0) - b)^T r_0 = 0. \tag{5.2.15}$$

Rearranging terms

$$\alpha(Ar_0)^T r_0 = (b - Ax_0)^T r_0 \tag{5.2.16}$$

and considering $r_0 = b - Ax_0$

$$\alpha(Ar_0)^T r_0 = r_0^T r_0. \tag{5.2.17}$$

The equation for the step size is obtained by clearing α

$$\alpha = \frac{r_0^T r_0}{r_0^T Ar_0} \tag{5.2.18}$$

Therefore, the SD method consists of a loop made of three fundamental steps given an initial value x_0: first, the residual, i.e., the steepest descent direction is calculated by

$$r_i = b - Ax_i, \tag{5.2.19}$$

second, the step size α is computed with

$$\alpha_i = \frac{r_i^T r_i}{r_i^T Ar_i}, \tag{5.2.20}$$

and finally, the solution is updated by

$$x_{i+1} = x_i + \alpha_i r_i. \tag{5.2.21}$$

The SD cycle continues until convergence, which is reached when the residual equals to zero. Through the iterative process, the matrix A and the vector b remain unchanged. The SD process produces a zigzag path at the paraboloid surface because each gradient is orthogonal to the previous gradient. This characteristic path produced by the orthogonality of the steepest descent directions is known as the 'ping pong effect', and is illustrated in the contour map shown in Fig. 5.3. The contour map corresponds to the function $f(x)$ of Fig. 5.1.

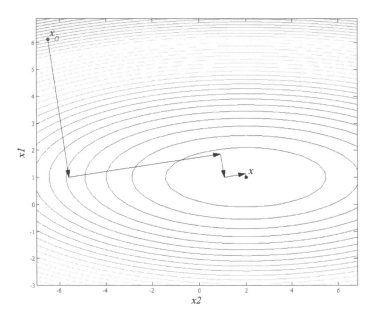

Fig. 5.3 Zigzag path produced by the SD method.

Color image of this figure appears in the color plate section at the end of the book.

The SD method as described, involves two matrix-vector products [Eqs. (5.2.19) and (5.2.20)]; however, the product Ax can be eliminated to reduce the computational load of the algorithm. As $e_i = x_i - x$ and $x_{i+1} = x_i + \alpha_i r_i$, then $e_{i+1} = e_i + \alpha_i r_i$. Substituting e_{i+1} in the residual

$$r_{i+1} = -Ae_{i+1},$$ (5.2.22)

an alternatively definition for r_{i+1} is obtained. Thus, Eq. (5.2.19) is used to compute r_0 outside the SD loop, and the equation

$$r_{i+1} = r_i - \alpha_i A r_i$$ (5.2.23)

is used to calculate r_i for every iteration thereafter. Accordingly, the vector Ar, which appears in Eqs. (5.2.20) and (5.2.23), need only be computed once per iteration (Jorge Nocedal 2006).

5.2.3 Conjugate directions method

As a consequence of the 'ping pong effect', many steps in the SD method are taken in the same direction as earlier steps (see Fig. 5.3). This characteristic is reflected in the slow convergence of the SD method.

A better approach is to propose a set of orthogonal search directions $d_0, d_1, ..., d_{N-1}$ in which for each search direction d_i, only one step is taken. This concept is illustrated in Fig. 5.4. The first step leads to the correct x_2 -coordinate and the second step reaches the target. It can be seen that d_0 and e_1 are orthogonals.

The expression to define the iterative steps is

$$x_{i+1} = x_i + \alpha_i d_i. \tag{5.2.24}$$

In order to define α_i, the fact that e_{i+1} should be orthogonal to d_i is considered, so that only one step will be taken for a given d_i direction (Edwin and Chong 2001). According to this condition,

$$d_i^T e_{i+1} = 0, \tag{5.2.25}$$

which is the orthogonality condition for the d_i and e_{i+1} vectors. Since $e_{i+1} = x_{i+1} - x$ and $x_{i+1} = x_i + \alpha_i d_i$, $e_{i+1} = e_i + \alpha_i d_i$ so

$$d_i^T (e_i + \alpha_i d_i) = 0. \tag{5.2.26}$$

Rearranging terms and clearing α_i, we get

$$\alpha_i = -\frac{d_i^T e_i}{d_i^T d_i}. \tag{5.2.27}$$

Equation (5.2.27) cannot be computed because the term e_i is unknown as it depends on the actual solution, so the scheme shown in Fig. 5.4 cannot be used. To overcome this problem, the search directions are considered A-orthogonal (or conjugated with respect to A) instead of orthogonal.

A set of nonzero vectors $\{d_0, d_1, ..., d_{N-1}\}$ is said to be A-orthogonal or conjugated with respect to the SPD matrix A if

$$d_i^T A d_j = 0 \tag{5.2.28}$$

for all $i \neq j$. Any set of vectors satisfying this property is also linear independent (Andreas Antoniou 2007).

So, the new requirement and better choice of search directions is to choose directions that are A-orthogonal instead of orthogonal. In other words, d_i must be A-orthogonal to e_{i+1}. This orthogonality condition is equivalent to finding the minimum point along the search direction d_i as in the steepest descent method.

Hence, the expression for α_i when the search directions are A-orthogonal could be derived from

$$d_i^T A e_{i+1} = 0. \tag{5.2.29}$$

As $e_{i+1} = e_i + \alpha_i d_i$
$$d_i^T A(e_i + \alpha_i d_i) = 0. \tag{5.2.30}$$

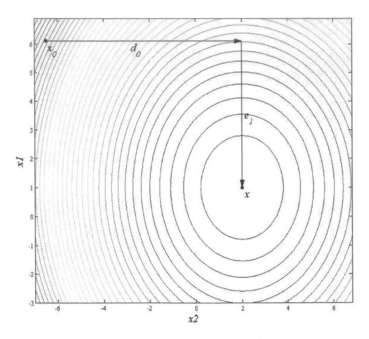

Fig. 5.4 Method of one-step per orthogonal direction.

Color image of this figure appears in the color plate section at the end of the book.

Rearranging the terms and clearing α, we get

$$\alpha_i = -\frac{d_i^T A e_i}{d_i^T A d_i}. \tag{5.2.31}$$

Replacing Eq. (5.2.6) in Eq. (5.2.31), we get

$$\alpha_i = \frac{d_i^T r_i}{d_i^T A d_i}. \tag{5.2.32}$$

Figure 5.5 shows the A-orthogonal vectors that minimize the problem presented in Fig. 5.3. If Fig. 5.5 is stretched until the ellipses appeared circular, the A-orthogonal vectors would then appear orthogonal, as in Fig. 5.4.

The contour map of Fig. 5.5, corresponds to the function $f(x)$ of Fig. 5.1. As illustrated, the method of conjugate directions converges in N steps. The first step is taken along a given direction d_0 and the point x_1 is chosen by the constraint that d_i must be A-orthogonal to e_{i+1}.

Unlike Eq. (5.2.27), α_i can be calculated with Eq. (5.2.32) because all terms are known. If the search vector d_i were the residual r_i, Eq. (5.2.32) turns into the formula used by the SD method, i.e., Eq. (5.2.20).

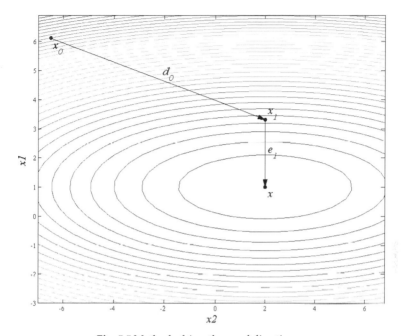

Fig. 5.5 Method of A-orthogonal directions.

Color image of this figure appears in the color plate section at the end of the book.

Considering their convergence rate, it is clear that the CD method is more efficient than the SD method. However, to implement the CD method, a way to generate a set of A-orthogonal search directions d_i is needed.

The conjugate Gram-Schmidt algorithm could be used to generate the A-orthogonals directions given an initial d_0, e.g., the coordinate axes (Andreas Antoniou 2007).

Let $\{u_0, u_1, ..., u_{N-1}\}$ be a set of linearly independent vectors. In order to construct d_i, the parts of u_i that are not A-orthogonal to any previous search directions need to be removed. The process is illustrated in Fig. 5.6, where it is shown the method to obtain d_1 from the linearly independent vectors u_0 and u_1.

First, set $d_0 = u_0$. The vector u_1 is composed of two components: $u_{||}$ which is parallel to vector u_0 and u^* which is A-orthogonal to u_0. After conjugation, only the A-orthogonal part remains, so $d_1 = u^*$. In other words, $d_1 = u_1 - u_{||}$.

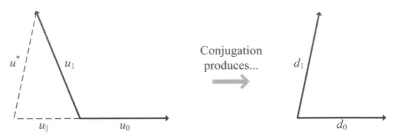

Fig. 5.6 Gram-Schmidt conjugation of two vectors.

Therefore, after setting $d_0 = u_0$, the search directions can be represented as

$$d_i = u_i + \sum_{k=0}^{i-1} \beta_{ik} d_k \tag{5.2.33}$$

where β is the length of the parallel component $u_{\parallel}$. The value β is found multiplying both sides of Eq. (5.2.33) by $d_j A$

$$d_i^T A d_j = u_i^T A d_j + \sum_{k=0}^{i-1} \beta_{ik} d_k^T A d_j. \tag{5.2.34}$$

Solving Eq. (5.2.34) for β_{ij}

$$\beta_{ij} = -\frac{u_i^T A d_j}{d_j^T A d_j}. \tag{5.2.35}$$

Therefore, the main drawback of the conjugate Gram-Schmidt method is that all the old search vectors must be stored to construct each new one, affecting directly the computational cost of the algorithm (Jorge Nocedal 2006).

5.2.4 Linear conjugate gradient method

To overcome the problem of the computational cost of the Gram-Schmidt conjugation but still use it as the method to generate the A-orthogonal directions, the conjugation of the residuals is used to construct the search directions d_i, i.e., $u_i = r_i$. In fact, the conjugation of the residuals gives the name to the CG method, as the residual is the gradient of $f(x)$, see Eq. (5.2.6) (Jorge Nocedal 2006).

The residual r_i is a good choice for u_i because it is orthogonal to the previous search directions. Thus, in the CG method each new residual r_i is orthogonal to all previous residuals, and each new search direction constructed from r_i is A-orthogonal to all previous search directions (Andreas Antoniou 2007).

Therefore, after computing r_0, the search directions can be represented as

$$d_i = r_i + \sum_{k=0}^{i-1} \beta_{ik} d_k. \tag{5.2.36}$$

Multiplying both sides of Eq. (5.2.36) by Ad_j, we get

$$d_i^T A d_j = r_i^T A d_j + \sum_{k=0}^{i-1} \beta_{ik} d_k^T A d_j. \tag{5.2.37}$$

Clearing β_{ij}

$$\beta_{ij} = -\frac{r_i^T A d_j}{d_j^T A d_j}. \tag{5.2.38}$$

From Eq. (5.2.22) and $e_{j+1} = e_j + \alpha_j d_j$,

$$r_{j+1} = r_j - \alpha_j A d_j. \tag{5.2.39}$$

Multiplying Eq. (5.2.39) by r_i

$$r_i^T r_{j+1} = r_i^T r_j - \alpha_j r_i^T A d_j \tag{5.2.40}$$

and clearing $r_i^T A d_j$

$$r_i^T A d_j = \frac{r_i^T r_j - r_i^T r_{j+1}}{\alpha_j}. \tag{5.2.41}$$

If $i = j$, then

$$r_i^T A d_j = \frac{r_i^T r_i}{\alpha_i} \tag{5.2.42}$$

due to the orthogonality of the residual vectors. If $j = i - 1$, then

$$r_i^T A d_j = \frac{-r_i^T r_i}{\alpha_{i-1}}. \tag{5.2.43}$$

For other cases, $r_i^T A d_j = 0$.

Replacing Eq. (5.2.41) with Eq. (5.2.38) and setting $j = i - 1$, we get

$$\beta_{ij} = \frac{r_i^T r_i}{\alpha_{i-1} d_{i-1}^T A d_{i-1}}. \tag{5.2.44}$$

Considering Eq. (5.2.32),

$$\beta_{ij} = \frac{r_i^T r_i}{d_{i-1}^T r_{i-1}}. \tag{5.2.45}$$

As $d_i^T r_i = u_i^T r_i$, and $u_i = r_i$, then for the CG method

$$\beta_i = \frac{r_i^T r_i}{r_{i-1}^T r_{i-1}} \tag{5.2.46}$$

because the sub-indexes i,j can be replaced by i. The Eq. (5.2.36) for search direction is defined now as

$$d_{i+1} = r_{i+1} + \beta_{i+1} d_i. \tag{5.2.47}$$

Therefore, the CG method consists in a loop made of five fundamental steps given an initial value x_0 (Edwin and Chong 2001): first, the residual, i.e., the search direction is calculated outside the loop by

$$r_0 = d_0 = b - Ax_0, \tag{5.2.48}$$

then, inside the loop, α, x, r, β and d values are computed respectively with

$$\alpha_i = \frac{r_i^T r_i}{d_i^T A d_i}, \tag{5.2.49}$$

$$x_{i+1} = x_i + \alpha_i d_i, \tag{5.2.50}$$

$$r_{i+1} = r_i - \alpha_i A d_i, \tag{5.2.51}$$

$$\beta_{i+1} = \frac{r_{i+1}^T r_{i+1}}{r_i^T r_i}, \tag{5.2.52}$$

$$d_{i+1} = r_{i+1} + \beta_{i+1} d_i. \tag{5.2.53}$$

Summarizing, by means of conjugacy, the CG method makes the SD direction have conjugacy, thus increasing the efficiency and reliability of the algorithm. If A is not symmetric, the CG will find a solution for the system

$$\frac{1}{2}(A^T + A)x = b. \tag{5.2.54}$$

5.2.5 Non-linear conjugate gradient method

The non-linear CG method generalizes the CG algorithm to nonlinear optimization. The Fletcher-Reeves version (Fletcher 1964) is considered the first non-linear CG method.

There are three main changes for the linear CG method: the computation of the residual r, the step size α and β.

In the non-linear CG method, the residual r is set to the negative gradient

$$r_i = -f(x_i). \tag{5.2.55}$$

Also, the line search parameter α that minimizes

$$f(x_i + \alpha_i d_i) \tag{5.2.56}$$

is found by ensuring that the gradient is orthogonal to the search direction. So, the Newton-Raphson method or secant method could be used to found α. By Newton-Raphson method

$$\alpha_i = -\frac{f'^T d}{d^T f' d}. \tag{5.2.57}$$

For non-linear methods, different definitions for β_i produce different nonlinear CG methods with significantly different properties (Hager 2006). The Fletcher-Reeves version considers (Fletcher 1964)

$$\beta_{i+1} \frac{\|\nabla f\,(x_{i+1})\|^2}{\|\nabla f\,(x_i)\|^2}. \tag{5.2.58}$$

5.3 Phase Field Model

Phase field models are used to describe the behavior of multiphase and multi-component flows. In a phase field model, the interfaces are replaced by extremely thin transition regions. The main idea is to introduce an order parameter ϕ that varies sharply but continuously across the thin interfacial region and has an almost uniform value on the bulk phases (Badalassi 2003). This concept is illustrated in Fig. 5.7, in which ϕ changes its value from phase 1 to phase 2 in a continuous manner across the interface region. In fact, the interface region is a kind of a linear interpolation of ϕ, which is used to characterize the phases and could be density, concentration or mass fraction among others parameters.

Assume a binary fluid made of A and B particles in which diffusion is the transport mechanism. Denote by $\phi = -1$ a phase completely made up of A particles and by $\phi = 1$ a phase completely made up of B particles. A free energy can be defined for times when the system is not in equilibrium. The system evolution is driven by the minimization of this free energy, which is given by the functional (Junseok 2012)

$$E\,[\phi] = \int \left[\varepsilon\,|\nabla\phi|^2 + F\,(\phi)\right] d\Omega, \tag{5.3.1}$$

where Ω is the volume of the system under consideration. The term $F(\phi)$ is the bulk energy potential defined as a classical double-well potential function with two minima for $\phi = -1$ and $\phi = 1$ corresponding to the two phases of the system, the function shape is shown in Fig. 5.7. The term $|\nabla\phi|^2$ is the gradient energy, called capillary term, which acts as a penalty for sharply varying concentration of ϕ; ε is its coefficient. The bulk energy,

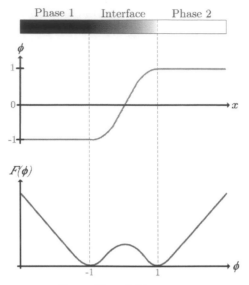

Fig. 5.7 Phase field concept.

Color image of this figure appears in the color plate section at the end of the book.

also called Helmholtz free energy, describes the entropy of the system. The gradient energy term describes the energy of the interactions between particles of type A and B (Badalassi 2003, Junseok 2012).

Variation of E with respect to ϕ is quantifying how the energy changes when particles change position, i.e., the chemical potential of the system (Badalassi 2003). This chemical potential is found by applying variational calculus to Eq. (5.3.1)

$$\frac{\delta E}{\delta \phi} = \mu = \varepsilon \nabla^2 \phi - F'(\phi). \tag{5.3.2}$$

The evolutionary equation for the system is obtained by making the chemical potential (5.3.2) a time-dependant system. The result is the Allen-Cahn equation (Allen 1979).

$$\frac{\partial \phi}{\partial t} = \varepsilon \nabla^2 \phi - F'(\phi) \tag{5.3.3}$$

where $F(\phi) = \frac{1}{4}(1 - \phi^2)^2$. Replacing $F'(\phi)$ in Eq. (5.3.3) we obtain

$$\frac{\partial \phi}{\partial t} = \varepsilon \nabla^2 \phi + \phi - \phi^3. \tag{5.3.4}$$

To completely specify the model of Eq. (5.3.4), it is assumed that the boundary and initial conditions are known (Huang 2011).

The Allen-Cahn Eq. (5.3.4) is used to model the separation process of a binary fluid, in which the fluid is decomposed into a fine-grained mixture of particles. This process is known as spinodal decomposition (Badalassi 2003).

5.3.1 Discretization of the allen-cahn equation

In order to discretize Eq. (5.3.4), the finite difference method is used. The discretization is done through the Crank-Nicholson scheme to obtain a second-order accuracy in time. The finite difference method replaces the derivatives in the partial differential equation by numerical differentiation formulas.

For the 1-D case, the domain is $[-1; 1]$, N is the number of points of the spatial grid and T is the number of iterations. The boundary conditions are constant throughout all the iterations.

Equation (5.3.4) is discretized in time with the forward Euler method (Ken F. Riley 2006)

$$\frac{\partial \phi}{\partial t} = \frac{\phi_i^{t+1} - \phi_i^{'T}}{\Delta t} + O(\Delta t), \tag{5.3.5}$$

and is discretized in space using the centered finite difference equation for the second-order derivative

$$\nabla^2 \phi = \frac{\phi_{i-1} - 2\phi_i + \phi_{i+1}}{h^2} + O(h^2). \tag{5.3.6}$$

Replacing Eqs. (5.3.5) and (5.3.6) in Eq. (5.3.4)

$$\frac{\phi_i^{t+1} - \phi_i^t}{\Delta t} = \varepsilon \frac{\phi_{i-1} - 2\phi_i + \phi_{i+1}}{h^2} + \phi_i - (\phi_i)^3. \tag{5.3.7}$$

applying the Crank-Nicholson scheme, we get

$$\frac{\phi_i^{t+1} - \phi_i^t}{\Delta t} = \frac{\varepsilon}{2} \left[\frac{\phi_{i-1}^{t+1} - 2\phi_i^{t+1} + \phi_{i+1}^{t+1} + \phi_{i-1}^t - 2\phi_i^t + \phi_{i+1}^t}{h^2} \right]$$
$$+ \frac{1}{2} \left[\phi_i^{t+1} - \left(\phi_i^{t+1}\right)^3 + \phi_i^t - \left(\phi_i^t\right)^3 \right] \tag{5.3.8}$$

and defining the constants

$$k_1 = \frac{\varepsilon \Delta t}{2h^2}, k_2 = \frac{\Delta t}{2} \tag{5.3.9}$$

the final discrete equation is

$$-k_1 \phi_{i-1}^{t+1} + l_1 \phi_i^{t+1} - k_1 \phi_{i+1}^{t+1} = k_1 \phi_{i-1}^t + l_2 \phi_i^t + k_1 \phi_{i+1}^t, \tag{5.3.10}$$

where

$$l_1 = 1 + 2k_1 - k_2 + k_2(\phi_i^{t+1})^2, \tag{5.3.11}$$

$$l_2 = 1 - 2k_1 + k_2 - k_2(\phi_i^t)^2. \tag{5.3.12}$$

Equation (5.3.10) gives rise to a non-linear system of equations which can be solved by the non-linear CG method implemented in a GPU.

The scheme for the 2-D numerical solution is similar to the 1-D case, except that the space discretization is done through a bidimensional mesh; the time discretization is done in the same way as in the 1-D case. The domain is an $N \times N$ square shaped region, where the interval for both the x and y axes is $[-1; 1]$, N is the number of points in each direction and T is the number of iterations. The initial value for all the points of the mesh except the ones at the boundaries are given by the initial condition; the values of the points at the boundaries are given by the boundary conditions through all the iterations.

The 2-D AC equation is discretized in time the same way as the 1-D case. The space discretization is done with Eq. (5.3.6), but for the 2-D case, the Laplacian is calculated with respect to the x and y axes. This means that another subindex is needed, unlike Eq. (5.3.10), so the respective derivatives of a point can be denoted.

Expanding Eq. (5.3.10) to 2-D, we get

$$\frac{\phi_{i,j}^{t+1} - \phi_{i,j}^t}{\Delta t} = \frac{\varepsilon}{2} \left[\frac{\phi_{i-1,j}^{t+1} - 2\phi_{i,j}^{t+1} + \phi_{i+1,j}^{t+1} + \phi_{i-1,j}^t - 2\phi_{i,j}^t + \phi_{i+1,j}^t}{\Delta x^2} \right]$$

$$\frac{\varepsilon}{2} \left[\frac{\phi_{i,j-1}^{t+1} - 2\phi_{i,j}^{t+1} + \phi_{i,j+1}^{t+1} + \phi_{i,j-1}^t - 2\phi_{i,j}^t + \phi_{i,j+1}^t}{\Delta y^2} \right] \tag{5.3.13}$$

$$+ \frac{1}{2} \left[\phi_{i,j}^{t+1} - \left(\phi_{i,j}^{t+1} \right)^3 + \phi_{i,j}^t - \left(\phi_{i,j}^t \right)^3 \right]$$

Setting the constants

$$c_1 = \frac{\varepsilon \Delta t}{2\Delta x^2}, c_2 = \frac{\varepsilon \Delta t}{2\Delta y^2}, c_3 = \frac{\Delta t}{2} \tag{5.3.14}$$

$$-c_1\phi_{i-1,j}^{t+1} - c_2\phi_{i,j-1}^{t+1} - c_1\phi_{i+1,j}^{t+1} - c_2\phi_{i,j+1}^{t+1} + l_1\phi_{i,j}^{t+1} = c_1\phi_{i-1,j}^t + c_2\phi_{i,j-1}^t$$
$$+ c_1\phi_{i+1,j}^t + c_2\phi_{i,j+1}^t \tag{5.3.15}$$
$$+ l_2\phi_{i,j}^t$$

where

$$l_1 = 1 + 2c_1 + 2c_2 - c_3 + c_3(\phi_{i,j}^{t+1})^2 \tag{5.3.16}$$

and

$$l_2 = 1 - 2c_1 - 2c_2 + c_3 - c_3(\phi_{i,j}^t)^2. \tag{5.3.17}$$

Like Eq. (5.3.10), Eq. (5.3.15) can be represented as a non-linear system of equations. For the 2-D case, Dirichlet boundary conditions are used at the bottom and top of the square domain, i.e.,

$$\phi(i, -1) = -1, \phi(i, 1) = 1. \tag{5.3.18}$$

For the right and left sides of the domain, homogeneous Neumann boundary conditions are used.

The physical interpretation of these boundary conditions and domain implies a container in which the bottom and top sides correspond to different phases or components of a fluid, and the right and left sides represent the walls of the container.

5.3.2 CUSPARSE and CUBLAS

The NVIDIA CUSPARSE library contains a set of basic linear algebra subroutines designed for sparse matrix operations. The library can be called from C or C++. These subroutines can be classified into four main categories: (1) conversion routines for different sparse matrix formats, (2) routines for operations between a sparse matrix and a set of vectors in dense format, (3) routines for operations between a sparse matrix and a vector in dense format, and (4) functions for operations between a vector in sparse format and a vector in dense format (NVIDIA 2012b).

The library takes advantage of the CUDA parallel programming model as well as of the computational resources of the NVIDIA graphics processor unit (GPU). The CUDA programming model lets the use of a GPU as a co-processor of the CPU. For a detailed description of CUDA see (NVIDIA 2012c).

The CUBLAS library is an implementation of BLAS (Basic Linear Algebra Subprograms) for dense matrices on top of the CUDA runtime. The library includes matrix-vector and matrix-matrix products. The CUBLAS library also provides functions for writing and retrieving data from the GPU (NVIDIA 2012a).

To use the functions of both libraries, the data must be transferred to the GPU memory and must be converted to the corresponding format. Also, the corresponding libraries need to be initialized.

5.3.3 GPU-based solution for the A-C equation

Algorithm 5.1 shows the pseudocode of the nonlinear CG method that solves Eq. (5.3.4). Due to the several linear algebra operations involved in the algorithm, such as sparse matrix-vector multiplication, dot product and vector-vector addition, the functions of CUBLAS and CUSPARSE libraries are a good choice for the parallel implementation.

The first step to use the GPU is to transfer the necessary data from the CPU memory to the GPU memory. Next, the CUSPARSE and CUBLAS libraries are initialized. The instructions of Algorithm 5.1 are performed sequentially but the linear algebra operations are performed in parallel, i.e., each element of a sparse matrix or vector is computed by a single thread within the GPU. This is possible because in a CUDA program, the sequential code sections are performed in the CPU, generally, with a single thread and the parallel code sections are computed in the GPU by thousands of threads that are executed concurrently in order to reduce the computation time.

Algorithm 5.1 Parallel implementation of the nonlinear CG method.

---

allocate CPU and GPU memory
CPU – GPU data transfer
CreateSparseMatrix() A
initialize CUBLAS library
initialize CUSPARSE library
for t = 0 to T
　give an initial solution $u^{(0)}$
　$r^{(0)} = F(u^{(0)}) - Au^{(0)}$
　$p^{(0)} = r^{(0)}$
　for i = 0, 1, 2 . . . until convergence
　　for k = 1 to M
　　　$num = (\alpha_i^{(k)} < Ap^{(i)}, p^{(i)}> - < r^{(i)}, p^{(i)}> + <F(u^{(i)}) - F(u^{(i)} + \alpha_i^{(k)}p^{(i)}), p^{(i)}>$
　　　$den = < Ap^{(i)}, p^{(i)}> - <F(u^{(i)} +\alpha_i^{(k)}p^{(i)})'p^{(i)}, p^{(i)}>$
　　　$d^{(k)} = num / den$
　　　$\alpha_i^{(k+1)} = \alpha_i^{(k)} - d^{(k)}$
　　end for
　　$u^{(i+1)} = u^{(i)} + \alpha_i^{(k+1)}p^{(i)}$
　　$r^{(i+1)} = r^{(i)} - F(u^{(i)}) + F(u^{(i+1)}) - \alpha_i^{(k+1)}Ap^{(i)}$
　　Convergence test
　　$\beta^{(i+1)} = - < r^{(i+1)}, r^{(i+1)}>/< r^{(i)}, r^{(i)}>$
　　$p^{(i+1)} = r^{(i+1)} - \beta^{(i+1)}p^{(i)}$
　end for
end for

To compute a given linear algebra operation, the corresponding function is called by the CPU thread, and the execution is performed in the GPU by thousands of simultaneous threads. If the operation involves a sparse matrix, a CUSPARSE function is called; otherwise, a CUBLAS function is called.

The CPU thread has the flow control of the instructions and loops of Algorithm 5.1, whereas the GPU threads compute the linear algebra operations.

5.4 Results

In this section, we present the numerical results obtained with the parallel solution of the A-C equation [Eq. (5.3.4)].

5.4.1 1-D case

Figure 5.8 shows the process of phase separation in 1-D. Figure 5.8a shows the initial condition, which is made of random values −1 and 1. Intermediate steps of the evolution are shown in Fig. 5.8b and c. The final shape of the phase separation, which is shown in Fig. 5.8d, resembles the interface shape of Fig. 5.7. The boundary conditions are $\phi(-1) = -1$ and $\phi(1) = 1$, so the initial random values can be grouped with their respective component.

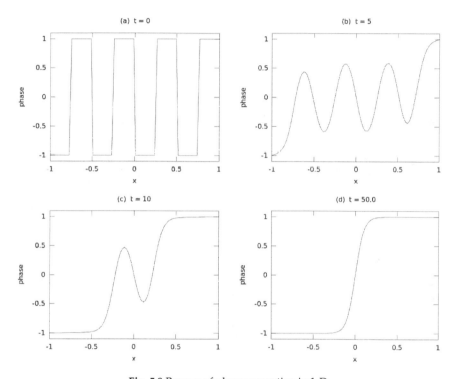

Fig. 5.8 Process of phase separation in 1-D.

5.4.2 2-D case

Figure 5.9 shows the process of phase separation of a binary fluid in 2-D. Fig. 5.9a shows the initial condition, which is made of values −1 and 1 and represents a non-homogeneous fluid made of two components, one represented by the value −1 and the other represented by 1. Intermediate separation steps are shown in Figs. 5.9b and c. The final fluid separation is shown in Fig. 5.9d, where the two components are completely separated,

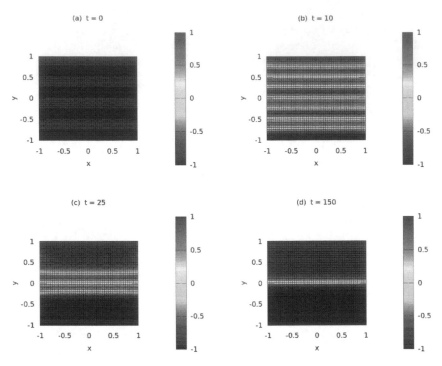

Fig. 5.9 Process of phase separation in 2-D.
Color image of this figure appears in the color plate section at the end of the book.

and the interface area is represented by the horizontal line in the middle of the domain. As in the 1-D case, the interface thickness of the fluids separation line is determined by the ε value of Eq. (5.3.4).

5.5 Conclusions and Future Work

The discretized A-C equation can be represented as a non-linear system which can be solved by different methods. Among these methods, the non-linear CG method is suitable for parallelization in a GPU due to its involved operations. Although these operations can be implemented with kernels, see (NVIDIA 2012c), the CUSPARSE and CUBLAS functions are used because their relatively easy implementation and their optimal design from the GPU hardware point of view.

As we stated in Section 3, ε controls the interface thickness of the phase field. However, when we use very small ε values, the regular grid for the numerical solution is no longer numerically accurate. At this point, is necessary the use of an adaptive mesh refinement method in order to get a good numerical solution over the interface area. This is a future step of our investigation. In particular, we will seek for Conjugate Gradient methods for the adaptive mesh function, because, as we have demonstrated in this work, the CG method can be parallelized in a GPU.

A comparison between the execution time of an algorithm implemented in a CPU and in a GPU is not the best way to evaluate a CUDA application, although it is the most common way. It is not the best choice, because it is not objective, fair nor qualitative. Many factors are involved such as the capacity of the processors, the available resources among many others. That is why we think that an execution time comparison, as well as an objective measurement of the GPU and CPU resources utilization, is the best way to measure a GPU application. For this reason, the objective measurement of the kernels and C code that are used in this work is the next step of the project. To accomplish it, we need to establish the adequate metrics and find the tools for debugging and profiling the CUDA programs and the C programs, e.g., Compute Visual Profiler and CUDA-gdb.

Acknowledgments

The authors would like to thank the "Instituto Politécnico Nacional (IPN)", "Comisión de Operación y Fomento de Actividades Académicas del IPN (COFAA-IPN)" and the Mexico's entity "Consejo Nacional de Ciencia y Tecnología (CONACYT)" for supporting our research activities.

References

Allen, S.M. and J.W. Cahn (1979). A microscopic theory for antiphase boundary motion and its application to antiphase domain coarsening. Acta Metall. 27(6): 1085–1095.
Andreas, A. and W.-S. Lu (2007). Practical Optimization: Algorithms and Engineering Applications. Springer.
Badalassi, V.E., H.D. Ceniceros and S. Banerjee (2003). Computation of multiphase systems with phase field models. J. Comput. Phys., 190(2): 371–397.
Edwin K.P. and S.H. Chong (2001). An introduction to optimization (2nd edn.). Springer.
Fletcher, R. and C.M. Reeves (1964). Function minimization by conjugate gradients. Comput. J. 7(2): 149–154.
Galiano V., H. Migallón, V. Migallón and J. Penadés (2012). GPU-based parallel algorithms for sparse nonlinear systems. J. Parallel. Distrib. Comput., 72(9): 1098–1105.
Garcia, N. (2010). Parallel power flow solutions using a biconjugate gradient algorithm and a Newton method: A GPU-based approach. IEEE Power and Energy Society General Meeting, Minneapolis, MN, USA, pp. 1–4.
Hager, W.H. and H. Zhang (2006). A survey of nonlinear conjugate gradient methods. Pac. J. Opt., 2(1): 35–58.
Huang, P., and A. Anduwali (2011). A numerical method for solving Alen-Cahn equation. J. Appl. Math. and Informatics, 29(5-6): 1477–1487.
Nocedal, J. and S. Wright (2006). Numerical Optimization (2nd edn.). Springer.

Junseok, K. (2012). Phase-Field Models for Multi-Component Fluid Flows. Comm. Comput. Phys., 12(3): 613–661.

Riley, K.F., M.P. Hobson and S.J. Bence (2006). Mathematical Methods for Physics and Engineering (3rd edn.). Cambridge University Press.

Kirk, D.B. and W.-m.W. Hwu (2013). Programming Massively Parallel Processors A Hands-on Approach (Second edn.). Elsevier.

M.R. Hestenes, E.S. (1952). Methods of conjugate gradients for solving linear systems. J. Res. Nat. Bur. Stand., 49: 409–436.

Michels, D. (2011). Sparse-matrix-CG-solver in CUDA. Proceedings of the 15th Central European Seminar on Computer Graphics. Viničn, Slovakia.

NVIDIA. (2012a). CUDA Toolkit 4.2 CUBLAS Library. Available at: http://docs.nvidia.com/cuda/pdf/CUBLAS_Library.pdf.

NVIDIA. (2012b). CUDA Toolkit 4.2 CUSPARSE Library. Available at: http://docs.nvidia.com/cuda/pdf/CUSPARSE_Library.pdf.

NVIDIA. (2012c). NVIDIA CUDA C Programming Guide. Available at: http://docs.nvidia.com/cuda/pdf/CUDA_C_Programming_Guide.pdf.

6

Parallel Computing Applied to a Diffusive Model

Roberto Salas, Juan José Tapia* and Fernando Villalbazo

ABSTRACT

A Markovian Random-Walk approach is used to obtain the partial differential equation of a diffusive one-dimensional and bi-dimensional process, modeled as particles passing walls with certain probability of reflecting or passing along. An analytical solution is deduced and compared with a finite difference approach solved with MATLAB; further considerations of numerical solution under the CUDA architecture are presented in this chapter.

6.1 Introduction

Diffusive problems appear in many contexts not just in Physics, but even in Social Sciences too. We distinguish between the dynamic or evolutionary component of the diffusion process and the particular characteristics of media along which particles or any other objects are to travel. Communication theory considers also diffusive problems, especially regarding to noise distribution, or strong versus weak components of signals competing to reach a device (Lehner 2008).

In general terms, diffusion is a dynamic process, which depends on time, that implies that at least an elementary unit of matter, information or any measurable entity travels along a particular environment (so called media)

Instituto Politécnico Nacional, CITEDI, Tijuana, México.
  E-mail: jtapiaa@ipn.mx
* Corresponding author

along time. In the present chapter we start the discussion from problems in a discrete environment, constituted of homogeneous cells, the particles are of mechanical type (therefore having mass and velocity).

One of the most famous ways to treat the problem is the theory or Random Walk. From the diverse approaches to apply Random Walk we will take the Markovian style, justifying below the plausibility of the proposal in terms of the one-dimensional comparison with a classical, well studied model, like Fick's approach (Bird, Stewart and Lightfood 2007).

6.2 Random Walk

Let us consider a material particle embedded in a one-dimensional infinite array of identical cells. This particle will have a definite mass. The possible motions of the particle could be to the right or to the left under the following rules:

a) If the particle moves to the right or to left in any step, motion in the same direction in the next step will be called transmission.
b) If the particle moves to the right or to left in any step, motion in the opposite direction in the next step will be called reflection.
c) The particle will be reflected by probability R.
d) The particle will be transmitted by probability T (obviously $T = 1-R$).
e) According to the characteristics of the walls of the cells there is an initial probability of motion to the left or to right.

An agent following this rule could take pieces of information (for example e-mail notes) and could return the note to the original sender or could forward the note to another. If the example sounds unreal (forwarding process is frequently done to more than one receiver) surely information passing from one to another person shall be better adjusted.

An additional assumption is in order: the agents participating in this hypothetical experiment must take decisions regarding pass or retain information subject to the same probability. So, the process, defined in this way, adopts a Markovian character: motion in any step depends just in what happened at the previous one. We are aware that beyond the feasibility in a practical case, in mechanical conditions (transport of mass), we start and finish with the same number of particles (continuity), while in the case of information spreading, an agent is informed, increasing the number of informed individuals (supposing that we establish a binary condition). The concept of density changes to a percentage of informed agents.

From now on we continue the discussion based on the mechanical (material particles) model. Let us use N for the succession counter, being the time elapsed an integer multiple of the time to pass from one cell to another (and assuming this time of passing as constant). We will label every cell with the corresponding coordinate x.

Let us describe in terms of this consideration three successive jumps. First, we state (as initial condition) that the particle is able to move with identical probability to the right and left, being therefore the starting probability $1/2$. The probability for the particle to exert two transmissions and one reflection is given by:

$$\frac{T^2 R}{2} = \frac{R(1-R)^2}{2}$$

(6.2.1)

If we define a particular position, x_0, the particle could reach this point by several possible combinations of jumps. We summarize the possible results of the process defined in (6.2.1) in a very simple scheme in Table 6.1.

Henin presents some important aspects of this phenomenon (Henin 2011), based on the original works of Chandrasekhar, who considers the problem represented by (6.2.1) with a great difference: independent steps at all times (Chandrasekhar 1943). A recent work from Lawler applies a similar approach to the heat equation (Lawler 2010). We can extend the discussion to a single particle or to a stream (ensemble), probabilities then represents percentage of the stream to be found in each cell. Particles far from origin will late some finite time to receive the stream. According to Fick's model the stream would arrive immediately to all points.

Table 6.1 A scheme for diffusion applying a Random Walk approach. One particle inside symmetrical cells.

	X = -3	X = -2	X = -1	X = 0	X = 1	X = 2	X = 3
N = 0				1/2 → 1/2 ←			
N = 1			T/2←	R/2← R/2→	T/2→		
N = 2		TT/2←	TR/2→ RT/2←	RR/2→ RR/2←	TR/2← TR/2→	TT/2→	
N = 3	TTT/2←	TRR/2← RTT/2→	TRR/2← RTR/2→ RRT/2←	RRR/2← RRR/2→ TRT/2← TRT/2→	TRR/2→ RTR/2← RRT/2→	TTR/2← RTT/2→	TTT/2→

For this process we make the following definitions:

$P_1(X, N) = P_+$ Probability for the particle to be found at the coordinate X at step N, or, which is the same, at time $t = N\tau$ where τ is the time to complete a jump and moves to right.

$P_2(X, N) = P$ Probability for the particle to be found at the coordinate X at step N, or, which is the same, at time $t = N\tau$ where τ is the time to complete a jump and moves to left.

$P_1 + P_2 = \rho(X, N)$. The density function, which measures the probability of the particle of reach coordinate X at jump N.

$P_1 - P_2 = (1/c)\, J(X, N)$. The current density, here c is a measure of speed: δ/τ where δ is the size of the cell.

It follows immediately that under these conditions the values of P must fulfill the restrictions of a probability function of density, namely:

$$\left. \begin{array}{c} 0 \leq P_1, P_2 \leq 1 \\ 0 \leq \rho \leq 1 \\ 0 \leq |J| \leq 1 \\ \displaystyle\sum_{x=-\infty}^{\infty} \rho(X, N) = 1. \end{array} \right\} \tag{6.2.2}$$

The restriction above of course makes sense of counting integrity, giving bounds to the final result. Of course (6.2.2), would be an obvious consequence of the random character of the model.

The Markovian character of the process will be assumed if we postulate

$$\bar{P}(X, N) = A\, \bar{P}(X, N-1) \tag{6.2.3}$$

Here A is a squared matrix and

$$\bar{P} = \begin{pmatrix} P_1 \\ P_2 \end{pmatrix} \tag{6.2.4}$$

We can state the probability for a particle to reach the cell labeled as X at jump N with two components: a transmission from the cell $X–1$ or a reflection to maintain the same position (first dynamic equation, consequence of (6.2.3) viewed as vector (6.2.4)):

$$P_1(X, N) = TP_1(X-1, N-1) + RP_2(X, N-1) \tag{6.2.5}$$

And in an analogous form we state the second dynamic equation:

$$P_2(X, N) = RP_1(X, N-1) + TP_2(X+1, N-1) \tag{6.2.6}$$

The reader will easily check that, for example, taking $X = 0, N = 3$.

$$P(0,3) = \begin{pmatrix} P_1(0,3) \\ P_2(0,3) \end{pmatrix} = \begin{pmatrix} P_1(-1,2) & P_2(0,2) \\ P_2(1,2) & P_2(0,2) \end{pmatrix}\begin{pmatrix} T \\ R \end{pmatrix} = \begin{pmatrix} TR/2 & RR/2 \\ RT/2 & RR/2 \end{pmatrix}\begin{pmatrix} T \\ R \end{pmatrix} = \begin{pmatrix} \frac{RT^2}{2} + \frac{R^3}{2} \\ \frac{RT^2}{2} + \frac{R^3}{2} \end{pmatrix} \tag{6.2.7}$$

These results (6.2.5)–(6.2.6) for successive steps of the process and (6.2.7) for its matrix representation are in good agreement with our virtual experiment, expressing the dynamic character on sequence order if time counting is not available. Successive application of the process will predict any state.

If we apply the Fourier Transform to the dynamic equations, a solution is found:

$$r = \sqrt{|T^2 - R^2|}$$

$$
\begin{aligned}
P_{1,2}(x,N) = \frac{T}{2}r^{N-1} \sum_{m=0}^{\left\|\frac{N-1}{2}\right\|} (-1)^m \frac{(N-1-m)!}{m! \left\|\frac{N-1-2m\pm1-x}{2}\right\|! \left\|\frac{N-1-2m\mp1+x}{2}\right\|!} \left(\frac{T}{r}\right)^{N-1-2m} \\
+ \frac{R}{2}r^{N-1} \sum_{m=0}^{\left\|\frac{N-1}{2}\right\|} (-1)^m \frac{(N-1-m)!}{m! \left\|\frac{N-1-2m-x}{2}\right\|! \left\|\frac{N-1-2m+x}{2}\right\|!} \left(\frac{T}{r}\right)^{N-1-2m} \\
- \frac{1}{2}r^N \sum_{m=0}^{\left\|\frac{N-1}{2}\right\|} (-1)^m \frac{(N-2-m)!}{m! \left\|\frac{N-2-2m-x}{2}\right\|! \left\|\frac{N-2-2m+x}{2}\right\|!} \left(\frac{T}{r}\right)^{N-2-2m}
\end{aligned}
\tag{6.2.8}
$$

A parallel computing tool can be developed to assign calculation in threads of parts of the summation in (6.2.8), instead of generating a step-by-step solution. The disadvantage of the description is of course that we must solve the model previously. Applying an approximate model to solve (6.2.5) and (6.2.6) is necessary to use parallel computing in a more effective way.

6.3 Continuous Distributions

In the study of a very large number of particles, or ensembles, counting each of them shall be foolish. However some physical considerations could be taken into account to have a plausible formulation.

Let us state a limiting process for τ and δ, as decrease to zero. Then the ratio δ/τ will assume a constant value c, giving a measure for the velocity of the process. The first approximation for the dynamic equations is then:

$$P_1(x,t) + \tau \frac{\partial P_1}{\partial \tau} \approx T\left(P_1(x,t) - \delta\frac{\partial P_1}{\partial x}\right) + RP_2(x,t) \tag{6.3.1}$$

We now divide by t and make an additional assumption: the transmission (T) will be a dominating effect, so it will tend to unit, and therefore:

$$\frac{\partial P_1}{\partial t} + \frac{\delta}{\tau}\frac{\partial P_1}{\partial x} \approx \frac{R}{\tau}[P_2(x,t) - P_1(x,t)] \tag{6.3.2}$$

Relationships (6.3.1) and (6.3.2) synthesize the nature of the physical approximation, we are working in an intermediate scale between classical and quantum approaches (because considerations about momentum and energy exchanges are not included in the description). We also state that, as T tends to unit, R tends to zero and the following ratio has a definite value in the limit:

$$\frac{R}{\tau} = \frac{1}{2\theta} = \lambda. \tag{6.3.3}$$

Here θ will be denoted as the characteristic time. In addition $c^2\theta = D$ will be stated as the diffusion coefficient of the process, so:

$$\left.\begin{aligned}
\frac{\partial P_1}{\partial t} &= -c\frac{\partial P_1}{\partial x} + \frac{1}{2\theta}P_2 - \frac{1}{2\theta}P_1 \\
\frac{\partial P_2}{\partial t} &= -c\frac{\partial P_2}{\partial x} + \frac{1}{2\theta}P_1 - \frac{1}{2\theta}P_2
\end{aligned}\right\} \tag{6.3.4}$$

On the other hand, we can require from the process the following conditions as a consequence of taking the limit:

$$\left.\begin{aligned}
0 &\leq P_1, P_2, \rho, |J| \leq 1 \\
\rho &= P_1 + P_2 \\
J &= P_1 - P_2 \\
\int_{x=-\infty}^{\infty} &\rho(x,t)dx = 1
\end{aligned}\right\} \tag{6.3.5}$$

In other words, the description (6.3.4), with the simplification in (6.3.3) and the approximations (6.3.1)–(6.3.2) and the stochastic character (6.3.4) constitute a plausible description in intermediate scale: neither quantum nor classic, this is the so-called mesoscopic approach (Horsthemke et al. 2010).

Under these conditions, after a short manipulation we finally obtain:

$$\frac{1}{c^2}\frac{\partial^2 P_2}{\partial t^2} + \frac{1}{D}\frac{\partial P_2}{\partial t} - \frac{\partial^2 P_2}{\partial x^2} = 0 \tag{6.3.6}$$

An identical equation is verified by P_1. By superposition we deduce that the density of particles also satisfies (6.3.6), which is the Wave Equation with damping, better known as the Telegrapher's equation. Regarding this equation, an analytical interpretation to solve problems in electrodynamics comes as far as Stratton in his treatment of the Electromagnetic theory (Stratton 1941). New updated versions of the problem and solutions for simplest boundary conditions are considered by Griffiths (2012).

6.4 Solution via Finite Difference Approach

The final expression for the density of particles is:

$$\frac{1}{c^2}\frac{\partial^2 \rho}{\partial t^2} + \frac{1}{D}\frac{\partial \rho}{\partial t} - \frac{\partial^2 \rho}{\partial x^2} = 0 \qquad (6.4.1)$$

We can take, without losing of generality, c and δ as unitary quantities, implying a simple change of scale in a real situation. For a more complex problem in three dimensions, we would expect that the generalized expression for (6.4.1) gives the following formulation:

$$\frac{\partial^2 \rho}{\partial t^2} + \frac{\partial \rho}{\partial t} - \nabla^2 \rho = 0 \qquad (6.4.2)$$

Let us solve the analytic solution of (6.4.1), considering an infinite available one-dimensional space and the initial conditions:

$$\rho(x,0) = \delta(x); \quad \frac{\partial \rho}{\partial t}(t=0) = 0; \quad \rho(|x| > t) = 0 \qquad (6.4.3)$$

Here δ represents the Dirac Delta function. Regarding the boundary conditions, we must make an important consideration. The fastest particles (those which passed without a single reflection) move as $x = ct$, which could be called a "cauda effect", but beyond this extreme position density must be zero, therefore, as $c = 1$, the following condition must prevail (indeed are two conditions, for each branch of real axis):

The analytical solution is the following:

$$\rho = e^{-\frac{t}{2}}\left\{\delta(t+x) + \delta(t-x) + \frac{1}{2}I_0\left(\frac{\sqrt{t^2-x^2}}{2}\right) + \frac{t}{2}\frac{I_1(\sqrt{t^2-x^2}/2)}{\sqrt{t^2-x^2}}\right\}H(t-x) \quad (6.4.4)$$

In Fig. 6.1 we represent this solution and observe that the distribution of particles is similar to a classical spreading of particles, having this one as the limit for a long time.

Here we are studying an "infinite" amount of particles starting at the origin and spreading along the x axis. Among the many mathematical tools to solve the problem we can use whatever in terms of computing time, error handling and convergence. We could consider several ways to solve the problem, but sometimes an analytic determination could be, if not cumbersome, extremely difficult or impossible, above all, under diverse boundary or initial conditions. A numerical approach should then be considered. Between many possibilities one of the simplest is a Finite Difference approach, essentially consisting of expressing the partial derivatives as difference quotients.

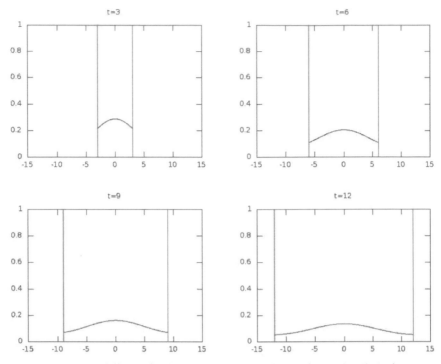

Fig. 6.1 Mesoscopic diffusion for the problem (6.4.2) under conditions (6.4.3). As far as time increases the general trend is the classical description.

Color image of this figure appears in the color plate section at the end of the book.

The differences mentioned above could take into account successive or nearest points (Strikwerda 2004), referring to the one which has an associate value we are looking for, that suggests the creation of a template or stencils, whose points represent locations, and have associated the value of the variable in question.

To describe the finite difference approximations, we state the following notation:

$$\rho_{i,j}^{k}$$

This is the density of particles at the discrete location labeled as i, j in two dimensions and just i for one dimension. The super index k represents the time step.

The problem we are going to consider now is exactly that described by (6.4.1) under the following boundary and initial conditions:

$$\rho(0,t) = \psi_1(t) = 0 \tag{6.4.5}$$

$$\rho(L = 200, t) = \psi_2(t) = 0 \tag{6.4.6}$$

$$\rho(x,0) = \eta_1(x) = (x-a)(b-x)\sin x\,[H(x-a) - H(x-b)]; a < b \tag{6.4.7}$$

$$\left.\frac{\partial\rho}{\partial t}\right|_{t=0} = \eta_2(x) = 0 \tag{6.4.8}$$

Conditions (6.4.5) and (6.4.6) refer to no available density of particles at the extreme points of a finite linear distribution. On the other hand, conditions (6.4.7) and (6.4.8) refers to an initial pulse in static condition.

A centered finite difference approach will be briefly considered in (6.4.1) giving the following result:

$$c^2 \frac{\rho_{i+1}^k - 2\rho_i^k + \rho_{i-1}^k}{\Delta x^2} = a \frac{\rho_{i+1}^k - \rho_i^k}{\Delta t} + \frac{\rho_i^{k+1} - 2\rho_i^k + \rho_i^{k-1}}{\Delta t^2} \tag{6.4.9}$$

From this, successive solutions for each discrete time step follow $(\gamma{=}c\Delta t/\Delta x)$:

$$\rho_i^{k+1} = \frac{\gamma^2}{1+a\Delta t}\rho_{i+1}^k + \frac{2-2\gamma^2+a\Delta t}{1+a\Delta t}\rho_i^k + \frac{\gamma^2}{1+a\Delta t}\rho_{i-1}^k - \frac{1}{1+a\Delta t}\rho_i^{k-1} \tag{6.4.10}$$

Some comments are in order. First, the problem will be reduced in any way, to generate successive sets of points as a recurrence relation, and then the method is so called explicit [modifying the general expression (6.4.9) to give a recursive formulation like (6.4.10) states]. On the other hand, if a system of linear equations were generated, we would had an implicit method. So we must solve a system of linear equations at each step of time, the unknowns being the values of ρ.

A criterion proposed by Courant, Friederichs and Lewy for the formulations in (6.4.9) and (6.4.10) (LeVeque 2007), imposes an absolute convergence condition:

$$\gamma = c\frac{\Delta t}{\Delta x} \leq \frac{1}{2} \tag{6.4.11}$$

Under the constraint (6.4.11) we can state (6.4.10) as a matrix and a more compact form of the description is:

$$\rho^{k+1} = A\rho^k - \rho^{k-1} + \frac{\gamma^2}{1+a\Delta t}\vec{b_j} \tag{6.4.12}$$

The quantities A and $\vec{b_j}$ are represented by

$$
A = \begin{pmatrix}
\dfrac{2(1-\gamma^2)}{1+a\Delta t} & \dfrac{\gamma^2}{1+a\Delta t} & 0 & & 0 \\[2mm]
\dfrac{\gamma^2}{1+a\Delta t} & \dfrac{2(1-\gamma^2)}{1+a\Delta t} & \dfrac{\gamma^2}{1+a\Delta t} & \cdots & 0 \\[2mm]
0 & \dfrac{\gamma^2}{1+a\Delta t} & \dfrac{2(1-\gamma^2)}{1+a\Delta t} & & 0 \\[2mm]
 & \vdots & & \ddots & \vdots \\[2mm]
0 & 0 & 0 & \cdots & \dfrac{2(1-\gamma^2)}{1+a\Delta t}
\end{pmatrix}
\tag{6.4.13}
$$

$$
\vec{b_j} = \begin{pmatrix}
\eta_0(t_j) \\
0 \\
0 \\
\cdots \\
\eta_L(t_j)
\end{pmatrix}
\tag{6.4.14}
$$

At each step we need information of two previous values of ρ, therefore we have to use the initial condition to determine the first group of those values.

6.5 High Performance Computing Resources

Handling the matrix formulation in (6.4.9) and (6.4.10), taking advantage of the characteristics of the matrix, a high performance computing approach is in order.

The problem at hand requires not so complex considerations in one dimension. We can take advantage of an easy-access resource such as MATLAB, for which high-performance resources are fully embedded. This is the case in handling the concept of sparse matrix, for which only the diagonal differs from zero. The corresponding treatment for any Linear Algebraic usual operations is really simplified. For this simple problem we present a proposal now.

The solution of the equation can be represented for successive times, considering also reflecting walls, as represented in Fig. 6.2.

Program 6.1 generates successive solutions for the problem, with the advantage that a nonlinear system must be solved. However, in implicit methods the necessity of the solution of such a system will be unavoidable. Another interesting strategy to solve these kind of problems (with increasingly complexity) and obtain graphic representations similar to the previous solution follows this consideration: we must first identify the characteristics of the matrix that describe the problem, to treat the sparse character when it is convenient.

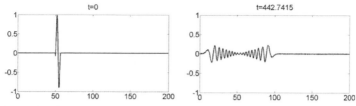

Fig. 6.2 (a) Initial pulse subject to damped displacement with reflecting walls. (b) Reflecting effect of walls after three interactions.

Let us now consider the corresponding MATLAB solution code represented in Program 6.1. One more time, even when the Finite Difference description has the principal disadvantage of uniform division (and identical computational cost to analyze all the points inside the domain), a very quick method of generating the successive set of points representing densities is necessary.

Program 6.1 Explicit solution of telegraphers equation.

```
clear all; clc; hold off;
g2 - 0.2;c - 1;N - 400;L = 200;atten - .0010; t=0;
dx = L / N;
dt = sqrt(g2)*dx/c;
x = dx : dx : L-dx;
n = length(x);
n1 = fix(n/4);
nr = fix(n/3.5);
x1 = x(n1);
xr = x(nr);
Y0 = zeros(n,1);
I = find(x >= x1 & x <= xr);
Y0(I) = (x(I)-x1).*(xr-x(I)).*(sin(x(I)));
Y0 = Y0 / max(Y0);
plot(x,Y0); hold on; axis([0 L -1 1]); plot([0 L],[0 0]);
title(strcat('t=',num2str(t)))
pause()
A = sparse(diag(2*(1-g2+atten*dt/2)/(1+atten*dt)*...
    ones(n,1))+g2/(1+atten*dt)*diag(ones(n-1,1),1)+...
    g2/(1+atten*dt)*diag(ones(n-1,1),-1));
Y1 = Y0; hold off
for t = 2*dt:dt:2000*dt
Y2 = A * Y1 - Y0/(1+atten*dt);
    Y0 = Y1;
    Y1 = Y2;
    if(fix(t/10/dt) == t/10/dt)
        plot(x,Y1)
        axis([0 L -1 1]);
title(strcat('t=',num2str(t)))
        pause(0.1)
    end
end
hold off
```

Let us now make the description of the diffusive problem in two dimensions, but within a bounded rectangular domain. Also, the initial distribution of particles will be

$$\rho(x, y, 0) = \frac{1}{2\pi\sigma^2} e^{-\frac{(x^2+y^2)}{2\sigma^2}} \tag{6.5.1}$$

with boundary conditions

$$\rho(x, L) = \rho(x, 0) = \rho(a, y) = \rho(0, y) = 0. \tag{6.5.2}$$

Program 6.2 can be used for the solution of the problem. Let us bear in mind that the initial distribution (6.5.1) represents a pulse in two dimensions at the beginning of the experiment. As this is a static pulse, we assume a temporal derivative equals zero. Again, the restriction of zero density along all the boundaries is represented by (6.5.2). As we will see, a file with the solutions generated will be created, and then these values can be plotted via Program 6.3. We recommend here Barret's book to find the formal and useful additional templates for other applications (Barret et al. 1987).

Program 6.2 CUDA solution of Telegrapher's equation in two dimensions.

```
#include <string.h>
#include <stdio.h>
#include <stdlib.h>
#include <math.h>
# define N 34
texture<float,2,cudaReadModeElementType> Z1ref;
__global__ void wave(float *Z, float *Z0, float r, float dt)
{
        int idx=blockIdx.x*blockDim.x+threadIdx.x+1;
        int idy=blockIdx.y*blockDim.y+threadIdx.y+1;
        int id=idy*N+idx;
        Z[id]=(2-dt)*tex2D(Z1ref,idx,idy)+(dt-
1)*Z0[id]+r*(tex2D(Z1ref,idx+1,idy)+tex2D(Z1ref,idx-
1,idy)+tex2D(Z1ref,idx,idy+1)+tex2D(Z1ref,idx,idy-1)-4*tex2D(Z1ref,idx,idy));
        Z0[id]=tex2D(Z1ref, idx, idy);
}
int main()
{
        int bloquesize=32,i,j=0, t;
        float L=20.0, Zmax=0.0, dtx=0.0, dty=0.0, sigma=3, sigma2=sigma*sigma,
pi=3.14159265;
        float* x = (float*)malloc(N*sizeof(float));
        float* I = (float*)malloc(N*sizeof(float));
        float* Z0 = (float*)malloc(N*N*sizeof(float));
        float* Z1 = (float*)malloc(N*N*sizeof(float));
        float* Z = (float*)malloc(N*N*sizeof(float));
        float* Z0_d;
        float* Z_d;
```

Program 6.2 contd....

Program 6.2 contd.

```
    cudaMalloc((void**)&Z0_d,sizeof(float)*N*N);
    cudaMalloc((void**)&Z_d,sizeof(float)*N*N);
    cudaChannelFormatDesc desc = cudaCreateChannelDesc<float>();
    cudaArray* Z1tex;
    cudaMallocArray (&Z1tex, &desc, N, N);
    dim3  numBlocks(N/bloquesize,N/bloquesize);
    dim3  threadsperblock(bloquesize, bloquesize);
    float c=10, ix=-L/2, iy=-L/2;
    float dx=L/(N-1);
    float dy=L/(N-1);
    float dt=0.9*dx/(2*c);
    float dx2=dx*dx;
    float dt2=dt*dt;
    float r=(c*c)*dt2/dx2;
    for (i=0; i<N; i++){
          Z0[i]=0.0;
          Z1[i]=0.0;
          Z[i]=0.0;
    }
    for (i=0; i<N; i++){
          ix=-L/2;
          for (j=0; j<N; j++){
                Z0[i*N+j]=(1/(2*pi*sigma2)*exp(-(abs(ix)*abs(ix)+
                abs(iy)*abs(iy))/(2*sigma2)));
                ix=ix+dx;
          }
          iy=iy+dy;
    }
    for (i=0; i<N; i++){
          Z0[i]=0.0;
          Z0[N*(N-1)+i]=0.0;
          Z0[N*i]=0.0;
          Z0[N*i-1]=0.0;
    }
    Zmax=Z0[0];
    for (i=1; i<N*N; i++)
          if (Z0[i]>Zmax)
                Zmax=Z0[i];
    for (i=0; i<N*N; i++)
          Z0[i]=Z0[i]/Zmax;
    cudaMemcpyToArray(Z1tex, 0, 0, Z0, sizeof(float)*N*N,
cudaMemcpyHostToDevice);
    Z1ref.addressMode[0]=cudaAddressModeClamp;
    Z1ref.filterMode = cudaFilterModePoint;
    cudaMemcpy(Z0_d, Z0, sizeof(float)*N*N, cudaMemcpyHostToDevice);
    cudaMemcpy(Z_d, Z, sizeof(float)*N*N, cudaMemcpyHostToDevice);
    cudaBindTextureToArray(Z1ref, Z1tex, desc);
    for (t=0; t<300; t++) {
          wave<<<numBlocks,threadsperblock>>>(Z_d, Z0_d, r, dt);
          cudaMemcpyToArray(Z1tex, 0, 0, Z_d, sizeof(float)*N*N,
cudaMemcpyDeviceToDevice);
    }
```

Program 6.2 contd....

Program 6.2 contd.

```
    cudaMemcpy(Z, Z_d, sizeof(float)*N*N, cudaMemcpyDeviceToHost);

    /*File generation*/
    FILE *p;
    p = fopen("plot3d.dat", "w");
                    printf("Preparing file...\n");
    for (i=0; i<N; i++){
            dtx=0.0;
            for (j=0; j<N; j++) {
                    dtx=dtx+dt;
                    fprintf(p, "%f ",Z[i*N+j]);
            }
    fprintf(p,"\n");
    dty=dty+dy;
    }
    fclose(p);
    cudaUnbindTexture(Z1ref);
    cudaFree(Z0_d);
    cudaFree(Z_d);
    printf("\nConcluded succesfully.\n");
return 0;
}
```

Program 6-3: MATLAB interpreter for the data generated by program 6-2.

```
clc
close all
data = load('plot3d.dat');
surf(data)
 print -djpg figure.jpg;
```

The corresponding plot is then represented in Fig. 6.3. It is remarkable to note that the results were obtained at a notebook with a GeForce GT 445M GPU, under Linux lasting just 304 ms. The CUDA compiler was version 5.0, V0.2.1221. Less important is the fact that plotting was made using Octave 3.2.4.

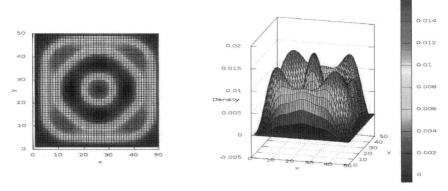

Fig. 6.3 Solution of Telegrapher's equation in two dimensions.

Color image of this figure appears in the color plate section at the end of the book.

6.6 Conclusions

Future work in numerical analysis, combined with the correct and pertinent application of computational architectures is being developed to handle huge matrix formulations. Resources available today offer strong working platforms to solve relatively complex problems. The possibility of improving the efficiency of the process depends also on a computational parallel effective method to generate successive recursive solutions or to solve linear systems generated in implicit methods. Extending the work to three dimensions would generate even bigger matrices (coming from a Kronecker tensor product for the numerical description) whose treatment will be delegated to a GPU, whose competitive advantages are subject to formal consideration nowadays.

Acknowledgments

The authors would like to thank the "Instituto Politécnico Nacional (IPN)", "Comisión de Operación y Fomento de Actividades Académicas del IPN (COFAA-IPN)" and the Mexico's entity "Consejo Nacional de Ciencia y Tecnología (CONACYT)" for supporting our research activities.

References

Barret, Richard, Michael Berry, Tony F. Chan, James Demmel, June Donato, Jack Dongarra, Victor Eijkhout, Roldan Pozo, Charles Romine, Henk van der Vorst (1987). Templates for the Solution of Linear Systems: Building Blocks for Iterative Methods (Miscellaneous Titles in Applied Mathematics Series No 43). London: Society for Industrial and Applied Mathematics.

Bird, R. Byron, Warren. Stewart, Edwin Lightfoot (2007). Transport Phenomena, 2nd. Edition. Danvers, MA, USA: John Wiley & Sons.

Chandrasekhar, S. (1943). Stochastic Problems in Physics and Astronomy. REV MOD PHYS, vol. 15, no. 1, pp. 1–89, 1943. DOI: 10.1103/RevModPhys.15.1

Griffiths, David (2012). Introduction to Electrodynamics, 4th Ed. New York: Addison Wesley.

Pignedoli, A. (2011). Some Aspects of Diffusion Theory: Lectures given at a Summer School of the Centro Internazionale Matematico Estivo (C.I.M.E.) held in Varenna (Como), Italy. New York: Springer.

Mendez, Vicenc; Sergei Fedotov, Werner Horsthemke. Reaction-Transport Systems: Mesoscopic Foundations, Fronts, and Spatial Instabilities (Springer Series in Synergetics). Berlin: Springer.

Lawler, Gregory F. (2010). Random Walk and the Heat Equation. Rhode Island, USA: American Mathematical Society. Student Mathematical Library, Vol. 55.

Lehner, Günter. (2008). Electromagnetic Field Theory for Engineers and Physicists. Berlin: Springer.

LeVeque, Randall J. (2007). Finite Difference Methods for Ordinary and Partial Differential Equations: Steady-State and Time-Dependent Problems (Classics in Applied Mathematics). Philadelphia: Society for Industrial and Applied Mathematics.

Stratton, J.A. (1941). Electromagnetic Theory. London: McGraw Hill.

Strikwerda, John C. (2004). Finite Difference Schemes and Partial Differential Equations., 2nd Ed. Philadelphia, PA: Society of Industrial and Applied Mathematics.

7

Ant Colony Optimization (Past, Present and Future)

Oscar Montiel,* Roberto Sepúlveda and
Nataly Medina Rodríguez

ABSTRACT

This chapter provides a survey of the Ant Colony Optimization (ACO) metaheuristics, the original proposals that make flourish this natural computing optimization method are included, as well as other proposals that have reported good results to the present time. Some of the included approaches are the Simple ACO (SACO), Ant systems, Elitist Ant systems, Ant-Q, Ant Colony systems, parallelization of AS, and others. We also have included an application of SACO in mobile robotics.

7.1 Introduction

Ant colonies are social, like termites, bees and many other insects; they live in large communities, and work together. Ants feed and share their food with other ants in the colony so each member in the colony has a job to do, from laying eggs to gathering food for the colony.

Collective behaviors that emerge from insects have intrigued humans, and there have been many studies aimed to get a better understanding of this social interaction within an insect colony. Within an ant colony, there are three types of individuals: the queen, workers and soldiers; these

Instituto Politécnico Nacional, CITEDI, Tijuana, México.
  E-mail: oross@ipn.mx
* Corresponding author

individuals have to attend to a number of tasks such as reproduction by the queen, defense task for a soldier, food collection and nest building, which are done by worker ants.

7.1.1 Introduction to swarm intelligence

When a designing problem–solving is inspired by nature's examples of collective behaviors such as social insects, the study of self-organization in these colonies is called Swarm Intelligence (Eric Bonabeau 1999).

Using Swarm Intelligence, computational models can be solved by using an adaptive algorithm; however, this algorithm does not need to be inspired on animal communities; collective and distributed intelligence can be seen at almost any level, by using the biological foundation of cells, organs, immune and nervous systems or a human interaction model.

Nowadays, Swarm Intelligence refers to the design of distributed control systems that displays some "intelligence", and control is fully distributed among the agents or individuals of a community, by using the interaction between agents.

The interaction between agents can be done by using a Stochastic decision (Eric Bonabeau 1999), this means that decisions are often made without a precise knowledge of their impact on future behavior of a random process; this can be achieved by using appropriate actions that influence future evolution.

In deterministic models, inputs and initial conditions determine output processes, and sometimes they are known as systems with complete information, where the current value of state and input uniquely determine the future state (Carlo A. Furia 2012). Example of a deterministic system can be a light switch, where pressing the button yields the unique possible future state of light ON; other examples of deterministic systems are computer, classical physics, etc. Systems that are not deterministic are called non–deterministic, stochastic or random systems, they can change to different future states from the same present state and input by making arbitrary choices; examples of non-deterministic systems are quantum physics, lotteries, radioactive decay and some metaheuristics, which are approximate and usually non-deterministic. A metaheuristic can be seen as a general algorithm which can be applied to different optimization problems, some of them are Genetic Algorithms, Simulated Annealing and Ant Colony Optimization (Yang 2010).

A stochastic system involves uncertainty, whose states are non-deterministic; future states are determined by the system's predictable actions and by a random element, where the chance of occurrence of the variable is considered by introducing the concept of probability this classification is illustrated in Fig. 7.1, according to it, probabilistic models

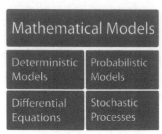

Fig. 7.1 General classification of mathematical models.

are related to stochastic processes; deterministic models to differential equations, where the output will be the same with a given starting condition or initial state, these models can be represented by using differential equations with fixed initial conditions (Alexandre Shapiro 2009).

Applications of stochastic programming can be presented in the finance area, data transmission on a network, transportation of energy and general routing problems; for routing problems, we are focused on Ant Colony Optimization (ACO) as a metaheuristic to find the shortest path between nodes with minimal costs (Nataly Medina-Rodríguez 2011).

7.1.2 Biological inspiration

The behavior of ant colonies has inspired the development of a large number of ant-based algorithms used to solve in most cases, combinatorial optimization problems; these colonies are used as a computational intelligence technique for solving these kinds of problems.

One of the first behaviors studied by ethologists (study of animal behavior) was the ability of ants to find the shortest path between their nest and a food source; Pierre–Paul Grassé (Marco Dorigo 2004) observed that some insects react to what he called "significant stimuli", and it means the interaction of some species through the environment; some species respond to changes in the environment. Grassé used the term stigmergy (Marco Dorigo 2004) to describe this particular type of communication in which the workers are stimulated by the performance they have achieved.

An example of stigmergy can be observed in ant colonies; ants walking to and from a food source depositing on the ground a substance called pheromone, so other ants can perceive the presence or absence of pheromone, and they tend to follow paths where pheromone concentration is higher. This mechanism allows an ant to be able to transport food from a source to their nest through an optimal path with a shortest distance.

In an experiment known as the "double bridge experiment" (Marco Dorigo 2005) as shown in Fig. 7.2, the nest of a colony of ants was connected to a food source by two bridges of equal lengths; ants started to explore the possible paths and eventually reached the food source. Initially, each ant randomly chose one of the two bridges, however, after some time one of the two bridges presented a higher concentration of pheromone; the ants started to follow this path with higher concentration of pheromone.

A variant of the double bridge experiment in which one bridge is longer than the other one, see Fig. 7.3. In this case, stochastic fluctuations in the initial choice of a bridge are much reduced: the ants choosing by chance the short bridge are the first to reach the nest. The short bridge has a higher level of pheromone than the longer bridge; this fact increases the probability that further ants select it rather than the long one.

Deneubourg (E. Foundas 2006) and colleagues proposed a simple stochastic model that adequately describes the dynamics of the ant colony as observed in the double bridge experiment; in this model, m ants per second cross the bridge in each direction at a constant speed; suppose that at a given moment in time, m_1 ants have used the first bridge, and m_2 the second one, the probability p_1 for an ant to choose the first bridge is given by Eq. (7.1.1):

$$p_1 = \frac{(m_1+k)^\alpha}{(m_1+k)^\alpha + (m_2+k)^\alpha} \qquad (7.1.1)$$

where α is an experimental value, identified on trail–following experiments, and k is the trail update. The probability of choosing the second bridge is $p_2 = 1-p_1$.

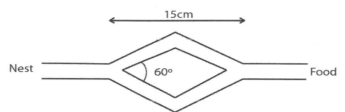

Fig. 7.2 Experimental setup for the double bridge experiment where branches have equal lengths.

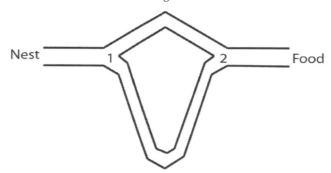

Fig. 7.3 Experimental setup for the double bridge experiment where branches have different lengths.

This model proposed by Deneubourg and colleagues for explaining the foraging behavior of ants was the main source of inspiration for the development of several Ant Colony Optimization (ACO) Algorithms; a list of successful ACO algorithms are shown in Table 7.1.

The original ACO algorithm was proposed by Marco Dorigo and colleagues, and it is known as the Ant System (M. Dorigo 1996), and a number of ACO algorithms were introduced since the original ACO algorithm was proposed and accepted as a new metaheuristic for solving combinatorial problems. All ACO algorithms share the same purpose, and they are inspired by the same idea. The following sections in this chapter present a brief introduction for each ACO algorithm in chronological order, and we describe how a generic ACO algorithm can be applied to the well–known traveling salesman problem.

Table 7.1 Ant Colony Optimization Algorithms (M. Dorigo 2004).

Algorithm	Authors	Year
Ant System (AS)	Dorigo et al. (M. Dorigo 1996)	1991
Elitist AS	Dorigo et al. (Marco Dorigo 2004)	1992
Ant – Q	Gambardella and Dorigo (Gambardella 1995)	1995
Ant Colony System	Dorigo and Gambardella (M. Dorigo 2004)	1996
$\mathcal{MAX}$-$\mathcal{MIN}$ AS	Stützle and Hoos (Thomas Stützle 2000)	1996
Rank – based AS	Bullnheimer et al. (Bernd Bullnheimer 1997)	1997
ANTS	Maniezzo (Maniezzo 1998)	1999
BWAS	Cordon et al. (O. Cordón 2000)	2000
Hyper – Cube AS	Blum et al. (Christian Blum 2004)	2001

7.2 The Simple ACO Algorithm

The Simple Ant Colony Optimization algorithm (SACO) was originally proposed as a didactic tool to explain the fundamental mechanisms of ACO algorithms (M. Dorigo 2004). SACO addressed four important issues, they are:

1. Probabilistic forward ants and solution construction.
2. Deterministic backward ants and pheromone updates.
3. Pheromone update based on solution quality.
4. Pheromone evaporation.

Moreover, the ants have two working modes:

1. *Forward mode.* In this mode, the ants move from the nest to the food, they do not deposit any pheromone while moving; an ant constructs the solution by choosing probabilistically the next node to move in the graph, being the available node those in the neighborhood of the

node where it is located. The probabilistic selection is biased by the pheromone trials previously deposited on the graph by other ants. In general, on each iteration or epoch all ants build a path to the destiny node, the probabilistic formula for the next node selection is given by (7.2.1),

$$
p_{ij}^k = \begin{cases} \dfrac{\tau_{ij}^\alpha}{\sum_{j \in N_i^k} \tau_{ij}^\alpha} & \text{if } j \in N_i^k \\ 0 & \text{if } j \notin N_i^k \end{cases}
\tag{7.2.1}
$$

where N_i^k is the set of feasible nodes connected to the node i with respect to ant k; τ_{ij}^α is the total pheromone concentration of the link (i, j), and α is a positive constant used as gain for the pheromone concentration influence.

2. *Backward mode*. In this case, when an ant has arrived at its destination, it switches to backward mode; the ant moves from the food back to the nest. Since the forward path of each ant was memorized, before starting to move backward on the path, the ants eliminate any loop that might exist. In the process of returning to the nest, the ant leaves pheromone on the arc it traverses, the amount may depend on the quality of the solution found to direct future ants more strongly toward better solutions. The pheromone evaporation is calculated using Eq. (7.3.2), and the pheromone update by (7.3.3), for each iteration.

Other particularities of the SACO are in (M. Dorigo 2004, Porta Garcia et al. 2009).

7.3 Ant System

Ant Colony Optimization was introduced by Marco Dorigo in the early 1990s (M. Dorigo 2004). Using very simple communication mechanisms, an ant group is able to find the shortest path between any two points by choosing the paths according to pheromone levels. ACO metaheuristics can be applied to the TSP, where the pheromone trails are associated with arcs and therefore τ_{ij} refers to the desirability of visiting city j directly after city i. The heuristic information is chosen as $\eta_{ij} = \frac{1}{d_{ij}}$, that is, the heuristic desirability of going from city i to city j is inversely proportional to the distance between the two cities. For implementation purposes, pheromone trails are collected into a pheromone matrix whose elements are the τ_{ij}'s.

Tours are created by applying the following simple constructive procedure to each ant:

1. Each ant chooses, according to some criterion, a start city at which the ant is positioned.
2. Each ant uses a pheromone and heuristic values to probabilistically construct a tour by iteratively adding cities that the ant has not visited yet, until all cities have been visited.
3. Each ant goes back to the initial city.
4. After all ants have completed their tour, they may deposit pheromone on the tours they have followed.

7.3.1 Tour construction in ant system

In AS, m ants concurrently build a tour of the TSP; initially, ants are put on randomly chosen cities. At each construction step, ant k applies a probabilistic action choice rule, called random proportional rule, to decide which city is going to visit next. The probability with which ant k, currently at city i, chooses to go to city j is shown in Eq. (7.3.1):

$$p_{ij}^k = \frac{[\tau_{ij}]^\alpha [\eta_{ij}]^\beta}{\sum_{l \in N_i^k} [\tau_{ij}]^\alpha [\eta_{ij}]^\beta}, \quad if\, j \in N_i^k \tag{7.3.1}$$

where $\eta_{ij} = \frac{1}{d_{ij}}$ is a heuristic value, α and β are two parameters which determine the relative influence of the pheromone trail and the heuristic information, and N_i^k is the feasible neighborhood of ant k when being at city i, that is, the set of cities that ant k has not visited yet. The probability of choosing a city outside N_i^k is 0.

- If $\alpha = 0$, the closest cities are more likely to be selected, it regulates the influence of τ_{ij}.
- If $\beta = 0$, only pheromone amplification is used, without any heuristic bias, it regulates the influence of η_{ij}.

7.3.2 Update of pheromone trails

After all the ants have constructed their tours, the pheromone trails are updated. This is done by first lowering the pheromone value on all arcs by a constant factor, and then adding pheromone on the arcs the ants have crossed in their tours. Pheromone evaporation is implemented by Eq. (7.3.2):

$$\tau_{ij} \leftarrow (1-\rho)\tau_{ij}, \quad \forall (i,j) \in L \tag{7.3.2}$$

where $0 \leq \rho \leq 1$ is the pheromone evaporation rate. After evaporation, all ants deposit pheromone on the arcs they have crossed in their as it is shown in Eq. (7.3.3):

$$\tau_{ij} \leftarrow \tau_{ij} + \sum_{k=1}^{m} \Delta\tau_{ij}^{k}, \quad \forall(i,j) \in L \tag{7.3.3}$$

where $\Delta\tau_{ij}^{k}$ given by (7.3.4), is the amount of pheromone deposited on the visited arcs by ant k.

$$\Delta\tau_{ij}^{k} = \begin{cases} \frac{Q}{L_k} & \text{if ant } k \text{ used edge } (i,j) \text{ in its tour} \\ 0 & \text{otherwise} \end{cases} \tag{7.3.3}$$

where Q is a constant, and L_k is the length of the tour constructed by ant k.

7.3.3 ACO for the traveling salesman problem

Given an undirected graph $G = (V,E)$ connected with edge–weights, each vertex V of this graph represents the cities, and the edge weights represent the distance between cities. The main purpose in this problem is to find a closed path in G that contains each node exactly once, called a tour, and whose length is minimal. Thus, the search space S consists of all tours in G. The optimized objective function value $f(s)$ of a tour $s \in S$ is defined as the sum of the edge–weights of the edges that are in s (Kumar 2010). Under this context, the TSP problem (M. Dorigo 1997, M. Dorigo 1997) is the problem of a salesman who, starting from his hometown, wants to find a shortest tour that takes him through a given set of costumer cities and then back home, visiting each costumer city exactly once. The TSP can be represented by a completed weighted graph $G = (N,A)$ with N being the set of nodes representing the cities, and A being the set of arcs. Each arc $(i,j) \in A$ has assigned a value (length) d_{ij} which is the distance between cities i and j; each path becomes a directed graph.

In ACO, the problem is solved by simulating a number of artificial ants moving on a graph as shown in Fig. 7.4, in which each vertex represents a city and each edge represents a connection between two cities, associated with a level of pheromone to ensure visibility on ants.

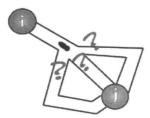

Fig. 7.4 Ant decision for choosing the best edge with high level of pheromone.

Algorithm 7.1 Ant Colony Optimization general Metaheuristic.

```
procedure ACO Metaheuristic
        begin
        Initialize Parameters
        for each cycle
            for each ant
                for each city
                    Build a solution k
                    Evaluate solution k
                end for
                Save Best solution
                Update Trail
            end for
            Print Best Solution
        end for
end Procedure
```

ACO is an iterative algorithm and at each iteration, a number of artificial ants can be considered as constant or variable; each ant builds a solution by following the path from vertex to vertex of the graph with the constraint of not visiting any vertex that the ant has already visited, and it chooses the following vertex to be visited according to a stochastic mechanism influenced by the pheromone, see Eq. (7.3.1). The general ACO metaheuristic can be represented by the pseudocode shown in Algorithm 7.1 (M. Dorigo 1997).

To explain the ACO metaheurisitc, we present a simple general example of the solution construction for a TSP problem consisting of four cities shown through Fig. 7.5 to Fig. 7.8. The solution construction starts by randomly choosing a start node for an ant; in Fig. 7.5, the start node is labeled with the number 1. In this case, the ant has to make a decision to choose the best edge for all these three possible solutions, marked by thick lines.

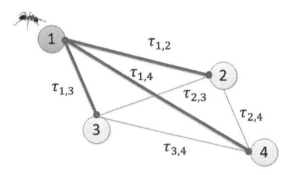

Fig. 7.5 Example of the solution construction for a TSP problem, starting from node 1.

Once the ant made a decision to choose the best edge, the next step needs to apply once again the probabilistic decision; this is the main reason why ACO algorithms are mainly iterative. Note that the ant has visited edge (1,3) and it is marked with dashed lines in Fig. 7.6; the next decision is represented in Fig. 7.7, choosing the edge (3,4).

Figure 7.8 shows the complete solution after the final construction step, this is because the ant in node 2 has to choose node 1 with two possibilities, but one of these edges cannot be chosen because other edges connected to this node were visited before, therefore the edge (2,1) with pheromone $\tau_{1,2}$ is the final edge to complete the solution construction for a TSP.

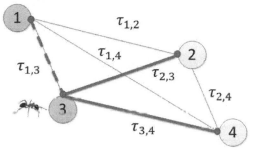

Fig. 7.6 Example of the solution construction for a TSP, once the ant has made a decision.

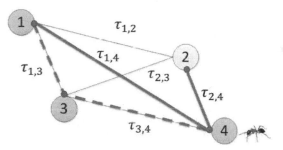

Fig. 7.7 Example of the solution construction for a TSP, once the ant has made again a decision.

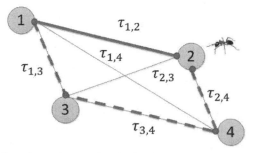

Fig. 7.8 Complete solution after the final construction step for a TSP.

7.4 Elitist AS: An Extension of AS

With an *Elitist Strategy* (Marco Dorigo 2004), there is a reinforcement to the arcs belonging to the best tour found since the beginning of the algorithm which is denoted as T^{bs} (*best–so–far* tour), and it is applied with a feedback to the *best–so–far* tour, and it consists in additional pheromone deposited by an additional ant called *best–so–far* ant.

The additional reinforcement of tour T^{bs} is achieved by adding a parameter e/C^{bs} to its arcs, where e is a parameter that defines the weight given to the *best–so–far* tour, and C^{bs} is its length, see Eq. (7.4.1),

$$\tau_{ij} \leftarrow \tau_{ij} + \sum_{k=1}^{m} \Delta\tau_{ij}^{k} + e\Delta\tau_{ij}^{bs} \qquad (7.4.1)$$

where $\Delta\tau_{ij}^{bs}$ is defined as (7.4.2):

$$\Delta\tau_{ij}^{bs} = \begin{cases} \dfrac{1}{C^{bs}}, & \text{if arc } (i,j) \text{belongs to } T^{bs} \\ 0, & \text{otherwise} \end{cases} \qquad (7.4.2)$$

Marco Dorigo et al. (M. Dorigo 1996, Dorigo 1992) reported that an appropriate value for e allows the algorithm to find better solutions in a few iterations than in AS.

7.5 Ant-Q: Introduction to Q-Learning

Q-Learning is an extension of reinforcement learning, which was an important improvement for these kinds of algorithms. To explain how Q-learning was an improvement, let us first suppose that we have a model of the environment with some states and actions that an agent has to follow, and each arc has a reward to choose the best path as shown in Fig. 7.9.

In this example, we set an agent located in node 3 as a start node, and we want the agent to learn to reach node 6, which is the end node (the goal); each arrow contains an instant reward value, so the nodes that lead immediately to the goal have an instant reward of 100 in this case. Other nodes not directly connected to the target node have zero reward.

Suppose that the agent can learn through experience to find the target, choosing the best path in this case. The terminology in Q-Learning includes the terms "state" and "action" to learn the next state with some action and some kind of reward. We can put the state diagram and the instant reward values into the following reward table shown in Fig. 7.10, commonly named as the "matrix R".

If the agent starts at node 3, from this state, it can go to state 2 directly because state 2 is connected and there is a reward value higher than other paths; however, the agent cannot directly go to state 6 because there is no

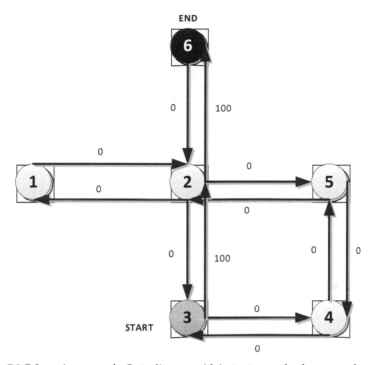

Fig. 7.9 Q-Learning example: State diagram with instant reward values on each arrow.

$$R =$$

Action State	1	2	3	4	5	6
1	-	0	-	-	-	-
2	0	-	0	-	0	100
3	-	100	-	0	-	-
4	-	-	0	-	0	-
5	-	0	-	0	-	-
6	-	0	-	-	-	-

Fig. 7.10 Reward table example with states and actions.

direct connection. In Fig. 7.10, the dash symbol '-' represents null values, where there is no link between nodes, for example, state 6 cannot go to state 1 directly.

The learning process consists of adding a similar matrix "Q" to the brain of the agent, representing the memory of what the agent has learned through experience, where the rows of the matrix represent the current state of the agent and the columns represent the possible actions leading to the next state. The agent starts out knowing nothing, the matrix Q is initialized to zero. The transition rule of Q learning is defined as (7.5.1):

$$Q(state, action) = R(state, action) + \gamma * Max[Q(\text{next state, all actions})] \quad (7.5.1)$$

Where γ is the learning parameter, with a range of $0 \leq \gamma \leq 1$. If r is closer to zero, the agent will tend to consider only immediate rewards. If γ is closer to one, the agent will consider future rewards with higher value of weights.

Ant–Q is the first attempt to enhance an ACO algorithm with a reinforcement using Q-learning; this algorithm is proposed by Marco Dorigo et al. (Gambardella 1995) to improve the performance of Ant System with an important modification in the pheromone update (7.3.2) from Ant System, adopted from Q–learning as it is shown in (7.5.2):

$$\tau_{ij} \leftarrow (1 - \rho)\tau_{ij} + \alpha \left(\Delta\tau_{ij} + \gamma \cdot \max_{l \in N_j^k}(\tau_{jl}) \right), \quad \forall(i, j) \in L \quad (7.5.2)$$

The pheromone update rule is now implemented by using the techniques in Q-learning; it updates the pheromone value of the transition (i,j) based on the pheromone value of the successive transition (j,l) with a learning rate α and here the decay rate is γ. The value of $\Delta\tau_{ij}$ is computed based on the calculation on the best tour.

7.6 Ant Colony System (ACS)

The Ant Colony System differs from Ant System (AS) in three main aspects (Miagkikh 1999):

1. ACS uses a transition rule that gives a higher priority to exploitation than exploration.

   When an ant is located at city i, ant k moves to a city j using the following tour construction called *pseudorandom proportional* rule (M. Dorigo 2004), given by (7.6.1):

$$j = \begin{cases} argmax_{l \in N_i^k}\{\tau_{il}[\eta_{il}]^\beta\}, & if \ q \leq q_0 \\ J, & \text{otherwise} \end{cases} \quad (7.6.1)$$

   where q is a random variable uniformly distributed in the range of [0, 1]; according to this transition rule, ant k at city i chooses to move to the city j with the best edge in terms of pheromone trail and heuristic value with q_0 probability, therefore it exploits the current knowledge of the colony; otherwise, it selects a random city in the neighborhood. J is a random variable selected according to the probability distribution given by equation (7.3.1) with $\alpha = 1$; notice that there is no explicit weight parameter α for τ_{il}.

Tuning the parameter q_0 allows modulation of the degree of exploration for exploring other tours or to concentrate the search in an optimal path.

2. The pheromone update is only applied to the best–so–far tour.

3. After all the ants in the colony construct their solutions, only the tour corresponding to the best solution receives a pheromone update, implemented by Eq. (7.6.2):

4. $\tau_{ij} \leftarrow (1 - \rho)\tau_{ij} + \rho\Delta\tau_{ij}^{bs}, \quad \forall(i,j) \in T^{bs}$ \hfill (7.6.2)

where $\Delta\tau_{ij}^{bs} = 1/C^{bs}$; T^{bs} stands for the best solution so far, and C^{bs} stands for the cost of that solution. The parameter ρ is not the pheromone evaporation rate as in AS, it is now a parameter that weights the pheromone to be deposited in ACS (Pournos 2004), this means that the new pheromone trail becomes a weighted average of all the pheromone trail, and the pheromone to be deposited. This algorithm is important, because the computational complexity of the pheromone update at each iteration, is reduced from $O(n^2)$ to $O(n)$, where n is the size of the instance being solved (M. Dorigo 2004).

Local pheromone update is done after each transition on any edge, it means that when each time an ant uses an arc (i, j) to move from city i to city j, it removes a pheromone concentration from the edge to increase the exploration of alternative paths, implemented by Eq. (7.6.3):

$\tau_{ij} \leftarrow (1 - \xi)\tau_{ij} + \xi\tau_0,$ \hfill (7.6.3)

where ξ, $0 < \xi < 1$ is a parameter that decreases the desirability of choosing an edge, and τ_0 stands for the initial value for the pheromone trails; τ_0 can be defined as $1/nC^{nm}$, where n is the number of cities in the TSP instance and C^{nm} is the length of a nearest–neighbor tour.

7.7 *MAX-MIN* AS

In *MAX-MIN* AS, pheromone evaporation occurs on all edges after all ants construct their solutions, however, pheromone deposit is only done for the edges of the best-so-far solution (Thomas Stützle 2000). This algorithm differs in three aspects from AS:

- After each iteration, only one ant adds pheromone; this ant may be the one which found the best solution in the current iteration or the one which found the best solution from the beginning of the trial.
- The pheromone trails on each solution are constrained to the interval $[\tau_{min}, \tau_{max}]$.
- The pheromone trails are initialized to the upper pheromone trail limit with a small pheromone evaporation rate, causing an increase on the exploration of tours at the start of the search.

After all ants have constructed a tour, pheromone trails are updated by applying evaporation as in AS, followed by the deposit of new pheromone implemented by the Eq. (7.7.1):

$$\tau_{ij} \leftarrow \tau_{ij} + \Delta\tau_{ij}^{best} \tag{7.7.1}$$

where, $\Delta\tau_{ij}^{best} = 1/C^{best}$, and C^{best} denotes the solution cost of either the iteration-best or the global-best solution; while in ACS typically only global-best solution is used, $\mathcal{MM}$ AS focuses on the use of the iteration-best solutions.

In $\mathcal{MM}$ AS, lower and upper limits τ_{min} and τ_{max} on the possible pheromone values on any arc, are imposed in order to avoid search stagnation, these values have the effect of limiting the probability p_{ij} of choosing a city j when an ant is in city i to the interval $[p_{min}, p_{max}]$, with $0 < p_{min} \leq p_{ij} \leq p_{max} \leq 1$ (M. Dorigo 2004).

7.8 Rank-based AS

In rank-based AS (AS_{rank}), after all m ants have generated a tour, the ants are sorted by tour length ($L_1 \leq L_2 \leq \cdots \leq L_m$), and the contribution of an ant to the trail level update is weighted according to the rank μ of the ant; only the ω best ants are considered (Bernd Bullnheimer 1997).

AS_{rank} is a modification of Elitist AS. It uses only the best $\omega-1$ tours of the iteration and the best–so–far tour. Experimentally ω is set equal to 25% of the number of ants (Bernd Bullnheimer 1997, M. Dorigo 1997).

The amount of deposited pheromones depends on the tours rank, where T^{bs} in this case, has a weight factor of ω, the iteration-best tour a weight factor of $\omega-1$, the second ranked tour a weight factor of $\omega-2$, and so on. The pheromone update rule in this algorithm is set as (7.8.1)

$$\tau_{ij} \leftarrow (1-\rho) \cdot \tau_{ij} + \sum_{k=1}^{\omega-1}(\omega-r)\Delta\tau_{ij}^{r} + \omega\Delta\tau_{ij}^{bs}, \qquad \forall(i,j) \in L \tag{7.8.1}$$

where ω is the amount of ranked ants including T^{bs}, and the lower bound of $(\omega-r)$ is 0, to ensure that no pheromones are subtracted. Experimental results by Bullneheimer et al. (Bernd Bullnheimer 1997) suggest that AS_{rank} performs slightly better than Elitist AS and significantly better than AS.

7.9 ANTS

The *Approximate Nondeterministic Tree Search* (ANTS) was proposed in 1999 by (Maniezzo 1998); the name derives from the fact that the proposed algorithm can be extended in a straightforward way to a branch & bound procedure (Bertsekas 1995).

ANTS is based on the use of heuristic values derived from domain knowledge a priori, but computes lower bounds on completing a *partial solution* after temporarily adding a node (i,j_1), and it uses all these lower bounds of (i,j_1) to (i,j_n), with n the number of neighbors, as heuristic values. The expected *lower bound* (LB) on the expected result of a tour, is the sum of the lengths of the partial solution, the chosen edge (i,j_k) and an estimate of the edges needed to complete the solution from node j_k. The solution construction and pheromone update have also been changed, with solution construction as Eq. (7.9.1):

$$p_{ij}^k = \frac{\xi \tau_{ij} + (1-\xi)\eta_{ij}}{\sum_{l \in \mathcal{N}_i^k} \xi \tau_{il} + (1-\xi)\eta_{il}}, \quad if \ j \in \mathcal{N}_i^k \tag{7.9.1}$$

where α and β in Eq. (7.3.1) have been replaced by the parameter ξ and multiplication has been replaced by addition; in the pheromone update rule, pheromone evaporation has been removed, the equation is now changed to (7.9.2)

$$\tau_{ij} \leftarrow \tau_{ij} + \sum_{k=1}^{m} \Delta \tau_{ij}^k \tag{7.9.2}$$

where $\Delta \tau_{ij}^k$ is given by (7.9.3),

$$\Delta \tau_{ij}^k = \begin{cases} \vartheta (1 - \frac{C^k - LB}{L_{avg} - LB}, & if \ edge(i,j) \ belongs \ to \ T^k, \\ 0, & otherwise \end{cases} \tag{7.9.3}$$

where C^k is the length of the tour of ant k, L_{avg} the average length of the last l iterations and ϑ is a parameter usually set to τ_0; in this equation, the paths for tours better than L_{avg} are reinforced, while if the path is worse than L_{avg}, pheromones will be subtracted from the tour.

7.10 BWAS: 2000

The Best-Worst Ant System (BWAS) (O. Cordón 2000), is an algorithm which uses elements from Evolutionary Computation such as the term *mutations,* but also some elements from ACS with the pheromone update mechanism; and from $\mathcal{MM}$ AS, the restart and reset of pheromones have been included.

The BWAS model improves the performance of ACO models using evolutionary algorithm concepts. The proposed BWAS uses the AS transition rule as in Eq. (7.3.1); besides, the usual AS evaporation rule is changed in BWAS. In ACS, T^{bs} receives pheromone updates at each iteration, in addition the *iteration–worst* solution (T^{tw}) removes pheromones from edges that it contains but are not in T^{bs}. The pheromone update equation is changed to (7.10.1):

$$\tau_{ij} \leftarrow (1 - \rho) \cdot \tau_{ij} + \rho \cdot \Delta\tau_{ij}, \quad \forall(i,j) \in L \tag{7.10.1}$$

where

$$\tau_{ij} = \begin{cases} f\left(C^{bs}\right), & \text{if}(i,j) \in T^{bs} \\ 0, & \text{otherwise} \end{cases} \tag{7.10.2}$$

here, $f(C^{bs})$ is a function for the length of T^{bs}, usually $f(C^{bs}) = 1/C^{bs}$. To penalize T^{tw}, it is evaporated one more time after the global evaporation given by (7.10.3)

$$\tau_{ij} \leftarrow (1 - \rho) \cdot \tau_{ij}, \forall(i,j) \in T^{iw} \text{and}(i,j) \notin T^{bs} \tag{7.10.3}$$

When the difference between T^{bs} and T^{tw} becomes less than a pre-defined percentage, the algorithm converges too much to $\mathcal{MM}$ AS; in this case, the difference is defined as the amount of different edges.

Implementing evolutionary concepts such mutations to the pheromone values, which increase the explorative behavior, there exists a probability P_m for an edge to be mutated, as shown in the Eq. (7.10.4):

$$\tau'_{ij} = \begin{cases} \tau_{ij} + mut(k, \tau_{threshold}), & \text{if } a = 0 \\ \tau_{ij} - mut(k, \tau_{threshold}), & \text{if } a = 1 \end{cases} \tag{7.10.4}$$

where a is a random value between $\{0, 1\}$; k is the current iteration of the main loop and $\tau_{threshold}$ is the average of the pheromone values of the edges in the best-so-far solution as (7.10.5):

$$\tau_{threshold} = \frac{\Sigma_{(i,j)\in T^{bs}} \tau_{ij}}{|T^{bs}|} \tag{7.10.5}$$

must calculates the size of the mutation, which slowly will get bigger during the execution of the algorithm, see Eq. (7.10.6)

$$mut(k, \tau_{threshold}) = \frac{k - k_r}{N_k - k_r} \cdot \rho \cdot \tau_{threshold} \tag{7.10.6}$$

kr represents the last iteration of a restart, N_k is the maximum number of iterations and ρ is the mutation power which defines how fast the mutation reaches $\tau_{threshold}$, and how much higher it can go.

As we can see, in this algorithm, there are many parameters that have to be adjusted properly, and if it is done incorrectly, could greatly decrease the algorithms performance; for this reason, BWAS algorithm has its variants, such as the BWASC (BWAS with Ant Colony System) and many other algorithms with or without mutation, trail updating, etc.

7.11 10 Hyper–Cube AS (HC-AS): 2001

In this algorithm, it is not a change on the rules for the ants or the update mechanisms, but it is a remarkable change on the representation of the construction graph such that solutions can be made by binary vectors.

A binary vector, $\vec{v} = (v_1, ..., v_n)$ with every vector component representing the pheromone value of an edge, represents a solution by having every variable v_1 to v_n take a value in $\{0, 1\}$, and each unique ordering of 0's and 1's is a solution. In this case, an n–dimensional hyper–cube is generated, where each corner is a solution, but if values in the entire interval $[0, 1]$ are used instead of values in $\{0, 1\}$ this hyper–cube represents the entire search–space equivalent to the other Ant System algorithms (Christian Blum 2004).

HC-AS has a set of decision variables, corresponding to the solution components with values of $\{0, 1\}$. A 1 is indicating that the component is a member of the solution defined by the vector. Feasible solution s is a subset of the set of corners of the n–dimensional hyper–cube as it is shown in Fig. 7.11.

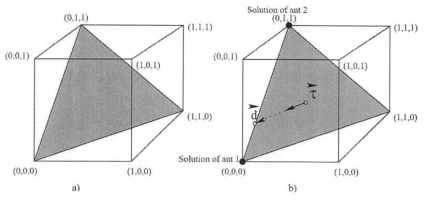

Fig. 7.11 A set of feasible solutions S consists of the three vectors $(0, 0, 0)$, $(0, 1, 1)$ and $(1, 1, 0)$ (Christian Blum 2004) shown in a). In b) are represented two solutions created by two ants, and $\vec{d}$ is the weighted average of these two solutions, so $\vec{\tau}$ will be shifted towards $\vec{d}$.

7.12 The Present of ACO: High Performance Computing

The availability of parallel architectures at low cost has widened the interest for the parallelization of Ant Colony Optimization algorithm. In order to parallelize the ACO, it is more important to modify the structure of ACO to get better optimization effect rather than to transfer the sequential ACO into a parallelization schema (L. Chen 2008).

The main purpose of parallel implementation of ACO is to obtain a high speed up and efficiency while the convergence and the ability of optimization are maintained or even improved. Some results on parallel Ant Colony algorithms have been reported, Bullnheimer (B. Bullnheimer 1998) proposed two parallelization strategies of synchronous and asynchronous for ACO using the TSP. Marcus Randal introduced a synchronous parallel strategy which assigns only one ant on each processor (Randall 2002). Marco Dorigo and Christian Blum introduced a parallel ACO on the hyper–cube architecture by modifying the rule of updating the pheromone so as to limit the pheromone values within the range of [0,1] (Christian Blum 2004).

Piriyakumar introduced an asynchronous parallel Max-Min ACO associated with the local search strategy (Douglas Antony Louis Piriyakumar 2002), and in (Tan 2012) a MapReduce Max–Min Ant System based on parallel programming model for the TSP was presented.

7.12.1 Introduction to parallel ant colony optimization

The sequential algorithm contains a high degree of natural parallelism (B. Bullnheimer 1998); the behavior of a single ant, during one iteration, is totally independent of the behavior of all other ants during that iteration. We will discuss a strategy called "synchronous (fork–join) algorithm".

A straight forward parallelization strategy for ACO is to compute the TSP tours in parallel; this would result in a fork–join structure, as shown in Fig. 7.12.

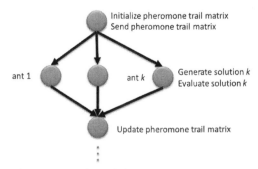

Fig. 7.12 Parallel implementation of Ant Colony Optimization in a message passing model.

In Fig. 7.12, an initial process (master) would spawn a set of processes, one for each ant. After distributing initial information about the problem, each process can generate a single solution for each ant k. After finishing this procedure, the result is sent from each process back to the master process. The master process updates the trail levels by calculating the intensity of the trails and checks for the best tour found so far. A new iteration is initiated by sending out the updated trail levels.

Bullnheimer (B. Bullnheimer 1998) states that ignoring any communication overhead, this approach would imply optimum (asymptotic) speedup as (7.12.1), assuming that an infinite number of processing elements (workers) is available, i.e., one process is assigned to one worker. In (7.12.1), m is the quantity of ants of the colony.

$$S_{asymptotic}(m) = \frac{T_{sec}(m)}{T_{par}(m,\infty)} = \frac{O(m^3)}{O(m^2)} = O(m) \qquad (7.12.1)$$

where $T_{seq}(m) = O(m^3)$ the computational complexity of the sequential algorithm for the problem is size m and $T_{par}(m,\infty) = O(m^2)$ is the computational complexity of the parallel algorithm for the problem size m and for infinite system size.

Communication overhead certainly cannot be disregarded and it has to be taken into account, further the system size (number of processing elements N) is restricted and it is typically smaller than the problem size (number of ants m).

Balancing the load among the workers is easily accomplished by assigning to worker $(j = 1...N)$ the processes (ants) m_i for $(i = 1...m)$ according to $m_i : j = i \ mod \ N$, thus each worker holds about the same number of processes and each process is of the same computational complexity (Nataly Medina-Rodríguez 2011, Oscar Montiel 2012).

When considering communication overhead, the ratio of the amount of computation assigned to a worker and the amount of data to be communicated has to be balanced. After each iteration, all completed tours and their lengths have to be sent to a central process (master). Then the new trail levels need to be computed and then broadcasted to each worker after which only then can start a new iteration (Nataly Medina-Rodríguez 2011).

7.13 Application of SACO in Mobile Robotics

Originally, SACO was proposed for didactical purpose; however, SACO combined with fuzzy logic was implemented in an experimental platform focus to teach how SACO algorithm is used to solve the problem of path planning for autonomous navigation for virtual and real environments with obstacle avoidance (Montiel-Ross et al. 2010). The software was called

ACO Test Center (ACP-TC), which is a global planner with two algorithms, the Simple Ant Colony Optimization Meta-Heuristic (SACO-MH), and a modification of the original proposal named SACOdm, where 'd' stands for distance and 'm' for memory. In this version, the algorithm is able to influence the decision making process based on the existing distance between the source and target nodes, the ants can remember the visited nodes in order to avoid stagnation. The new characteristics added to the SACO a performance increase (speed up) of 10x most of the times.

In the SACOdm, the original formula (7.2.1) was modified to accelerate the decision-making process in the free space path optimization. The transition formula for the SACOdm is given by (7.13.1)

$$
p_{ij}^k = \begin{cases} \dfrac{\tau_{ij}^\alpha}{\sum_{j \in N_i^k} \tau_{ij}^\alpha \varepsilon^\beta} & \text{if } j \in N_i^k \\[2ex] 0 & \text{if } j \notin N_i^k \end{cases} \tag{7.13.1}
$$

where ε is the Euclidian distance between the source and target nodes, and β is a value that amplifies the influence of ε. The valid range of β is $[0, \infty)$.

For the case of path planning, the SACOdm algorithm includes a fuzzy cost function based on heuristic knowledge that can be easily adjusted to improve performance using the Simple Tuning Algorithm (STA) (Gómez-Ramírez and Chavez-Plascencia 2004, Montiel et al. 2007).

7.14 The Future of ACO: A Final Conclusion

Ant Colony Optimization can be an algorithm with a high performance level implemented in several parallel architectures; ACO is therefore theoretically well–suited for implementation on Graphics Processing Units (GPUs) (Francisco Javier Diego 2012). Recently, GPU computation has become popular with great success in scientific fields (Ryoo 2008).

In Ch. 4, it was explained that processors in CUDA are grouped into multiprocessors in which each one consists of thread processors to exchange data via fast–squared memory. In a CUDA program, threads form two hierarchies: the *grid* and *thread blocks,* which is a set of threads and a grid is a set of blocks with the same size.

Each thread executes the same code specified by the *kernel function* (Tsutsui 2012). This allows to compute ACO algorithms faster than in parallel architectures with a fewer number of processors.

Moreover, in general, Ant Colony Optimization algorithms are extremely suitable to be implemented on FPGAs platforms because of its intrinsic parallel structure. FPGAs allow exploiting this feature providing several levels of parallelism, from the higher at the circuit level by parallelizing all the operations, to a user defined parallelism level.

The future of ACO is not mainly addressed to implementations on GPUs or in other parallel architectures; there are other proposals, for example, it can be implemented with a single processor by using ACO with memory–based immigrants, which is a new scheme that performs well on different variations of the dynamic travelling salesman problem (Mavrovouniotis 2012). In this approach, old environments will re–appear in the future, this might be useful to maintain the diversity within the population.

It is a great task for all researchers to find new proposals that improve the efficiency of metaheuristic algorithms. Implementation for real applications does not guarantee to find the optimal solution, or even the best near–optimal solution, just because this kind of algorithm will converge in theory, but in practice, it may take a significant number of iterations before the optimal solutions can be reached (Montiel and Diaz Delgadillo 2013).

Acknowledgments

The authors would like to thank the "Instituto Politécnico Nacional (IPN)", "Comisión de Operación y Fomento de Actividades Académicas del IPN (COFAA-IPN)" and the Mexico's entity "Consejo Nacional de Ciencia y Tecnología (CONACYT)" for supporting our research activities.

References

Alexandre Shapiro, D.D. (2009). Lectures on Stochastic Programming: Modelling and Theory. USA: Society for Industrial and Applied Mathematics.

B. Bullnheimer, G. Kotsis and C. Strauss (1998). Parallelization strategies for the ant system, in: R. De Leone, A. Murli, P. Pardalos, G. Toraldo (eds.), High Performance Algorithms and Software in Nonlinear Optimization; Series: Applied Optimization, 24, Kluwer, Dordrecht, pp. 87–100.

Bernd Bullnheimer, R.F. (1997). A New Rank Based Version of the Ant System—A Computational Study. Central European Journal for Operations Research and Economics, pp. 25–38.

Bertsekas (1995). Dynamic Programming and Optimal Control. Belmont, MA: Athena Scientific.

Carlo A. Furia, D.M. (2012). Modeling Time in Computing. Berlin Heidelberg: Springer-Verlag.

Christian Blum, A.R. (2004). The hyper-cube framework for ant colony optimization. IEEE Transactions on Systems, Man, and Cybernetics, Part B: Cybernetics, pp. 1161–1172.

Dorigo, M. (1992). Optimization, Learning and Natural Algorithms. PhD. Thesis, Dipartamento die Elettronica. Politecnico di Milano, Milan, Italy.

Douglas Antony Louis Piriyakumar, P.L. (2002). A new approach to exploiting parallelism in Ant Colony Optimization. Proceedings of 2002 International Symposium on Micromechatronics and Human Science, pp. 237–243.

E. Foundas, A.V. (2006). Pheromone models in ant colony optimization (ACO). Journal of Interdisciplinary Mathematics, pp. 157–168.

Eric Bonabeau, M.D. (1999). Swarm Intelligence: From Natural to Artificial Systems. New York, NY: Oxford University Press.

Francisco Javier Diego, E.M.-M.-S. (2012). Parallel CUDA Architecture for Solving de VRP with ACO. Industrial Engineering: Innovative Networks-Springer.

Gambardella, L.M. (1995). Ant-Q: A Reinforcement learning approach to the traveling salesman problem. Proc. ML-95, 12th Int. Conference on Machine Learning, pp. 252–260.

Gómez-Ramírez, E. and A. Chavez-Plascencia (2004). How to tune fuzzy controllers. Proceedings of FUZZ'04, Budapest, pp. 1287–1292.

Kumar, R. (2010). Theory of Automata, Languages and Computation. New Delhi: McGrawHill.

L. Chen, H.-Y. (2008). Parallel Implementetation of Ant Colony Optimization on MPP. Proceedings of the Seventh International Conference on Machine Learning and Cybernetics. A. Prieditis and S. Russell (Eds.), Morgan Kaufmann, 1995, pp. 252–260.

M. Dorigo, L.G. (1997). Ant Colony System: a cooperative learning approach to the traveling salesman problem. IEEE Transactions on Evolutionary Computation, 1(1): 53–66.

M. Dorigo, L.M. (1997). Ant colonies for the traveling salesman problem. Biosystems Vol. 43.

M. Dorigo, T.S. (2004). Ant colony optimization. Cambridge, MA: MIT Press.

M. Dorigo, V.M. (1996). Ant System: Optimization by a colony of cooperating agents. IEEE Transactions on Systems, Man, and Cybernetics, pp. 29–41.

Maniezzo, V. (1998). Exact and approximate nondeterministic tree-search procedures for the quadratic assignment problem. Vittorio Maniezzo Scienze Dell'informazione. Università di Bologna, Via Sacchi.

Marco Dorigo, B.C. (2005). Ant colony optimization theory: A survey. Theoretical Computer Science, Vol. 344, Issues, 2–3, 243–278.

Marco Dorigo, T.S. (2004). Ant Colony Optimization. Cambridge, MA: MIT Press.

Mavrovouniotis, M. (2012). Ant colony optimization with memory-based immigrants for the dynamic vehicle routing problem. IEEE Congress on Evolutionary Computation (CEC), Conference Location:Brisbane, QLD, pp. 1–8.

Miagkikh, V.P. (1999). An Approach to Solving Combinatorial Optimization Problems Using a Population of Reinforcement Learning Agents. Genetic and Evolutionary Computation Conference, pp. 1358–1365.

Montiel, O. and F. Diaz Delgadillo (2013). Combinatorial complexity problem reduction by the use of artificial vaccines. Expert Syst. Appl., 40(5): 1871–1879.

Montiel, O., R. Sepúlveda, P. Melin, O. Castillo, M. Porta and M. Meza (2007). Performance of a Simple tuned fuzzy controller and a PID controller on a DC motor. Proceedings of the 2007 IEEE Symposium of Foundations of Computational Intelligence FOCI 2007, Hawaii, USA, pp. 531–537.

Montiel-Ross, O., R. Sepúlveda, O. Castillo and P. Melin (2010). Ant Colony Test Center for Planning Autonomous Mobile Robot Navigation. Computer Applications in Engineering Education, doi: http://dx.doi.org/10.1002/cae.20463.

Nataly Medina-Rodríguez, O.M. (2011). Toolpath Optimization for Computer Numerical Control Machines based on Ant Colony. International Association of Engineers. Engineering Letters.

O. Cordón, F.H. (2000). A new ACO model integrating evolutionary computation concepts: The best-worst ant system. Proc. of ANTS' 2000, Brussels, Belgium, pp. 22–29.

Oscar Montiel, N.M. (2012). Methodology to optimize manufacturing time for a CNC using high performance implementation of ACO. International Journal of Advanced Robotic System.

Porta Garcia, M.A., O. Montiel, O. Castillo, R. Sepúlveda and P. Melin (2009). Path planning for autonomous mobile robot navigation with ant colony optimization and fuzzy cost function evaluation. Applied Soft Computing, pp. 1102–1110.

Pournos, P.A. (2004). Mapping of Ant Algorithms to the Reinforcement Learning Framework. University of Bristol: MSc Thesis submitted to Department of Computer Science.

Randall, M. (2002). A parallel implementetation of Ant Colony Optimization. Parallel and Distributed Computing.

Ryoo, S.R. (2008). Program optimization carving for GPU computing. J. Parallel Distrib. Comput., pp. 1389–1401.

Tan, Q. a. (2012). Parallel Max-Min Ant System Using MapReduce. In: Y. a. Tan, Advances in Swarm Intelligence. Springer Berlin Heidelberg, pp. 182–189.

Thomas Stützle, H.H. (2000). MAX - MIN Ant System. Future Generation Computer Systems, pp. 889–901.

Tsutsui, S. (2012). ACO on Multiple GPUs with CUDA for Faster Solution of QAPs. In: Lecture Notes in Computer Science, C.C., PPSN 2012, Part II, LNCS 7492. Berlin: Springer-Verlag Berlin Heidelberg, pp. 174–184.

Yang, X.S. (2010). Nature-Inspired Metaheuristic Algorithms. United Kingdom: Luniver Press.

8

Tool Path Optimization Based on Ant Colony Optimization for CNC Machining Operations

Nataly Medina-Rodríguez, Oscar Montiel* and Roberto Sepúlveda

ABSTRACT

In this chapter, we present an efficient solution to determine the best sequence of G commands for a set of holes of a printed circuit board, in order to minimize the manufacturing time by the optimization of the travel path using Multi-Ant Parallel Colonies metaheuristic. Several experiments that demonstrate how this proposal can help to outperform solutions provided by commercial software are presented.

8.1 Introduction

Computer Numerical Control (CNC) refers to the automation of machine tools, which is of primordial importance in any automated industrial process for many manufacturing products. Today manual machine tools have been largely replaced by CNC machines where all movements of the machine tools are programmed and controlled electronically rather than by hand (Mattson 2010), reducing time and avoiding human errors.

Instituto Politécnico Nacional, CITEDI, Tijuana, México.
  E-mail: oross@ipn.mx
* Corresponding author

The productivity of CNC machine tools is significantly improved by using Computer-Aided Design (CAD) and Computer-Aided Manufacturing (CAM) systems for automated Numerical Control (NC) program generation. Currently, many CAD/CAM packages that provide automatic NC programming have been developed for various cutting processes, hole-cutting operation or drilling being one of those processes.

There are several studies that focus on reducing the cutting time by optimizing some parameters such as part geometry, material and tool type. This chapter analyzes the cutting time, which is the time that the cutting tool moves with cutting speed in air or in material. A survey of the literature shows that extensive research has been done on minimizing the cutting time (K. Castelino 2003, Radhakrishnan 2008); however, there is a lack of literature about the study of the travel time between operations. In order to minimize the travel time, the cutting tool travel path between operations should be minimized. This travel path can be formulated as a special case of the travelling salesman problem (TSP) (J.E.A. Qudeiri 2006).

8.2 Introduction to CNC Programming using CAD/CAM

Figure 8.1 shows a vertical milling machine; the feed movement is to be realized by the individual or simultaneous movement of X and Y axes. Thus the milling machine requires three slide movements (X, Y and Z).

In order to carry out the milling operation on the workpiece, the coordinate information (X, and Y coordinates) of the starting point and the ending point, direction of rotation, speed, use of coolant and the feed rate, has to be coded in the NC (Numerical Control) program. The CNC controller decodes the positioning information coded in the NC program, and the slide is moved to the programmed position at the required feed rate (Radhakrishnan 2008).

Let us assume that a hole is to be drilled at location $X = 150.0$ and $Y = 120.0$. The corresponding block of program is read by the control system and the necessary inputs are sent to the X- and Y-axes servomotors. These motors drive the respective slides to the commanded position. When the distance information is given from the feedback devices, and it equals the programmed values, the slide movement stops. The input is then given to the Z-axes servomotor to perform the drilling operation.

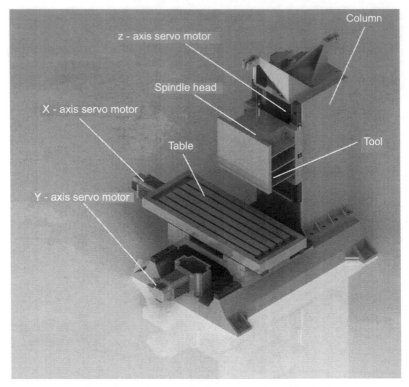

Fig. 8.1 Generic numerically controlled machine tool.

8.2.1 Numerical control programming

NC is control by information contained in a part program, which is a set of coded instructions given as numbers for the automatic control of a machine in a pre-determined sequence. NC is not a machine, it is an essential part of a machine in order to execute drilling/cutting operations, Computer Numerical Control (CNC) machines have more programmable features than older NC tape machinery; they are easier to program and can be programmed by loading an NC code into the machine's memory; once the NC program is loaded, a controller uses this program called an *executable program* to process the codes into the electrical pulses that control the machining operations (Seames 2002).

A simple program block is presented as:

```
N005 G01 U20 W-50 S1200 F0.2 M08;
```

N005   Block number (program line number)
G01   Linear interpolation
U20   X increment in slide movement
W-50   Z increment in slide movement
S1200   Spindle speed at 1200 rpm
F0.2   Feed rate at 0.2mm/rev
M08   Coolant on
;   End of block

Each of the above consists of a letter of an alphabet also known as a word address (N, G, U, W, S, etc.), and a numeric value (005, 01, 20, –50, 1200), which represents a function of a slide displacement position or machining data.

A typical movement line of a CNC program (called block) is given below:

```
N005 G01 X100.05 Y180.95 S450 M08;
```

Every block starts with a block number (three or four digits + N word), and a block that may have one or more G functions, for example, G01 for Linear Interpolation. The block may contain the X, Y and Z coordinates of the target point; the feed at which the slide movement is to be executed is specified in the feed value. If the feed is the same as specified in the previous block, we do not need to repeat this command.

G-codes are mainly NC functions; the following are common preparatory functions in a CNC system:

• Interpolation functions
   ○ Positioning (G00)
   ○ Linear interpolation (G01)
   ○ Circular interpolation (G02, G03)
   ○ Polar coordinate interpolation (G112, G113)
   ○ Cylindrical interpolation (G107)

• Thread cutting (G32, G34)
• Feed functions
   ○ Feed per minute (G98)
   ○ Feed per revolution (G99)
   ○ Dwell (G04)

• Reference point
   ○ Automatic reference point return (G28)
   ○ 2nd, 3rd, and 4th reference point (G30)

- Coordinate system setting (G50)
- Inch – metric conversion (G20, G21)
- Constant surface speed control (G96, G97)
- Canned cycles
  ○ Outer diameter cutting cycle (**G90**)
  ○ Thread cutting cycle (G92)
  ○ End face turning cycle (G94)
  ○ R level return in fixed cycle mode (**G99**)

- Multiple repetitive cycle
  ○ Stock removal in longitudinal turning (G71)
  ○ Contour parallel turning (G73)
  ○ Finishing cycle (G70)

- Canned cycles
  ○ Front drilling (**G83**)
  ○ Side drilling (G87)
  ○ Front tapping (G84)

As we mentioned earlier, an NC program will consist of a number of lines called blocks, and each block will consist of a number of words, as shown in Fig. 8.2. Each word will have two components: a word address and a numeric code representing information. The common word addresses are (Radhakrishnan 2008):

- **N**: Sequence number of instructions
- **G**: Preparatory function
- **XYZABC**: Coordinate and angular data
- **F**: Feed
- **S**: Spindle speed
- **T**: Tool code
- **M**: Miscellaneous function

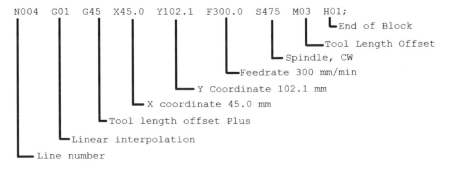

Fig. 8.2 Example of a NC program block.

8.2.2 The use of Computer Aided Manufacturing (CAM)

Commercial Software is available in the market, and this is capable of producing most CNC machine programs in a time much shorter than any manual programming method. Current CNC software, commonly known as CAM software (Smid 2003), has many features that translate the original part drawing into a CNC program, this means that a program generated by a computer has to be in format compatible with the CNC machine and its control system.

The starting point of CAM is the CAD file, by using solid models or surface modules; a CAD model represents the finished product. Solid models can be converted to IGES, VDA, DXF, CADL, and many more file extensions generated from a CAD Software. The following are the steps involved in creating a NC program using a CAM software package:

1. Create a manufacturing model from the design model and the workpiece.
2. Set up the tool database, defining tools for a manufacturing operation.
3. Select the set up for the machining operation.
4. Select parameters such as Spindle Speed and Feed Rate from the machinability database.
5. Create the manufacturing operations to generate the data.
6. Create a manufacturing route sheet at the end of the manufacturing session.
7. Post–process data to create the NC program.

8.2.3 Drilling holes

The hole making NC sequences include drilling, reaming, boring, tapping, etc; and these sequences are created by selecting the cycle type and specifying the holes to drill by defining the hole sets. A hole set includes one or more hole to be drilled; each hole set, will have a drilling depth specification (Radhakrishnan 2008).

To create an NC program, a number of tool motion parameters need to be considered as input, some of them are:

- Feed rate
- Cut Com (Tool compensation)
- Coolant
- Spindle Speed
- Speed sense
- Max spindle RPM
- Spindle range.

8.2.4 Absolute and incremental programming

The geometrical information contained in an NC file, defines the physical shape and size of the part to be made, and it is normally specified by a series of Cartesian coordinate points; these points can be expressed in two ways: Absolute or Incremental coordinates.

NC files include geometrical data relative to some suitable point, and this is the starting point for either method of specifying a sequence of coordinate points; for drilled parts, usually programmers choose the bottom left-hand corner of the component as a starting point, as is shown in Fig. 8.3.

Absolute coordinates involves specifying each point from the zero data point, as is shown in Fig. 8.3. In incremental dimensioning (also called chain dimensioning), each point is measured from the last position (Waters 1996), see Fig. 8.4.

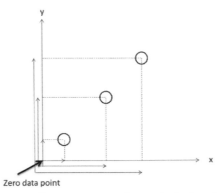

Fig. 8.3 For drilled parts, usually programmers choose the bottom left-hand corner of the component as a starting point as absolute coordinates.

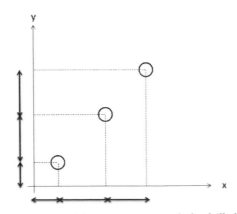

Fig. 8.4 Incremental dimensioning example for drilled parts.

Table 8.1 Absolute and Incremental dimensioning example.

Hole no.	Absolute dimensioning		Incremental programming	
	x	y	x	y
1	4	2	+4	+2
2	6	7	+2	+5
3	4	−3	−2	−10
4	8	−6	+4	−3
5	−10	−5	−18	+1
6	−6	−3	+4	+2
7	−6	+5	0	+8
0	0	0	+6	−5
			$\Sigma = 0$	$\Sigma = 0$

8.2.4.1 Example

A set of seven holes are going to be drilled in a component, following the pattern shown in Fig. 8.5. Table 8.1 lists the dimensions of the seven holes, using both absolute and incremental methods.

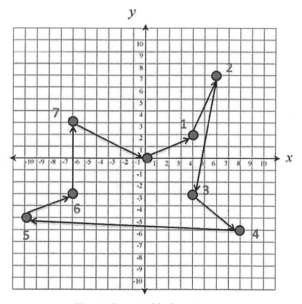

Fig. 8.5 Required hole pattern.

8.2.5 Canned cycles

Usually drilled parts involve a series of motions, these motions are repeated a number of times to common positions; for drilling operations, the tool has to position a little above the hole, next move to the required depth with the given feed rate and then the tool has to return to the top of the hole; for each hole, the same actions are repeated. The tool movement can be done by using a canned cycle (Rao 2009), it can replicate all the motions without having to repeat the same information for each of the hole. Table 8.2 shows the most common cycles that are useful for the hole–making operations, such as drilling, reaming, tapping, etc.

For the model shown in Fig. 8.6, the NC program for drilling the three holes without using canned cycles is shown in Program 8.1.

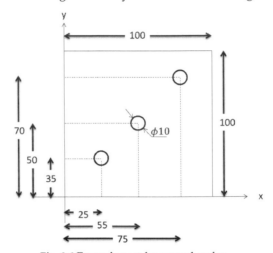

Fig. 8.6 Example part for canned cycles.

Program 8.1 NC program for drilling the three holes without using canned cycles.

```
N010 G00 X25.0 Y35.0 Z2
N015 G01 Z-18.0 F125
N020 G00 Z2.0
N025 X25.0 Y 35.0
N035 G01 X55.0 Y50.0 F125
N040 G00 Z2.0
N045 X25.0 Y35.0
N050 G01 X75.0 Y70.0 F125
N055 G00 Z2.0
N065 X0 Y0 X50
```

For the same model of Fig. 8.6, the NC program using canned cycles is shown in Program 8.2.

Program 8.2 NC program using canned cycles.

```
N010 G81 X25.0 Y35.0 X-18.0 R2.0 F125
N015 X55.0 Y50.0
N020 X75.0 Y70.0
N025 G80 X0 Y0 Z50
```

Table 8.2 Common canned cycles.

Desired Action	Command	Description of Variables	Example G Code
DRILL	G81	X, Y = location Z = depth R = clearance/rapid level F = Feed-rate G81 commands the machine to drill a hole at every subsequent X/Y location, to a depth Z, rapid traverse to a point, and feed to depth at the Feed-rate F.	G81 X3. Y1.5 Z-2. R.1 F3
DRILL – CHIP BREAKER	G73	X, Y = location Z = depth R = clearance/rapid level F = Feed-rate Q = depth of peck G73 commands the machine to do a high speed peck drilling cycle. This is not a full retract; the tool only retracts enough to break the chip.	G73X3.Y1.Z-2. R.1 Q.3F3
DRILL – PECK FULL RETRACT	G83	X, Y – location Z = depth R = clearance/rapid level F = Feed-rate Q = depth of peck G83 commands the machine to do a full retract peck to the R level.	G83X3.Y1.Z-2. R.1 Q.3F3
DRILL – COUNTER BORING	G82	X, Y = location Z = depth R = clearance/rapid level F = Feed-rate P = Dwell (seconds/1000) G82 commands the machine to stop at the final depth and dwell for amount of time specified by P.	G82X3.Y1.Z-2. P500F3.
TAPPING R. H.	G84	X, Y = location Z = depth R = clearance/ rapid level F = Feed-rate G84 commands the machine to feed to depth, reverse spindle direction then feed out to R level and return spindle to CW rotation.	G84X3.Y1.Z-2.R.3F31.25
BORING	G85	X, Y = location Z = depth R = clearance/rapid level F = Feed-rate G85 commands the machine to return to the R level at a Feed-rate instead of at a rapid traverse.	G85X3.Y1.Z-2F3.

8.3 Ant Colony Optimization for Hole Making NC Sequences: A Special Case of the Traveling Salesman Problem (TSP)

The TSP problem (M. Dorigo 1997a,b) is the problem of a salesman who, starting from his hometown, wants to find a shortest tour that takes him through a given set of customer cities and then back home, visiting each customer city exactly once. The TSP can be represented by a completed weighted graph $G = (N, A)$ with N being the set of nodes representing the cities, and A being the set of arcs. Each arc $(i,j) \in A$ has an assigned value (length) d_{ij} which is the distance between cities i and j.

Ant Colony Optimization (ACO), was introduced by Marco Dorigo (M. Dorigo 2004), by using an artificial ant group that can be able to find the shortest path between any two points according to pheromone levels. ACO metaheuristics can be applied to solve the TSP, where the pheromone trails are associated with arcs and therefore τ_{ij} refers to the desirability of visiting city j directly after city i. The heuristic information is chosen as $\eta_{ij} = \frac{1}{d_{ij}}$; that is, the heuristic desirability of going from city i to city j is inversely proportional to the distance between the two cities. For implementation purposes, pheromone trails are collected into a pheromone matrix whose elements are the τ_{ij}'s.

Tours are constructed by applying the following simple constructive procedure to each ant:

1. Each ant chooses, according to some criterion, a start city at which the ant is positioned.
2. Each ant uses a pheromone and heuristic values to probabilistically construct a tour by iteratively adding cities that the ant has not visited yet, until all cities have been visited.
3. Each ant goes back to the initial city.
4. After all ants have completed their tour, they may deposit pheromone on the tours they have followed.

8.3.1 Tour construction in ant system

In AS, m ants concurrently build a tour of the TSP. Initially, ants are put on randomly chosen cities. At each construction step, ant k applies a probabilistic action choice rule, called random proportional rule, to decide which city he is going to visit next. The probability with which ant k, currently at city i, chooses to go to city j is given by Eq. (8.3.1):

$$p_{ij}^k = \frac{[\tau_{ij}]^\alpha [\eta_{ij}]^\beta}{\sum_{l \in N_i^k} [\tau_{ij}]^\alpha [\eta_{ij}]^\beta}, if j \in N_i^k \tag{8.3.1}$$

where $\eta_{ij} = 1/d_{ij}$ is a heuristic value, α and β are two parameters which determine the relative influence of the pheromone trail and the heuristic information, and N_i^k is the feasible neighborhood of ant k when being at city i, that is, the set of cities that ant k has not visited yet. The probability of choosing a city outside N_i^k is 0.

- If $\alpha = 0$, the closest cities are more likely to be selected, it regulates the influence of τ_{ij}.
- If $\beta = 0$, only pheromone amplification is used, without any heuristic bias, it regulates the influence of η_{ij}.

8.3.2 Update of pheromone trails

After all the ants have constructed their tours, the pheromone trails are updated.

This is done by first lowering the pheromone value on all arcs by a constant factor, and then adding pheromone on the arcs the ants have crossed in their tours. Pheromone evaporation is implemented by:

$$\tau_{ij} \leftarrow (1 - \rho)\tau_{ij}, \quad \forall(i,j) \in L, \tag{8.3.2}$$

where $0 \le \rho \le 1$ is the pheromone evaporation rate. After evaporation, all ants deposit pheromone on the arcs they have crossed in their as shown in:

$$\tau_{ij} \leftarrow \tau_{ij} + \sum_{k=1}^{m} \Delta\tau_{ij}^k, \quad \forall(i,j) \in L, \tag{8.3.3}$$

where $\Delta\tau_{ij}^k$ is the amount of pheromone ant k deposits on the arcs it has visited.

8.3.3 NC for drilling operations using ant colony optimization: case study

To illustrate the application of the optimized NC sequence generation using ACO, a case of study is presented below; the application is based on a Printed Circuit Board (PCB) design, and the tool path for PCB drilling making, is generated by using our Parallel–ACO implementation.

To create a PCB design, we used a schematic program along with their board layout program; in this section, we describe how to export PCB files to Drawing Exchange Format (DXF) files, which can be used to generate an optimized G–code by using our parallel implementation of ACO. Our example is based on a 5V DC Power Supply PCB board, as shown in Fig. 8.7.

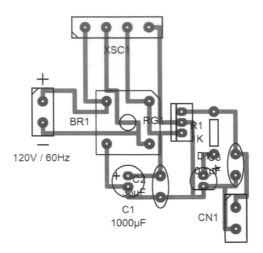

Fig. 8.7 5V DC Power Supply PCB Board.
Color image of this figure appears in the color plate section at the end of the book.

The DXF was created by Autodesk™ for its AutoCAD™ product, for saving a readable file containing all the information stored in a drawing. This information can be passed to, and read by the CAD/CAM system that supports the DXF system (Stephen F. Krar 2003) to generate NC operations. CAM systems have been specifically designed to use the information from CAD software such Solidworks™, AutoCAD™, or a PCB editor software, to assign tool paths. Once the tool paths and machining data have been defined, the information is run through a translator called a post processor to produce a CNC program that will machine the part (Stephen F. Krar 2003).

To export DXF files from a PCB editor, it is important to set the display in a PCB editor to show only the layers that we want to export into a DXF file, in this example, we must export the solder side to generate a tool path for drilling operations.

Our system reads an input DXF file to obtain a set of commands and coordinates for any part designed in a CAD/CAM commercial package; this input DXF file is now being processed by the parallel Ant Colony Optimization algorithm to generate a sequence of G commands for a CNC drilling tool path. This implementation can be represented, as shown in Fig. 8.8.

In order to minimize the cutting time, the Cutting Tool Travel Path (CTTP) (J.E.A. Qudeiri 2006) between operations should be minimized. We were working on a continuous travel path, in which the start point and the end point of each operation are the same, and it mainly appears in hole–cutting operations such as drilling, reaming and tapping.

The Travelling Salesman Problem (TSP) and Parallel Ant Colony Optimization (P-ACO) have been incorporated to find the shortest cutting tool travel path for drilling operations. Figure 8.9 shows our Graphics User (GUI) developed with C# .NET Framework.

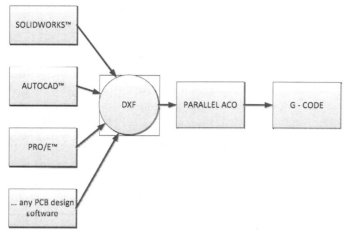

Fig. 8.8 CNC Tool path optimization for a drilling process implementation.

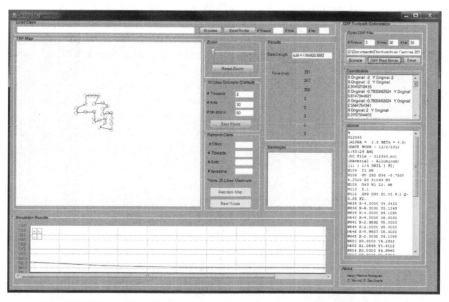

Fig. 8.9 GUI for a Parallel Ant Colony Optimization and NC code generator.

Color image of this figure appears in the color plate section at the end of the book.

The following procedure represents the generation of G code sequence for the cutting tool travel path:

- *Step 1.* Read the coordinate of each node of the optimized CTTP based on P-ACO.
- *Step 2.* Code the traverse motion command G00 and then the X and Y coordinates of the first hole in the CTTP.
- *Step 3.* For each coordinate in solution *k* generated by the P-ACO, code G00 rapid move and then X and Y coordinates for the next node.
- *Step 4.* If a change in tool is needed, then code M6 for a tool change and repeat steps 1 to 3. If no change in tool is needed, then proceed with steps 1 to 3 until the cutting tool reaches the last hole in the cutting travel tool path.

Our system generates this output file and then the user can save this file with .NC extension.

8.4 Parallel Implementation of Ant Colony Optimization

Recall that the main purpose of an operating system is to manage hardware and software resources on a computer system, and it also controls the allocation of memory of running programs and access to peripheral devices. When a user runs a program, the operating system creates a process, and most modern operating systems are multitasking, this means that the operating system provides support for a simultaneous execution of multiple process or programs.

Threading provides a mechanism to divide a program into more or less independent tasks with the property that when one thread is blocked, another thread can be run (Pacheco 2011). Threads are contained within processes, as shown in Fig. 8.10, so they can use the same executable unit,

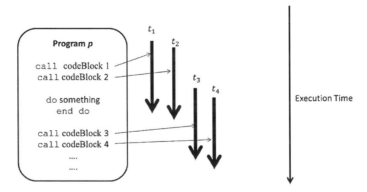

Fig. 8.10 Threads programming model.

and they usually share the same memory and the same I/O devices. When program *p* runs its code, it may create many threads and they are executed in parallel; each thread may have local data, and also it can share the entire resources of *p*, including global memory.

In Fig. 8.10, a main program *p* is scheduled to run by the operating system; when *p* runs its code, it may create many threads and they can run in a parallel mode; each thread may have local data, and also can share the entire resources of *p*, including global memory.

The ACO algorithm can be formulated by the following sequential implementation as described in Algorithm 8.1:

Algorithm 8.1 ACO extended sequential metaheuristic.

---

procedure ACO Sequential Metaheuristic
        begin
        Initialize Parameters
  for each cycle
           for each ant
                for each city
                      Build a solution *k*
                      Evaluate solution *k*
                end for
           end for
        Save Best Solution
        Update Trail
  end for
  Print Best Solution
  end

The availability of parallel architectures at low cost has widened the interest for the parallelization of ACO algorithm. In order to parallelize the ACO, it is more important to modify the structure of ACO to get a better optimization effect to change the sequential ACO into a parallelization schema (L. Chen 2008).

The main purpose of parallel implementation of ACO is to obtain a high speed up and efficiency, while the convergence and the ability of optimization are maintained or even improved. Some results on parallel ACO algorithms have been reported recently: Bullnheimer (B. Bullnheimer 1998) proposed two parallelization strategies of synchronous and asynchronous for ACO using the Traveling Salesman Problem; Piriyakumar in (Douglas Antony Louis Piriyakumar 2002) introduced an asynchronous parallel Max-Min ACO associated with the local search strategy; Marcus Randal in (Randall 2002) introduced a synchronous parallel strategy which assigns only one ant on each processor; Marco Dorigo in (C. Blum 2004) introduced a parallel ACO on the hyper–cube architecture by modifying the rule of updating the pheromone, so as to limit the pheromone values within the range of [0,1].

In this research, we present a parallel implementation of ACO for CNC Tool path optimization generating a set of G commands. The general procedure consists of the interpretation of a DXF file as input, detecting commands related to the coordinates of all points; then, this information is optimized by using ACO, and finally, our system generates the best route found by ants, using the fork–join model, so we interpret this route by generating a set of G commands.

8.4.1 Parallel implementation: problem formulation

The sequential algorithm contains a high degree of natural parallelism (B. Bullnheimer 1998); the behavior of a single ant in one iteration is totally independent of the behavior of all other ants during that iteration. We will discuss a strategy called "synchronous (fork–join) algorithm", a straight forward parallelization strategy for ACO is to compute the TSP tours in parallel; this would result in a fork–join structure, such as illustrated in Fig. 8.11.

An initial process (master) would spawn a set of processes, one for each ant; after distributing initial information about the problem, each process can generate a single solution for each ant k. Following the above procedure, the result from each process is sent back to the master process, which updates the trail levels by calculating the intensity of the trails, and it checks for the best tour found so far. A new iteration is initiated by sending out the updated trail levels.

According to Bullnheimer (B. Bullnheimer 1998) ignoring any communication overhead, since ants end up spending most of their time communicating the modifications they made to pheromone trails (M. Dorigo

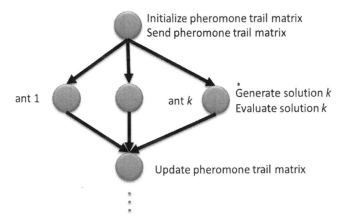

Fig. 8.11 Parallel implementation of ACO in a shared memory model.

2004), this approach would imply optimum (asymptotic) speedup, as it is explained with Eq. (8.4.1), assuming that an infinite number of processing elements (workers) is available, i.e., one process is assigned to one worker. In (8.4.1) m is the quantity of ants in the colony.

$$S_{asymptotic}(m) = \frac{T_{sec}(m)}{T_{par}(m, \infty)} = \frac{O(m^3)}{O(m^2)} = O(m) \tag{8.4.1}$$

where $T_{seq}(m) = O(m^3)$ is the computational complexity of the sequential algorithm for the problem is size m, and $T_{par}(m, \infty) = O(m^2)$ is the computational complexity of the parallel algorithm for the same problem of size m, considering an infinite system size.

Communication overhead certainly cannot be disregarded and it has to be taken into account, furthermore the system size (number of processing elements N) is restricted and it is typically smaller than the problem size (number of ants m).

To overcome the communication overhead, we need to balance the load among the workers or threads, this is easily accomplished by assigning to the worker j, the processes (ants) m_i ($i = 1...m$), thus each worker holds about the same number of processes and each process is of the same computational complexity (B. Bullnheimer 1998). In conclusion, when considering communication overhead, the ratio of the amount of computation assigned to a worker and the amount of data to be communicated has to be balanced (B. Bullnheimer 1998).

When considering communication overhead, the ratio of the amount of computation assigned to a worker and the amount of data to be communicated has to be balanced (B. Bullnheimer 1998).

After each iteration, all the completed tours and their lengths have to be sent to a central process (master). Then, the new trail levels need to be computed and broadcasted to each worker, afterwards a new iteration can be started.

In our parallel ACO, the ants are divided equally into P groups which are allocated into P processors. The ants in each group search for the best solution in its own processor independently as it is described in Program 8.3.

The main part of this implementation of an Ant System (AS) algorithm is of complexity $O(n^3)$, and the generation of one solution is of complexity $O(n^2)$, where n is the number of jobs; these two operations are independent for each ant of a given cycle, so they can be easily parallelized.

Figure 8.11 shows the behavior of our implementation of the ACO based on the parallel synchronous AS in a shared memory model. At the beginning of the algorithm, a master process initializes the information, spawns k processes (one for each ant m), and broadcasts the information. At the beginning of a cycle, the τ_{ij} matrix (the pheromone trail) is sent to each process and the computations of the generations and the evaluation of solutions, are done in parallel. Then, the solutions and their evaluations are sent back to the master, the τ_{ij} matrix is updated, and a new cycle begins by the broadcasting of the updated matrix.

Program 8.3 Parallel Ant Colony Optimization framework.

---

begin
>   An initial process initializes the pheromone matrix and some other control parameters.
>   **while** (not terminate) **do**
>>   **for each** processor **do** in parallel (P groups)
>>>   **for each** ant **do**
>>>>   Build a solution k
>>>>   Evaluate solution k
>>>   **end for**
>>   **end for**
>   **end while**

The parallel implementation of ACO is written entirely in C# .NET; a "Parameterized Thread Start" is used to start a thread with an argument, in this case, the argument is the Ant Cycle algorithm, initializing an array of threads, and implementing a `ParameterizedThreadStarts` for each element in the array; then the program starts them and joints them, improving performance and runtime for a program. With a `ParameterizedThreadStarts` we can pass a function name as the argument, as it is described below:

```
threads[a]=new Thread(newParameterizedThreaStart(Ant_Cycle));
threads[a].Start(tpACO);
```

where **tpACO** is an structure which contains parameters such as the number of ants, cities (holes), threads and a thread ID. When each thread job is done, a joint function must be implemented as it is described below:

```
foreach (Thread t in threads)
{
 t.Join();
}
```

Threads communicate with each other through global memory (updating addressed locations) (R 2009), so it requires synchronization constructs to ensure that more than one thread is not updating the same global address at any time, this is the main reason why a "join" method must be implemented in this case.

8.4.2 Performance of parallel computers

An important measure of a parallel architecture is the speedup, which is a measure of performance, such as it was explained in Chapter 1. In particular for this work, the speedup is defined using (8.4.2), as the ratio of the execution time of a single processor system, and the execution time of a parallel processing system.

$$S = \frac{T_s}{T_n} \tag{8.4.2}$$

where:
n: Number of processors
T_s: Single processor execution time
T_n: n processor execution time
S: Speedup

According to Speedup's Folk theorem (R 2009), for any algorithm of size N and any number of processors n we have $1 \leq S \leq n$.

The efficiency e of a parallel computer is defined as speedup S divided by the number of processors n:

$$e = \frac{s}{n} \tag{8.4.3}$$

8.5 Experimental Results

In this section, we show the test results of our parallel ACO algorithm implementation, it was applied to optimize G-codes of some PCB layouts, with different number of holes. The parameters used in the test were set as follows:

$\rho = 0.1$ for the evaporation coefficient.
$\alpha = 2.5$ for the parameter to regulate the influence of τ_{ij}.
$\beta = 4.5$ for the parameter to regulate the influence of η_{ij}.

We set the number of ants equal to the number of holes, and the number of available processors is six. Our experiment performs 30 trials for a test of 25 holes, the experiments were achieved using a single processor, as well as with many processors, in this case we have been working with six cores. The experimental results are shown in Fig. 8.12, where a DXF from a PCB design is tested using Parallel ACO and it is compared with a classical ACO algorithm which uses a single processor.

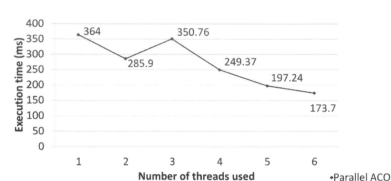

Fig. 8.12 CNC tool path optimization for a drilling process: 25 holes using P-ACO algorithm.

8.5.1 Analysis of the parallel implementation of ACO

Figure 8.12 shows a graph representing the execution time of a PCB design for a 5V DC Power Supply, consisting of 25 holes; the optimized G-code for drilling operations is generated by using the Parallel ACO.

We can observe that when the number of processors is increased, the computing time can be reduced because each processor has fewer ants assigned, but due to the overhead communication, which increases the total execution time, the speedup of our algorithm cannot increase linearly with the increasing of the number of processors exactly, but there is an excellent linear trend as it is shown in Fig. 8.13. This is in conformity with the Amdahl's Law (R 2009).

We can observe that speedup in these results is within a range of $1 \le S \le 6$ processors, so the Folk's theorem is satisfied.

And finally, the efficiency of the algorithm is stable after the use of two processors, and it maintains a value of 36% of efficiency, due to the overhead communication, this is illustrated in Fig. 8.14.

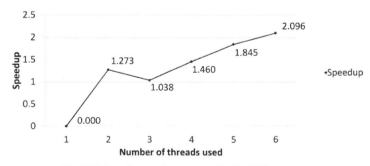

Fig. 8.13 Speedup analysis using the P-ACO program.

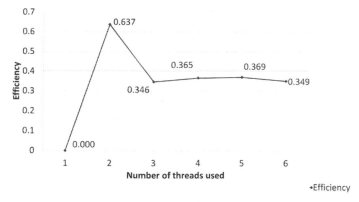

Fig. 8.14 Efficiency analysis using the P-ACO for different number of threads.

8.5.2 Tool path optimization analysis

Figure 8.15 shows a HAAS Automation™ CNC machine used for our experiments; the main purpose of the experiments is to compare the execution time of results, against those obtained by using commercial software such MasterCAM™.

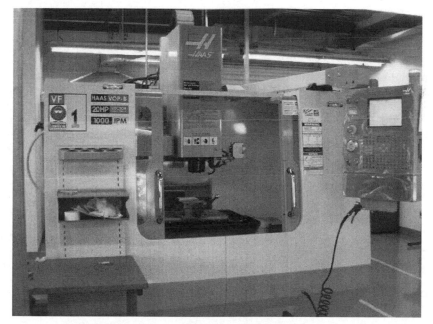

Fig. 8.15 HAAS Automation CNC machine used in this research.

Color image of this figure appears in the color plate section at the end of the book.

Figure 8.16 shows three different tool paths, the first two were obtained using two different options of the MasterCAM program, it is easy to note that they do not provide the optimal tour; meanwhile Fig. 8.16c) shows the optimal tour obtained by the application of the Parallel ACO to the DXF file sources.

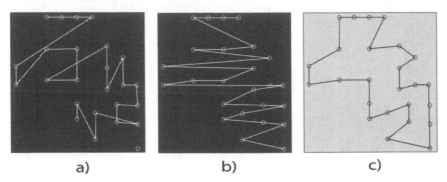

a)   b)   c)

Fig. 8.16 Tool path for a 25 holes PCB design. a) Option 1 from MasterCAM™, b) Option 2 from MasterCAM™ and c) Tool path using our Parallel ACO algorithm.

Color image of this figure appears in the color plate section at the end of the book.

We can observe that the determination of an optimal tool path is a good example of the well-known TSP, and it can be solved by using a metaheuristic. Figure 8.16a to c show two options created from commercial software, in this case using MasterCAM™; the total Cutting Tool Travel Path (CTTP) time for these three experiments are shown in Table 8.3, and they were implemented with the following machine parameters:

- Feedrate: 2
- Speed: 1200
- Tool selection: ¼ Drill
- Spindle Direction: CW
- Cut parameters: Peck drill

Table 8.3 Total Cutting Tool Travel Path (CTTP) Time.

Figure	Total CTTP Time
a	5 minutes and 41 seconds
b	6 minutes and 12 seconds
c	4 minutes and 52 seconds

8.6 Conclusions

We presented an efficient methodology to improve CNC G-code generation, which can be applied to free or paid software for industry applications; it is based on finding optimal time paths for CNC machines providing optimal G-code. The solution takes advantage of high-performance computing for implementing the P–ACO metaheuristics. A PCB design was used, here for drilling purposes, using commercial CAD/CAM software and our implementation of P–ACO. The results were verified using a professional CNC machine (see Fig. 8.15), and the analysis of numerical results and graphical trends utterly demonstrates that the use of this high-performance implementation of ACO can offer improvements around 62% (Oscar Montiel 2012) over solutions provided by application software, in this case commercial software. This methodology can be implemented straightforward, being no necessary to perform any modification to the application software, since the only thing to do is to feed the CNC machine with the optimal G-codes sequence, instead of the original.

Acknowledgments

The authors would like to thank the "Instituto Politécnico Nacional (IPN)", "Comisión de Operación y Fomento de Actividades Académicas del IPN (COFAA-IPN)" and the Mexico's entity "Consejo Nacional de Ciencia y Tecnología (CONACYT)" for supporting our research activities.

References

Bullnheimer, B., G. Kotsis and C. Strauss: Parallelization Strategies for the Ant System. Conference on High Performance Software for Nonlinear Optimization: Status and Perspectives (HPSNO), Ischia, Italy, 1997.
Blum, C. and M. Dorigo (2004). The hyper—cube framework for ant colony optimization", IEEE Transactions on SMC, 34(2): 1161–1172.
D. A. L. Piriyakumar and P. Levi (2002). A new approach to exploiting parallelism in ant colony optimization," in Micromechatronics and Human Science, 2002. MHS 2002. Proceedings of 2002 International Symposium on, pp. 237–243.
J. E. A. Qudeiri, A.-M. (2006). Optimization Hole-cutting Operations Sequence in CNC Machine Tools using GA. International Conference on Service Systems and Service Management, ICSSSM '06, 501, 506.
K. Castelino, P.K. (2003). Tool path optimization for minimizing airtime during machining. Journal of Manufacturing Systems, 22(3): 173–180.
L. Chen, H.-Y. (2008). Parallel Implementation of Ant Colony Optimization on MPP. Proceedings of the Seventh International Conference on Machine Learning and Cybernetics.
M. Dorigo, L.G. (1997a). Ant Colony System: a cooperative learning approach to the traveling salesman problem. IEEE Transactions on Evolutionary Computation.
Dorigo, M. and L.M. Gambardella (1997). Ant Colonies for the Traveling Salesman Problem. BioSystems, 43: 73–81.

M. Dorigo, T.S. (2004). Ant Colony Optimization. Cambridge, MA: MIT Press. Cambridge, MA.

Mattson, M. (2010). CNC Programming: Principles and Applications. Delmar Cengage Learning. Printed in Canada.

Oscar Montiel, N.M. (2012). Methodology to optimize manufacturing time for a CNC using high performance implementation of ACO. International Journal of Advanced Robotic System. Burlington, MA. USA.

Pacheco, P. (2011). An introduction to Parallel Programming. Elsevier Inc.

Moreshwar R. Bhujade (2009). Parallel Computing. New Age Science Ltd, UK.

Radhakrishnan, P.S. (2008). CAD/CAM/CIM. New Age International. Printed in India.

Randall, M. (2002). A parallel implementation of Ant Colony Optimization. Journal of Parallel and Distributed Computing, 62: 1421–1432.

Rao, P.N. (2009). Manufacturing Technology: Metal cutting and Machine Tools. New Delhi: McGraw Hill.

Seames, W.S. (2002). Computer Numerical Control: Concepts & Programming. USA: Thomson.

Smid, P. (2003). CNC programming Handbook: A comprehensive Guide to Practical CNC Programming. New York: Industrial Press Inc.

Stephen F. Krar, A.G. (2003). Machine Tool Technology basics. New York: Industrial Press.

Waters (1996). Fundamentals of Manufacturing for Engineers. Boca Raton, FL, USA: CRC Press.

9

A Compendium of Artificial Immune Systems

Oscar Montiel, * *Roberto Sepúlveda* and
Francisco Javier Díaz Delgadillo

ABSTRACT

In this chapter, we shall explore the field of Artificial Immune Systems (AIS), a series of algorithms that are inspired by the workings of the human immune system and their extrapolation to the world of mathematics, computer science and engineering, from their conception to what the future holds. Our goal is to give readers a notion of the history behind them, as well as the fundamental algorithms and applications of the AIS.

9.1 Introduction

Computer science has experienced over the course of the last 50 years, a veritable revolution in the way we look and solve some of the toughest problems one can imagine. A large part of this revolution is thanks to the relationship that has been created between the observations of nature, of what problems it faces and how it has elegantly solves them, and computer science itself. Scientists have come to the conclusion that some of the fundamental problems of computer science, such as optimization of

Instituto Politécnico Nacional, CITEDI, Tijuana, México.
  E-mail: oross@ipn.mx
* Corresponding author

resources, can also be found in nature. Because of this, the next logical step in computer science is to mimic nature, and to harness it as a metaphor to solve problems and improve existing algorithms.

Some of the most successful nature inspired algorithms are genetic algorithms (Holland 1975), neural networks (Haykin 1999), swarm systems (J. Kennedy 2001) and genetic programming (Banzhaf 1998). All of these works have spawned a great deal of derivative work. They have been applied to various fields of sciences; furthermore, they have found its way into industrial and commercial applications. Fields such as pattern recognition, optimization, machine learning, data validation and interpretation are just some of the success stories of these algorithms.

In recent years, a new proposal has emerged called Artificial Immune Systems (AIS). The premise behind these algorithms is to bring the knowledge about the human immune systems into the field of computer science and engineering.

The interest of these algorithms has grown significantly, because of the innate characteristics that the immune system (IS) presents, which can be very desirable properties to have in a computer systems; some of them are robustness, adaptability, diversity, scalability, dynamic and long lasting memory, etc.

Some of the early works of AIS saw the application of immune inspired algorithms to problems such as robotic control (R. Krohling 2002), network intrusion detection (Forrest 1997, Kim 2000), fault tolerance (RO Canham 2002, D.W. Bradley 2000), bioinformatics (Giuseppe Nicosia 2004) and machine learning (J. Kim 2002, T. Knight 2003, A. Watkins 2004).

Today, AISs have found a niche in the fields of pattern recognition, optimization and search, data mining and classification, machine learning, autonomous control and navigation with great success.

9.2 A Brief History of Natural and Artificial Immune Systems

The term "immunity" refers to a condition of the human body in which the organism can resist foreign pathogens and diseases. A broader definition establishes immunity as the reaction of the body towards foreign agents (pathogens) with the purpose of protecting itself. The Immune System (IS) cannot be classified as a specific organ or be pinpointed at a specific area of the body; instead, it is a series of organs, cells and molecules that have a great sense and ability to work in a harmonious and coordinated way in order to combat foreign agents that may harm us.

The origin of immunology is attributed to Edward Jenners in 1796, who discovered that when exposing an animal to small quantities of weakened samples of agents (vaccinia) responsible for causing the disease, the animal would generate defenses towards that particular disease that would otherwise cause death if infected. This is the origin of the word "vaccination", and in our modern times, we define it as the inoculation of healthy individuals with debilitated samples of illness causing pathogens, in such a way that the individual generates future protection against this particular disease (Leonardo N. de Castro 2002). In Table 9.1, we present a synthesis of the milestones in the field of immunology from a biological point of view until the year 1990 (Leonardo N. de Castro 2002).

In (J.D. Farmer 1986) was described one of the first paradigms that extrapolate the existing works of IS with adaptive systems and machine learning algorithms. All the pioneer works created an interest in the scientific community (Kephart 1994), and in the following years detailed algorithms regarding specific biological process such as Negative Selection (Stephanie Forrest 1994), Clonal Selection (L.N. de Castro 2002, Andrew Watkins 2004, Johnny Kelsey 2003) and Immune Networks (D. Cooke 1995, F.J. Varela 1988, J. Hunt 1999) emerged. In 1999 the first book devoted specifically to AIS was published (Dasgupta 1999).

The most-recent work in AIS has considered new areas in immunology such as Danger Theory and algorithms inspired in the Innate Immune System. Additionally, a great deal of work has been done exploring degeneracy models of the AIS (Paul S. Andrews 2006, M. Mendao 2007) because it is hypothesized that they play a very important role in learning and evolution of the IS and other biological systems (Edelman 2001, Whitacre 2010).

In the recent years, additional work has emerged in which hybrid intelligent systems have emerged. The most notable are the Quantum-Immune Hybrid models, where the concept of the "qubit antibody" appeared, a probabilistic representation that is based on the model of qubits, the building block of Quantum Mechanics, and a chain of them in order to produce the antibody. This concept is applied to the Clonal Selection algorithm in order to produce a set of operations between qubits and the standard procedures explained on the Clonal selection algorithm. Quantum-Inspired Immune Clonal Algorithm or QICA is one of such attempt applied to finding the optimal solution of test functions (Yangyang Li 2005) with successful results.

Even though originally AIS was proposed with the purpose of creating an abstraction of the IS, in order to model it for future medical study, it has recently generated great interest in the modeling of biological processes through these studies, and their application in the fields of mathematics, computing and engineering.

In 2008, Dasgupta and Nino published a book about immunological computation that provides a compendium of the state of the art techniques based on immunity and also presents some real world applications (Dipankar Dasgupta L.F. 2009).

Table 9.1 Milestones in the field of immunology (L.N. de Castro 2002).

Tendencies	Period	Pioneers	Concepts
Application	1796–1870	E. Jenner	Immunization
		R. Koch	Pathology
	1870–1890	L. Pasteur	Immunization
		E. Metchnikoff	Phagocytosis
Description	1890–1910	E. von Behring and S. Kitasato	Antibodies
		P. Ehrlich	Cell receptors
	1910–1930	J. Bordet	Specificity/Complement
		K. Landsteiner	Haptens/Blood types
Mechanisms	1930–1950	A. Breinl and F. Haurowitz	Antibody synthesis
		L. Pauling	Instructionism
	1950–1980	M. Burnet and Talmage	Clonal Selection
		N. Jerne	Network and cell co-operation
Molecular	1980–1990	S. Tonegawa	Structure and diversity of receptors

Despite Timmis's fear that AIS would fall short when compared to other more well established algorithms when it comes to industrial, medical and commercial applications, articles regarding this issue started to appear, being one of the first published, an immune algorithm for detecting the occurrence of tool breakage in milling machine operations (Dipankar Dasgupta 1999). Today, the International Conference on Artificial Immune Systems (ICARIS), is a forum dedicated for all research of AIS, and each year it proves the level of commitment in the scientific community to AIS, bringing us with new and promising algorithms, and many successful applications of AIS in a great variety of areas.

9.3 Interesting Properties of the Immune System

Some of the properties that makes the immune system a remarkable paradigm to study and to support computational algorithms, are the following (Jon Timmis 2007):

Self-organization: The IS lacks an organ. Its sole purpose is to direct the immune cells and molecules to what their role is, and what to do in the event of an attack. It is believed to function as the result of many local interactions, creating a complex system capable of self-organization. Computationally, this trait is very desirable as it promotes the idea of distributed and self-training adaptive systems.

Learning: Immune systems are able to learn and conserve knowledge of the types of antigens to which it has been exposed through the B cells, and their associated antibodies. Continuous exposure of the IS to antigens, also means a constant creation of new cells and mutation of existing ones to defend against these threats. This process, functions as a good metaphor in the development of systems capable of learning from repeated exposure (training), based lightly on the immune response.

Adaptation and diversity: The ability of IS to create cells capable of not only attacking a specific threat, but those that are similar to it, is achieved by the B cell clones through a process called *somatic hypermutation*. This concept has been one of the most important sources of inspiration for algorithms; specifically, when it is important to detect a specific data type, including similar data.

Classification: IS can classify antigens into self and non-self, thanks to the use of antigen receptors. Given the description of a system in its normal state, computationally, these receptors can monitor the system in search for something anomalous (non-self) that could occur.

Distributed system: The cells, molecules and organs that form the IS are distributed throughout the body. This allows IS to function in a non-centralized manner, in which the task of keeping the body healthy is shared by all of its parts.

Pattern recognition: The main mechanism that allows IS to identify threats to the body, is through its ability to recognize pathogens and foreign elements in the body, and also those that are part of itself and may cause harm. The ability to recognize patterns is of great value in computer science, being this specific task is one of the most critical in classification and decision making algorithms.

Although we know the IS has these properties, how it achieves these processes is still a main area of research for immunologists everywhere. The notion of a natural system with such characteristics is a good source of inspiration; as our understanding of the IS grows, extrapolation of the workings to our daily scientific and engineering lives will increase. Researchers everywhere are finding new ways and areas to apply AIS paradigms.

9.4 What Exactly is an Artificial Immune System?

AIS is defined as "adaptive systems, inspired by theoretical immunology and observed immune functions, principle and models, which are applied to problem solving" (Leonardo N. de Castro 2002).

Early pioneers of AIS, declared on their work on theoretical AISs, that the areas of opportunity or application would be a direct extrapolation of the workings of the natural human immune system. These areas are particularly good because their direct relationship between the natural function and a computational task. Some of the main applications of the AIS are (Leonardo N. de Castro 2002):

- Pattern recognition
- Fault detection and anomalies
- Data mining and classification
- Agent based systems
- Scheduling
- Machine learning
- Autonomous control and navigation
- Optimization and search
- Artificial life.

However, as time passed, and researchers started to explore and adopt AIS as a new way to solve classical problems, a tendency was clearly formed, and a lot of these areas of opportunity slowly started to fade, and then better areas emerged. Researches in the past decade has made it clear that AIS has found to be better suited to solve or to provide solutions and focus in the fields of pattern recognition, optimization and search, data mining and classification, machine learning and autonomous control and navigation.

This does not mean that efforts have completely been dropped on the previously stated areas, it just means that the scientific community has found that the AIS can provide useful solutions in these areas of application and efforts should be focused on improving them.

Furthermore, it is worth mentioning that the AIS paradigm is not a specific algorithm, but a series of methodologies inspired by different parts of the IS from which different algorithms have been created. Because of this, AIS cannot be classified into a single type of algorithm, since IS does not consist of a centralized system, thus many diverse algorithms exist inspired by different immune mechanisms.

9.5 Main Algorithms

In this section, we will describe the main biological functions and their extrapolations into computer science of the IS. They are the most known and utilized to this date; however, new and interesting ways to look at these same IS functions have given birth to new algorithms, such as the work of a hybrid quantum-immune algorithm.

9.5.1 Negative Selection Algorithm

The biological inspiration behind the Negative Selection Algorithm (NSA) comes from a natural immune process that prevents autoimmunity. T-cells, a type of lymphocyte, have the ability to bind to antigens via receptors that cover their surface. These receptors can only bind to a specific antigen, which means that a particular T-cell may only bind to one type of antigen. When T-cells are created the gene encoding of its antigen receptors goes through a process of random mutation which basically rearranges the genetic code of the receptors in order to generate T-cells capable of binding to foreign agents in the body, however, they may also mutate into receptors capable of binding to molecules of the body itself. Because of this, before the T-cells start to act in the IS, they undergo a screening process which removes those T-cells capable of binding to self-components (such as our cells) and they are destroyed (Aitkin et al. 2012, Parham 2000).

The original NSA algorithm was created with the idea of detecting the existence of a virus in a computer, through the constant monitoring of data in search for changes (Stephanie Forrest 1994). The NSA is a computational model that can discriminate the self and non-self strings; basically, it has two steps: detector generation and non-self detection. The way it works is that the algorithm produces a series of self-strings denoted S, which identify the normal state of the system, then a set of detectors are generated at random and compared with the self-strings. All the randomly generated detectors that can identify any self-strings from S are stored in a set called A. This results in a set of detectors for any self-strings patterns. The set A can then be used to compare with newly introduced data to execute the task of self and non-self classification. The pseudocode to implement this proposal is given in Algorithm 9.1.

Algorithm 9.1 Pseudocode for the Negative Selector Algorithm (Leonardo N. de Castro 2002).

```
function Negative_Selection (S,r,n)
returns a detector set A
inputs: S    set of strings that define the self
        r    cross-reactivity threshold
        n    numbers of detectors required
begin
        j ← 0
        while j ≤ n do
                m ← rand(1,L)
                for each s of S do
                        aff(p) ← match(m,s,r)
                        if aff ≤ r then
                                A ← insert(A,m)
                        endIf
                endFor
                j ← j+1
        endWhile
        return A
end
```

9.5.2 Clonal Selection

According to the Clonal Selection Immune theory (Burnet 1959), the cell repertoire of the IS is submitted to a selection mechanism during the course of the lifetime of each individual (Dipankar Dasgupta 2009). The theory states, that once a union with the antigen is established, lymphocyte activation is triggered causing the creation of clones of the lymphocytes, expressing identical receptors of the original lymphocytes that found the antigen; hence, a clonal expansion of the original lymphocyte is produced, effectively ensuring that only the specific lymphocytes responsible for the activation of the antigen are produced in large quantities.

The theory also states, that any lymphocyte capable of detecting molecules that belong to the body must be eliminated during the development of the lymphocytes. This is done in order to guarantee that only the antigens of a pathogen can cause a lymphocyte to produce a clonal expansion and produce a destructive adaptive immune response. Because of this mechanism, the IS can be characterized as an antigen classifier of self-antigens and non-self-antigens, eliminating from the body the non-self antigens assuming that they come from a pathogen.

During the clonal expansion of the B-cells, the average affinity of the antibodies is increased for that specific antigen. This phenomenon receives the name of affinity maturation, and it gives the IS more effective response to that antigen in the future, due to higher affinity for the antibodies. It is caused by a process called somatic hypermutation, and by the selection mechanism that produces the expansion of the B-cells. Somatic hypermutation changes the specificity of the antibodies during the introduction of random changes to the genes that code them (Aitkin et al. 2012, Parham 2000).

The theory of clonal selection has been a key factor in the development of AIS that deals with computational optimization and pattern recognition tasks. The most important source of inspiration has been taken from the process of the affinity maturation of the B-cells, somatic hypermutation mechanisms and memory cells in order to keep alive good solutions found in a given problem.

Two important B-cell affinity maturation characteristics are taken into consideration for the computational implementation of the theory (Leonardo N. de Castro 2002), they are:

1. Proliferation of B-cells is directly proportional to the affinity of the antigen it attaches itself to; therefore, the higher the affinity, more clones of the B-cell are created.
2. Mutation of the B-cells' antibodies are inversely proportional to the affinity of the antigen; therefore, the better the capacity of the B-cells antibodies to attach to the antigens is, the smaller the mutation they receive, as they are closer to the best possible form.

Algorithm 9.2 CLONALG Clonal Selection implementation pseudocode (Leonardo N. de Castro 2002).

```
function CLONALG (S, g, N, n₁, n₂)
returns set of memory individuals M
inputs:        S       patterns to be recognized
               g       number of iterations
               N       size of population
               n₁      number of high quality elements to be selected for cloning
               n₂      number of low quality elements to be replaced after
                       iterations
begin
j ← 0
P ← rand(N, L)
while j < g do
        for each s of S do
                for each p of P do
                        aff(p) ← match(s, p);
                endFor
                P ← sort(P, aff)
                P1 ← select(P, n₁)
                for i < n₁ do
                        C ← clone(P1, aff(P1))
                endFor
                for every c of C do
                        C1 ← hypermut(c, aff(P1))
                endFor
                for each c1 of C1 do
                        aff(c1) ← affinity(c1, s)
                endFor
                M1 ← sort(aff(C1))
                M(s) ← select(M1, 1)
                m ← rand(n₂, L)
                P ← replace(P, m, n₂)
        endFor
        j ← j + 1
endWhile
return M
end
```

One of the most widely available and successful implementations of the clonal selection is called CLONALG, although many other implementations exist (V. Cutello 2005). Algorithm 9.2 shows the pseudocode for the CLONALG implementation, which can be used in pattern search and multi-modal function optimization tasks (Leonardo N. de Castro 2002, Dipankar Dasgupta 2009).

Clonal Selection Algorithm (CSA) has also been implemented to tackle specific problems, which requires explicit changes in order to the particular problem's solution. For example, in the field of Combinatorial Optimization Problems (COPs) for the Traveling Salesman Problem (TSP) (Maoguo Gong 2006), the antibody definition and expression, operations such as affinity maturation, clonal expansion and somatic hypermutation, are only some changes required to adjust the CSA, in order to solve the TSP.

The CSA has become the most well-known algorithm of the AIS algorithm repertoire, this is mostly due to its similarities with Genetic Algorithms (GAs) and the fact that many of the optimization and technics that can be applied to GAs may also work on clonal selection, such as mutation operators and selection strategies. Moreover, the CSA can also find competitive solutions, for the same problem sets, benchmarks and applications where GAs have proved to work well finding solutions.

Figure 9.1 shows a multimodal function to be optimized by a CLONALG algorithm programmed by Leandro Nunes de Castro in Matlab programming language (Aitkin et al. 2012), and in Fig. 9.2 the result of the CLONALG optimization is shown, there are multiple local maximums. The specific parameters to test this program are:

- Number of generations: 25,
- Population size: 100,
- Mutation probability: 0.010
- Number of clones per candidate: 10.

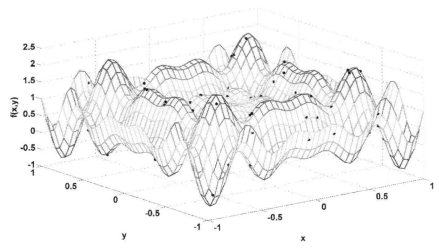

Fig. 9.1 Multimodal function to be optimized by a CLONALG algorithm.

Color image of this figure appears in the color plate section at the end of the book.

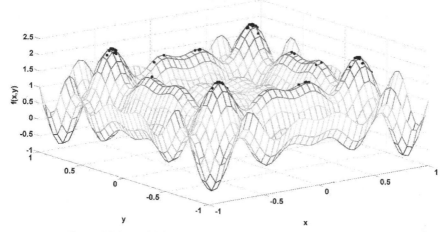

Fig. 9.2 Multimodal function optimized by a CLONALG algorithm.
Color image of this figure appears in the color plate section at the end of the book.

9.6 Immune Networks

With the aim to explain some properties of the biological immune systems, in 1974 the immune networks were proposed (Jerne 1974). In this theory, the IS is able to achieve immunological memory by means of a network of B-cells, which are constantly reinforcing themselves. The network functions by stimulating B-cells with binding matching between molecular portions of the antibody, called paratopes and idiotopes, located on the same cells. The stimulation is similar to what would happen if a pathogen is present in the body and its antigens were being bound to. Additional work must be done by the immune network in the form of suppression, between the B-cells in order to maintain it stability. It is believed that this process, gives to the IS the ability to retain memory of its cells, even in the absence of the antigen (Aitkin et al. 2012, Parham 2000).

In computer science, many implementations of the immune network theory have been developed, the aiNet (Leandro Nunes de Castro 2001) being one of the most important, which is an algorithm based on the CLONALG algorithm, but with important additions based on the immune network theory, specifically, the network suppression of cells with similar affinities. The pseudocode for the aiNet algorithm can be found in Algorithm 93. Initially the aim of the aiNet algorithm was to use it in data clustering and analysis operations, however, it has also found applications in multimodal optimization (De Castro 2002), and some hybridization with fuzzy systems in order to improve its performance (George Barreto Bezerra 2005).

Algorithm 9.3 Psuedocode for the aiNet algorithm (Leonardo N. de Castro 2002).

```
function aiNET (S, g, n₁, n₂, n₃, α, β)
returns set of memory individuals M
inputs:      S        patterns to be recognized
             g        number of iterations
             n₁       number of high quality elements to be selected for cloning
             n₂       number of high quality clones to be placed into the memory
                      set
             n₃       number of elements to be introduced
             α        threshold for clonal suppression
             β        affinity threshold
begin
j ← 0 ,
P ← rand(N, L)
while j < g do
        for each s of S do
                for each p of P do
                        aff(p) ← match(s, p);
                endFor
                P ← sort(P, aff) ,  P1 ← select(P, n₁)
                for i < n₁ do
                        C ← clone(P1, aff(P1))
                endFor
                for every c of C do
                        C1 ← hypermut(c, aff(P1))
                endFor
                for each c1 of C1 do
                        aff(c1) ← affinity(c1, s)
                endFor
                M1 ← sort(aff(C1))
                M2(s) ← select(M1, 1)
                for every m of M2 do
                        if aff(m) < β then
                                M2 ← remove(M2, m)
                        endIf
                        clonal_aff(m) ← affinity(M2, m)
                        if clonal_aff(m) < α then
                                M2 ← remove(M2, m)
                        endIf
                endFor
                M ← insert(P, M2)
        endFor
        for every m of M do
                aff(m) ← affinity(M, m)
                if aff(m) < β then
                        M ← remove(M, m)
                endIf
        endFor
        R ← rand(n₃, L)
        P ← insert(M, R)
        j ← j + 1
endWhile
return M
end
```

Figure 9.3 shows the training pattern to be used by the aiNet algorithm developed by Leandro Nunes de Castro in Matlab programming language (Aitkin et al. 2012). The specific parameters are as follows:

- Pruning threshold: 0.01
- Number of best matching cells to be selected: 4
- Percentage of the clones to be re-selected: 10%
- Number of generations: 20

Figure 9.4 shows the resulting dendrogram illustrating the arrangement of the clusters produced by hierarchical clustering of the aiNet algorithm. Finally, in Fig. 9.5 the correct clustering of the two spiral patterns is shown.

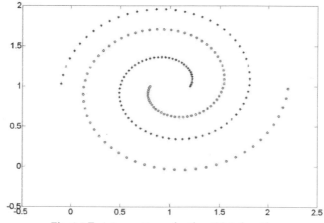

Fig. 9.3 Training patterns for the aiNet algorithm.

Color image of this figure appears in the color plate section at the end of the book.

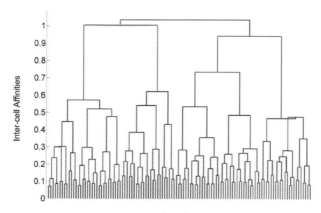

Fig. 9.4 aiNet Dendrogram.

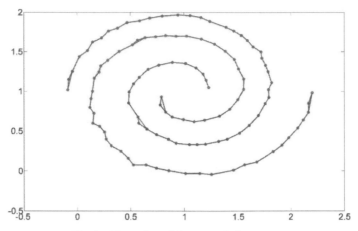

Fig. 9.5 Clustering of the two spiral patterns.
Color image of this figure appears in the color plate section at the end of the book.

9.7 Other Sources of Inspiration

Many other possible sources of inspiration exist in the world of immunology. For example, danger theory, which argues that the IS responds to danger signals rather than non-self substances, and makes it possible to explain why the IS is able to determine self-cells, and molecules that are harmful, and non-self cells and molecules which are not harmful. This extends one of the basic premises of immunity, in which, it was established that the IS can distinguish between self and non-self only. Danger theory has proven to be a great new research tool in the AIS community (U. Aicklin 2002).

The idea of utilizing vaccines, as an additional source of inspiration in the field of AIS, has been described as a potential source of new AIS algorithms in the future, however, no details of how this could be done, has been explored (Leonardo N. de Castro 2002). Biologically speaking, a vaccine consists of an inert foreign agent, which once introduced to the immune system, produces reinforcements against this particular foreign agent in future infections. As it was stated in the introduction, the concept of vaccination is what really created the interest in learning more about this phenomenon, and eventually the whole theory behind the IS was developed. It works as follows: Given to a host (algorithm), a vaccine is a piece of information that provides with meaningful insight into the solution of a problem, without having to modify the original host (algorithm). This means that vaccines act with the current algorithm, and they provide the means to produce better quality solutions by not changing the fundamental workings of an algorithm.

The concept of Artificial Vaccination was applied to COPs, because the performance of the vast majority of algorithms designed to solve them, decrease when the number of elements grows. Usually, this performance degradation is quantified in a substantial increase of the execution time needed to find a viable solution, or in degradation on the quality of the solutions themselves. This situation presents a very important challenge for all the developed algorithms, hence to find a methodology that helps to alleviate this problem is the focus of the Artificial Vaccination methodology (Oscar Montiel 2012).

9.8 Are Artificial Immune Systems Worth It?

Timmis proposed a rather interesting question that needs to be asked, but most importantly needs to be answered: "[…after all this research into AIS…] was it worth it?" (Emma Hart 2008). He goes into explaining that AIS has had success stories, no doubt; however, the issue is presented when AIS tackles some of the most important benchmarking problems, and finds out that the obtained results are inferior with those of more established and classical algorithms. The AIS community needs to take a step back and reevaluate what exactly AISs are good for, and continue the studies on those areas.

While novel paradigms may be extrapolated and new algorithms are created, at the end, it all comes down to know how effective this new concept is. In other words, finding its niche is the most important part of any new algorithm. Beyond simply telling us what it cannot do, ultimately the success of an algorithm is telling us what it can do, and how it does it better than anything else.

Proving that the AIS is noteworthy, and a paradigm that is worth exploring much more in depth, it has been and will continue to be a rough road to travel. No doubt this has to be with a relatively short maturation period that it has experienced, when it is compared to other algorithms, such as Evolutionary Algorithms, which have demonstrated to be useful in real-world applications in the industry.

One of the most important algorithms in the AIS network, the CSA, is also one of the most controversial. The CLA has been of great success in the scientific community, however due to its sharing too many similarities with other evolutionary algorithms, specifically Genetic Algorithms, its success is cut short, as one can argue that it is a special consideration of a Genetic Algorithm in which the Crossover operator is replaced by the Somatic Hypermutation.

9.9 Conclusions

AISs are computation models of IS, which extrapolate the workings of theoretical immunology and observed immune functions, principles and models, to problem-solving and engineering applications. Some of the main properties of the IS that are of particular interest to us, because to their direct connection to computational tasks are: self-organization, learning, adaptation and diversity, classification, distributed system, pattern recognition; most of these properties have a direct connection to computing related tasks.

From theoretical immunology, the main algorithms are based on the Negative Selection, Clonal Selection, Immune Networks and Danger theory; although constant research is still being done in this regard, meanwhile, new paradigms may be extrapolated and new algorithms are created.

Acknowledgments

The authors would like to thank the "Instituto Politécnico Nacional (IPN)", "Comisión de Operación y Fomento de Actividades Académicas del IPN (COFAA-IPN)" and the Mexico's entity "Consejo Nacional de Ciencia y Tecnología (CONACYT)" for supporting our research activities.

References

A. Watkins, J.T. (2004). Artificial Immune Recognition System (AIRS): An Immune Inspired Supervised Machine Learning Algorithm. Genetic Programming and Evolvable Machines, 5(3): 291–318.

Aitkin, J., P. Andrews, E. Clark, A. Greensted, R. Ismail, A. Knowles et al. (2012, November 25). AIS WEB. Retrieved March 8, 2013, from The Online Home of Artificial Immune Systems: http://www.artificial-immune-systems.org/index.shtml.

Banzhaf, W.N. (1998). Genetic Programming: An Introduction. San Francisco, USA: Morgan Kaufmann.

Burnet, F.M. (1959). The clonal selection theory of acquired immunity.

D.W. Bradley, A.M. (2000). Immunotronics: Hardware Fault Tolerance Inspired by the Immune System Evolvable Systems: From Biology to Hardware. Third International Conference, ICES 2000.

D. Cooke, J.H. (1995). Recognizing Promoter Sequences Using an Artificial Immune System. Proceedings of the Intelligent Systems in Molecular Biology Conference, pp. 89–97.

Dasgupta, D. (1999). Artificial Immune System and Their Applications. Springer-Verlag.

De Castro, L.N. (2002). opt-aiNet: An artificial immune network for multimodal function optimization. Proceedings of IEEE Congress on Evolutionary Computation, 1: 699–674.

Dipankar Dasgupta, L.F. (2009). Immunological Computation Theory and Applications. CRC Press.

Dipankar Dasgupta, S.F. (1999). Artificial Immune Systems in Industrial Applications. Intelligent Processing and Manufacturing of Materials. Proceedings of the Second International Conference on, pp. 257–267.

Edelman, G. (2001). Degeneracy and complexity in biological systems. Proceedings of the National Academy of Sciences, pp. 13763–13768.

Emma Hart, J.T. (2008). Application areas of AIS: The past, the present and the future. Applied Soft Computing, 8: 191–201.

F.J. Varela, A.C. (1988). Theoretical Immunology, Part Two. SFI Studies in the Science of Complexity, 3.

Forrest, S.e. (1997). Computer Immunology. Communications of the ACM, 40(10): 88–96.

George Barreto Bezerra, T.V. (2005). Adaptive Radius Immune Algorithm for Data Clustering. ICARIS 2005, pp. 290–303.

Giuseppe Nicosia, V.C. (2004). Artificial Immune Systems. Third International Conference - ICARIS 2004, 3239, pp. 13–16.

Haykin, S. (1999). Neural networks: a comprehensive fundation (2nd ed.). Upper Saddle Rever, New Jersey: Prentice Hall .

Holland, J.H. (1975). Adaptation in Natural and Artificial Systems. University of Michigan Press.

J. Hunt, J.T. (1999). The Development of an Artificial Immune System for Real World Applications. Artificial Immune System and Their Applications.

J. Kennedy, R.E. (2001). Swarm Intelligence. Morgan Kaufmann.

J. Kim, P.B. (2002). Immune memory in the dynamic clonal selection algorithm. Proceedings of the 1st International Conference on Artificial Immune Systems ICARIS, pp. 59–67.

J.D. Farmer, N.P. (1986). The immune system, adaptation and machine learning. Physica D, 2: 187–204.

Jerne, N.K. (1974). Towards a network theory of the immune system. Annales d'immunologie.

Johnny Kelsey, J.T. (2003). Immune Inspired Somatic Contiguous Hypermutation for Function Optimisation. Genetic and Evolutionary Computation (GECCO 2003), 202.

Jon Timmis, P.A. (2007). A Beginners Guide to Artificial Immune. Silico Immunology, pp. 47–62.

Kephart, J.O. (1994). A biologically inspired immune system for computers. Proceedings of Artificial Life IV: The Fourth International Workshop on the Synthesis and Simulation of Living Systems, pp. 130–139.

Kim, J.W. (2000). Integrating Artificial Immune Algorithms for Intrusion Detection. UCL, Department of Computer Science.

L. N. de Castro, F.J. (2002). Learning and Optimization Using the Clonal Selection Principle. IEEE Transactions on Evolutionary Computation, Special Issue on Artificial Immune Systems, 6(3): 239–251.

Leandro Nunes de Castro, F.J. (2001). Immune and Neural Network Models: Theoretical and Empirical Comparisons. International Journal of Computational Intelligence and Applications, 1(3): 239–257.

Leonardo N. de Castro, J.T. (2002). Artificial Immune Systems: A New Computational Intelligence Approach. Springer.

M. Mendao, J.T. (2007). The Immune System in Pieces: Computational Lessons from Degeneracy in the Immune System. Foundations of Computational Intelligence.

Maoguo Gong, L.J. (2006). Solving Traveling Salesman Problems by Artificial Immune Response. SIMULATED EVOLUTION AND LEARNING, Lecture Notes in Computer Science.

Oscar Montiel, F.J.-D. (2012). Combinatorial complexity problem reduction by the use of artificial vaccines. Expert Systems with Applications.

Parham, P. (2000). The Immune System. Routledge.

Paul S. Andrews, J.T. (2006). A Computational Model of Degeneracy in a Lymph Node. Lecture Notes in Computer Science, 4163.

R. Krohling, Y.Z. (2002). Evolving FPGA-based robot controllers using an evolutionary algorithm. In T. &. Bently.

RO Canham, A.T. (2002). A multilayered immune system for hardware fault tolerance within an embryonic array. Proc. 1st Int. Conf. on Artificial Immune Systems (ICARIS 2002).

Stephanie Forrest, A.S. (1994). Self-Nonself Discrimination in a Computer. In Proceedings of the 1994 IEEE Symposium on Research in Security and Privacy, pp. 202–212.

T. Knight, J.T. (2003). A Multi-layered Immune Inspired Machine Learning Algorithm. Applications and Science in Soft Computing, pp. 195–202.

U. Aicklin, S.C. (2002). The danger theory and its application to artificial immune systems. Proceedings of the First International Conference on Artificial Immune System, pp. 141–148.

V. Cutello, G.N. (2005). Clonal Selection Algorithms: A Comparative Case Study using Effective Mutation Potentials, optIA versus CLONALG. 4th Int. Conference on Artificial Immune Systems, ICARIS 2005, pp. 13–28.

Whitacre, J.M. (2010). Degeneracy: a link between evolvability, robustness and complexity in biological systems. Theoretical Biology and Medical Modelling 7.

Yangyang Li, L.J. (2005). Quantum-Inspired Immune Clonal Algorithm Institute of Intelligent Information Processing. ICARIS 2005, LNCS 3627, pp. 304–317.

10

Applications of Artificial Immune Algorithms

Francisco Javier Díaz Delgadillo, Oscar Montiel* and Roberto Sepúlveda

ABSTRACT

In this chapter, we present the main programming code to implement the Reduce-Optimize-Expand (ROE) methodology; it is based on the concept of Artificial Immune Systems (AIS). This method allows reducing problem size of Combinatorial Optimization Problems (COPs); hence, the complexity of the problem is reduced. To explain the methodology, the classical benchmarking problem known as the Travelers Salesman Problem (TSP) was used. A 711 cities problem was chosen to explain the methodology.

10.1 Introduction

There are two main ways to achieve better performance of algorithms, these are: adding the computational resources (more CPU frequency, RAM or even more CPU cores in the case of parallel computing) or we can handle the problem using efficient algorithms that allow optimizing the resources we already have. This work is framed in the Combinatorial Optimization Problem (COP) classification. To solve COPs efficiently is very important

Instituto Politécnico Nacional, CITEDI, Tijuana, México.
 E-mail: oross@ipn.mx
* Corresponding author

since they embrace a big diversity of real problems that can be found in different fields of science and engineering.

At this time, there is a big diversity of algorithms to solve COPs using different computational paradigms; because the problem of finding the optimal arrangement of elements grows exponentially with the problem size, the methods to solve them can be based on finding the exact solution such as the Dantzing–Fulkerson–Johnson algorithm (Dantizig et al. 1954), or in approximate approaches based on heuristics such as the "Polynomial Time Approximation Schemes for Euclidian Traveling Salesman and Other Geometric Problems" (Arora 1998); both branches of methods were conceived for a particular algorithm either exact or approximate. In this chapter, we present the main programming code to implement the Reduce-Optimize-Expand (ROE) method used to reduce the combinatorial complexity based on the reduction of the problem size, this method is more suitable to tackle large problems (Montiel et al. 2013). The particular characteristics of the ROE method is that it works over the problem itself instead of the solving algorithm. The aim is to reduce the problem complexity using the previous concept of artificial vaccines (Dasgupta and Nino 2009) that was inspired by the Artificial Immune System (AIS) paradigm (de Castro and Timmis 2002, Zhu et al. 2008, Zhao et al. 2008).

The Traveler Salesman Problem (TSP) is an important COP that is frequently used to test newly developed algorithms due to the easy comprehension of the problem description, but at the same time, it presents a high complexity since the amount of computational resources needed to solve a TSP that goes beyond just for few elements might be overwhelming; hence, great emphasis has been laid upon the generation of new algorithms that can provide faster and better solutions for the TSP, and in general for COPs, thus this has become one of the most important fields of research in optimization (Vazirani 2001).

In the following sections, first we explain the Reduction, Optimize and Expand stages of the ROE method. For the first stage, two main strategies were developed—Vaccine by Random Selector (VRS) and Vaccine by Elitist Selector (VES). For the second stage, we used a Genetic Algorithm (GA) as the main TSP solving algorithm. For the last stage, the algorithm named Expand was developed. Therefore, the ROE methodology at the present time consists of VRS, VES and Expand algorithms. Examples of generation of vaccines using VRS and VES algorithms are presented for a TSP problem with 711 cities. Here we show the main object-oriented classes of the ROE methodology, as well as the necessary programming code to implement the methods in the C# programming language using the .NET Framework.

10.2 Step 1. Reduction

In this section, we are going to explain the implementation of the artificial vaccination process in the object-oriented C# language in order of the problem size. To achieve this explanation, we will use the TSP hence the word City is used to denote the nodes of a graph $\mathcal{G}$. This step is based on five classes: `City`, `Tour`, `Vaccine`, and `Algorithms` that contains the main methods VRS, VES, and a Greedy search; additionally the `Algorithms` class contains methods used in other steps of the ROE Methodology such as the Expand method.

10.2.1 City representation using a Class

At the core of the implementation, we have the object `City`, which represents the basic element of the TSP to be manipulated by the optimization algorithm. `City` is also the main element with which artificial vaccines are created in the artificial vaccination process. Figure 10.1 shows the class diagram for `City`, which has the following attributes:

- A public field called `ID` that holds a unique identifier for each `City` in the form of an index.
- 2D space coordinates stored in the private fields named x and y. They can be accessed publicly through the corresponding X and Y properties.

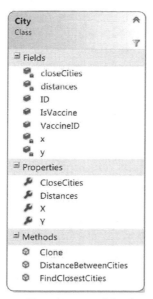

Fig. 10.1 Class diagram of the class City.

- Each City holds its relative Euclidean distance to the rest of the cities of the TSP in a private field array called distances accessible through the public property Distances.
- In order to facilitate the implementation of optimization algorithms such as Greedy Search, Nearest Neighbor and other algorithms, each City also contains the information of its nearest cities stored in a private field called closeCities accessible from the public property CloseCities. In order to obtain this information, a public method called FindClosestCities is provided.
- Utility methods are also implemented, such as calculating the Euclidian distance between cities with the public method DistancesBetweenCities and a way to clone City objects with the public method Clone.
- Additional public fields such as IsVaccine and VaccineID are added to help with the Expansion step and plotting purposes. The City object does not require these fields as part of the TSP solving algorithm.

10.2.2 Tour representation using a Class

The object Tour represents a possible solution to the TSP and depending on the TSP solving algorithm, anywhere from one to a very large number of Tour objects may exist at a given time. The main task of the TSP solving algorithm is to generate one Tour object which might be the optimal solution. At its core, the Tour class is a generic list of type Link and each Link contains the index of two City objects, effectively connecting them in such a way that they form a cycle that meets all of the TSP requirements. The Tour class is an important part of the process because it implements all the necessary methods to evaluate possible solutions, such as the total tour distance and the order in which a city comes after or before another given city. Figure 10.2 shows the class diagram for the Tour and Link objects which have the following attributes:

- Each Tour object provides information about the total distance of the tour (Hamiltonian cycle). In order to store this information the private field fitness is used, it is accessible from the public property Fitness. The field is named fitness, in order to stay consistent with the terminology used by GAs, the total tour length being the most appropriate measurement of fitness. Additionally, in order to calculate and update the fitness field, the public method named CalculateFitness is added. CalculateFitness uses the Euclidean distance formula.

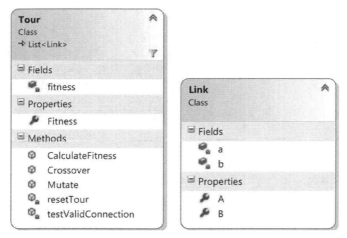

Fig. 10.2 Class diagram of the Tour class.

- Additional static methods that allow the GA to generate valid offspring tours are also implemented in this class. The GA uses the `Crossover` method to generate the offspring from two parent tours by exchanging sections of tours; with the `Mutate` method, mutation operation randomly swaps two cities of the same tour.
- Additional private methods that allow "sanity checks" and validation for `Tour` objects are included, such as a method called `testValidConnection` that allows to validate a `Tour` object after it has been modified, and `resetTour` which allows to empty a `Tour` object.
- The object `Link` holds two `City` object indexes, which allow retrieving the objects from a `City` list. The indexes are stored in private fields `a` and `b`, which are accessible from the public properties `A` and `B`.

10.2.3 Artificial Vaccines representation using the Vaccine Class

Artificial Vaccine is the main building block of the artificial vaccination methodology, it represents a group of `City` objects, which are removed from the original TSP in order to reduce its problem space. In Fig. 10.3, the fields, properties and methods to implement the `Vaccine` object are shown. In this proposal, a `Vaccine` object consists of a generic list of type `City` object, which has the following attributes:

- A public field called `VaccineID` that holds a unique identifier for each `Vaccine` object in the form of an index. This field is important in the systematical creation, optimization and expansion of the vaccinated TSP.

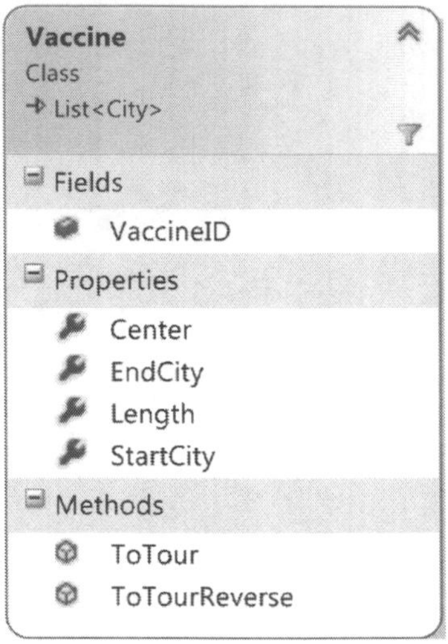

Fig. 10.3 Class diagram of a Vaccine.

- Being a list of `City` objects, the `Vaccine` has the properties `EndCity` and `StartCity`, where the first and last `City` objects of the vaccine are readily available for their later used for the Expansion step.
- With the property `Length`, the distance of the partial tour of the `City` objects included in the `Vaccine` object is obtained.
- The public property `Center` represents the location of the `Vaccine` on the map. It is required because the original `City` objects that are now part of the `Vaccine` are removed; therefore, it is indispensable to add a placeholder representation of the vaccine on the map. `Center` is an object of type `Point` which holds `X` and `Y` coordinates.
- The public methods `ToTour` and `ToTourReverse` are auxiliary methods used to review whether the current `Vaccine` is included in the vaccine repertoire already generated; with the method `ToTour`, a new `Tour` object, that represents the same order of the `Vaccine` object, is created; while `ToTourReverse` produces the `Tour` object in reverse order.

For a 2D map, the coordinates of the Center in x, y of a Vaccine is calculated using equations (10.2.1) and (10.2.2).

$$\text{Center in x: } X_c = \frac{\sum_{i=0}^{n} x_i}{n} \tag{10.2.1}$$

$$\text{Center in y: } Y_c = \frac{\sum_{i=0}^{n} y_i}{n} \tag{10.2.2}$$

The length of a `Vaccine` object (as a sub-path length) can be calculated by the Euclidian distance using (10.2.3). The implementation of the property

Length cycles through the entire list of **City** objects in the **Vaccine**, in order to calculate the respective distance; however, the loop is not closed as a vaccine because it is not a Hamiltonian cycle.

$$L = \sum_{i=0}^{n-1} \sqrt{(x_i - x_{i+1})^2 + (y_i - y_{i+1})^2} \tag{10.2.3}$$

Program 10.1 shows the C# code used to implement the vaccine described in Fig. 10.3 (Montiel et al. 2013).

Program 10.1 C# Implementation of an Artificial Vaccine.

```
public class Vaccine : List<City>
{
  public int VaccineID = -1;

  public Vaccine(int vaccineID)
  {
    this.VaccineID = vaccineID;
  }
  public City StartCity
  {
    get { return this[0]; }
    set { this[0] = value; }
  }
  public City EndCity
  {
    get { return this[Count - 1]; }
    set { this[Count - 1] = value; }
  }
  public Point Center
  {
    get
    {
      int x = 0;
      int y = 0;

      foreach (City c in this)
      {
        x += c.X;
        y += c.Y;
      }
      x = x / Count;
      y = y / Count;

      return new Point(x, y);
    }
  }
  public double Length
  {
    get
    {
      double length = 0;
      for (int i = 0; i < this.Count - 1; i++)
      {
```

Program 10.1 contd....

Program 10.1 contd.

```
    length +=
        Math.Round(Math.Sqrt(Math.Pow((double)(this[i].X - this[i + 1].X), 2D)
        + Math.Pow((double)(this[i].Y - this[i + 1].Y), 2D))));
    }

    return length;
    }
}
public Tour ToTour()
{
    Tour tour = new Tour(this.Count);

    for (int i = 0; i < this.Count - 1; i++)
    {
        Link link = Util.CreateLink(this[i], this[i + 1]);
        tour.Add(link);
    }
    return tour;
    }
    public Tour ToTourReverse()
    {
    Tour tour = new Tour(this.Count);
    for (int i = this.Count - 1; i > 0; i--)
    {
        Link link = Util.CreateLink(this[i], this[i - 1]);
        tour.Add(link);
    }
    return tour;
    }
}
```

10.2.4 Vaccine Generation by Random Selector (VRS)

To create Vaccines with the Random Selector (VRS) algorithm, we start providing the Original Node List (ONL) and randomly choosing a node that functions as the Initial Node (IN), from this point the rest of the vaccine is created by selecting the nearest neighbor until the required Number of Nodes (NN) per vaccine are completed, as it was indicated in (Montiel et al. 2013). The distance is calculated by Euclidean distance.

Program 10.2 shows the C# implementation code of this method, in this program, first we see that a clone of the original list is created. This is important because we want to create the vaccines in such a way that elements do not get repeated. In order to achieve this, we implemented a list from which already used elements are removed, so they cannot be selected again randomly. From there, as an element gets randomly selected, we proceed to calculate a greedy tour as long as it was defined in the variable numCitiesPerVaccine, this is made in a method called Greedy, which creates the Vaccine and also adds the unique identifier VaccineID to it. Before the Vaccine object is added to the list of created vaccines, an additional step is taken to confirm that this particular Vaccine is not already inserted in the list, the method Contains uses the public methods ToTour and ToTourReverse of the Vaccine to determine this. In case it is found, it is simply discarded, otherwise it is added. We continue with the iterations removing the randomly selected starting City from the availableCities repertoire, and creating and adding the new City based on the geometrical Center of the Vaccine. This process is repeated until we fulfill the needed quota stated at numVaccines.

Program 10.2 Generation of vaccines using VRS.

```
public List<Vaccine> VRS(int numVaccines,int numCitiesPerVaccine,List<City> cities)
{
    List<City> availableCities = new List<City>();
    availableCities.AddRange(cities);
    List<Vaccine> vaccines = new List<Vaccine>();

    for (int i = 0; i < numVaccines; i++)
    {
        while (true)
        {
            City temp = availableCities[new
                Random(Guid.NewGuid().GetHashCode()).Next(0,availableCities.Count)];
            Vaccine vaccine = Algorithms.Greedy(temp,availableCities,
            numCitiesPerVaccine - 1, -1-i);

            if (!vaccines.Contains(vaccine))
            {
                vaccines.Add(vaccine);
                availableCities.Remove(temp);
                foreach (City c in vaccine)
                {
                    cities.Remove(c);
                }
                cities.Add(new City(vaccine.Center,true,-1-i,-1-i));
                break;
            }
        }
    }
    return vaccines;
}
```

10.2.5 Vaccine Generation by Elitist Selector (VES)

The goal of the VES method to generate vaccines, is to choose the best vaccines for the specified requirements. The application of this algorithm requires calculating the matrix of distances, which corresponds to the distance from each one of the nodes to the rest, with the aim of sorting by the shortest distance. From this point, we construct the vaccines using the nearest neighbor algorithm. Those vaccines that have the shortest distance accumulated are chosen. This guarantees that the shortest route vaccines are picked. Program 10.3 shows the C# code that implements the method for this algorithm.

Similar to VRS, we provide the ONL and then we create a copy of it, which will be modified during the code execution. Additionally, since this methodology requires elitist selection criteria, we require to cycle through all of the cities and generate a `Vaccine` of the required length of `numCitiesPerVaccine` for each `City` by a greedy algorithm. Once this is calculated, we must sort the `Vaccine` list by the shortest distance, and select as many `numVaccines` as stated. After that, we must remove the selected starting `City` objects from the original `City` list `cities`, then create and add the new `City` object based on the geometrical `Center` of the `Vaccine` objects.

Program 10.3 Generation of vaccines using VES.

```
public List<Vaccine> VES( int numVaccines,int numCitiesPerVaccine,List<City> cities)
{
  List<City> originalCities = new List<City>();
  originalCities.AddRange(cities);
  List<Vaccine> vaccines = new List<Vaccine>();
  for (int i = 0; i < cities.Count; i++)
  {
    Vaccine vaccine = Algorithms.Greedy(originalCities[i], cities,
                           numCitiesPerVaccine - 1, -1 - i);
    if (!vaccines.ContainsCity(originalCities[i]) &&
        !vaccines.ContainsCity(originalCities[i]))
    {
      vaccines.Add(vaccine);
    }
  }
  var sorted = (from v in vaccines orderby v.Length select v).Take(numVaccines);
  List<Vaccine> result = new List<Vaccine>();
  int id = 0;
  foreach (Vaccine v in sorted)
  {
    v.VaccineID = -1 - id;
    result.Add(v);
    foreach (City c in v)
    {
      cities.Remove(c);
    }
    cities.Add(new City(v.Center, true, -1 - id, -1 - id));
    id++;
  }
  return result;
}
```

10.3 Step 2: Optimization

This step consists of optimizing the vaccinated city list of the reduced graph for a particular TSP. At this point the problem complexity was already reduced and it will be easier to any suitable optimization method to find the optimal tour. As mentioned earlier, at this stage, we can use any exact or approximate optimization method, in our case we used a GA (Montiel et al. 2013).

10.4 Step 3: Expansion

The expansion algorithm is required to transform the solution provided by the optimization method used as it was described in step 2 (Montiel et al. 2013). This process is critical, as the solution obtained so far is not valid, as it does not take into consideration the length of the removed vaccines.

The temporally removed nodes that represent the vaccines are recovered, to reconstruct the original problem; i.e., original set of nodes. Program 10.4 shows C# the implementation of this method.

Program 10.4 Algorithm to expand the Reduced Node List to the Original Node List.

```
publicList<City> Expand(List<City> cityList,List<Vaccine> vaccines,Tour bestTour)
{
  if (bestTour.Count != 0)
  {
    List<City> expandedCities = newList<City>();
    List<City> cityListCopy = cityList.Clone();
    List<Vaccine>vaccinesCopy = newList<Vaccine>();
    vaccinesCopy.AddRange(vaccines);

    int last = 0;
int next = bestTour[0].A;
    if (!cityListCopy[last].Clone().IsVaccine)
    {
    expandedCities.Add(cityListCopy[last].Clone());
    }
    else
    {
    Vaccine v = vaccinesCopy.Find(vv => vv.VaccineID ==
            cityListCopy[next].VaccineID);
      double d1= City.DistanceBetweenCities (v.StartCity, cityListCopy[last]);
    double d2= City.DistanceBetweenCities (v.EndCity, cityListCopy[last]);
    if (d1 <= d2)
      {
      expandedCities.Add(v.StartCity);
      for (int j = 1; j < v.Count - 1; j++)
        {
      expandedCities.Add(v[j]);
        }
        expandedCities.Add(v.EndCity);}
        else
        {
        expandedCities.Add(v.EndCity);
        for (int j = v.Count - 2; j >= 1; j--)
                expandedCities.Add(v[j]);
        expandedCities.Add(v.StartCity);
        }
    }
```

Program 10.4 contd....

Program 10.4 contd.

```
for (int i = 0; i < cityListCopy.Count; i++)
{
  if (!cityListCopy[next].IsVaccine)
    {
    expandedCities.Add(cityListCopy[next]);
    }
  else
    {
    Vaccine v = vaccinesCopy.Find(vv => vv.VaccineID
                 == cityListCopy[next].VaccineID);
        if (v != null)
```

10.5 Example of Vaccinated TSP

We have chosen TSP instance RBX711 from the available public repertoire of TSP called TSPLIB, as the main example to explain the process behind the application of vaccines. Figure 10.4 shows the graph of the TSP instance with no tours or any vaccination methods applied. This is the default instance with which the TSP solving algorithm would start its optimization process.

In Fig. 10.5 the same TSP instance is shown, but now it is vaccinated by the VRS algorithm. Notice especially the colored dots. The blue dots represent the edges of a vaccine, while the red dots represent the geometrical center. The blue dots are removed, and the red dots are added in order to reduce the TSP instance. The green line represents the path that the vaccine is removing, which will be reintroduced in the expansion step. Being a vaccination chosen by the random selector, we have vaccines of different

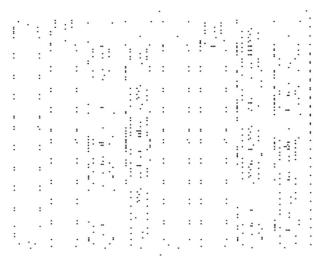

Fig. 10.4 TSP Instance RBX711 from TSPLIB.

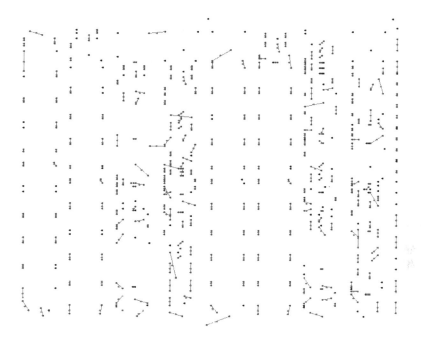

Fig. 10.5 TSP Instance RBX711 from TSPLIB vaccinated by VRS Method with NV=ONLSize×0.4, and NNV=2. Blue dots represent the Start and End Nodes and the red dot represents the Geometrical Central that represents the vaccine.

Color image of this figure appears in the color plate section at the end of the book.

lengths, some short but some long. Having longer length vaccines may produce unwanted results, as they could potentially be taking out of the TSP routes which are needed to optimize it correctly. This example also works with a Number of Vaccines (NV) equal to 40% of the totally amount of cities. Additionally, the Number of Nodes per Vaccine (NNV) is set to two, meaning that we are generating the smallest city count vaccines possible. Both properties can be changed in order to provide different reductions. Figure 10.6 shows VRS with NNV equal to four.

Figure 10.7 shows the same TSP instance vaccinated by VES. Compared to Fig. 10.5 we can notice that all the closest nodes have been grouped together, and this is precisely the performance we are looking for when speaking of an elitist behavior. This is mostly desirable because our hypothesis is that the optimal solution will most likely have a path that travels between the closest nodes.

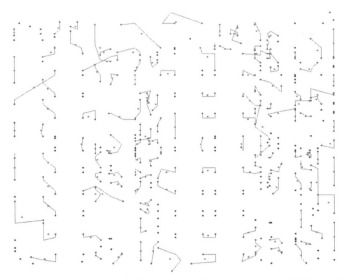

Fig. 10.6 TSP Instance RBX711 from TSPLIB vaccinated by VRS Method with NV=ONLSize×0.20, and NNV=4. Blue dots represent the Start and End Nodes and the red dot represents the Geometrical Central that represents the vaccine. Green lines are the vaccines.

Color image of this figure appears in the color plate section at the end of the book.

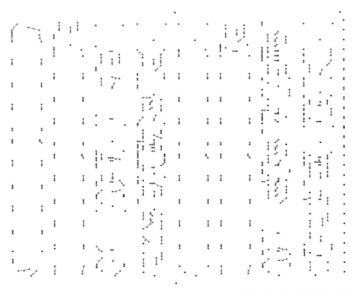

Fig. 10.7 TSP Instance RBX711 from TSPLIB vaccinated by VRS Method with NV=ONLSize×0.4, and NNV=2. Blue dots represent the Start and End Nodes and the red dot represents the Geometrical Central that represents the vaccine.

Color image of this figure appears in the color plate section at the end of the book.

10.6 Example: Application of the ROE Method and a GA

Even though the methodology can be applied to any TSP solving algorithm, we have chosen a Genetic Algorithm (GA) for the Artificial Vaccination process example. The main goal behind choosing a GA is its widely known importance and also of the ease of implementation due to the great amount of information that exists on them.

The main details of our GA implementation are as follows:

- The population was generated by a random tour generator with a greedy selector of 90% probability.
- The selection strategy of the parents is random.
- The crossover operation is a Partially Matches Crossover (PMX).
- The replace strategy consists of substituting the worse elements from the population with children, rather than parents, only if they are better (elitist).
- Mutation is implemented as city swap with a 3% chance of occurrence.

We have designed three experiments that embrace a big diversity of comparative situations that will help to conclude about the usefulness of the method.

Experiment 1—Obtaining benchmarks: This set of experiments consisted in obtaining the control data to be used as benchmarks in order to achieve comparisons. The behavior of the GA optimizing the TSP can be seen in Fig. 10.8.

Experiment 2—Vaccination with Random Selector (VRS): In this set of experiments, the vaccination was achieved using VRS on the TSP City list, as can be seen in Fig. 10.8.

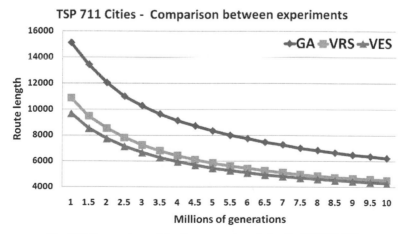

Fig. 10.8 Comparison of the three experiments for the TSP RBX711.

Experiment 3—Vaccination using the Elitist Selector (VES): Similar to Experiment 2, after reducing the set of cities, the GA was used to obtain the shortest path, and then the expansion of the vaccinated route was achieved. See Fig. 10.8.

For the three experiments, the GA ran 10 million of generations, and results are recorded every 500,000 generations, as is shown in Fig. 10.8. For the three experiments, we used a convergence value (stuck point). It was chosen such that the algorithm did not provide any significant advance (less than 1% of improvement compared the previous generation).

Continuing with the use of the TSP RBX711 example shown in Fig. 10.8, we present the example data of the average of 20 runs of each of the three experiments.

Performance wise we look forward to show the quality of the solution, the execution time and also the convergence values. By analyzing the graph in Fig. 10.8, it is clear that the vaccination algorithms, VRS and VES, will converge faster than the GA alone. The final route values for VRS and VES were better than the GA by 37.65% and 44.71% respectively.

Figure 10.9 is a graph that shows the execution time for each of the TSP solved by the three different algorithms GA, VRS and VES. From this graph we can see that the execution time was improved significantly in all the studied problems, independently of the number of cities used by the vaccination method.

The improvements in execution time and quality of the solutions are attributed to the effective reduction of nodes (cities) that the optimization algorithm has to work in order to provide the optimal (or suboptimal) solution.

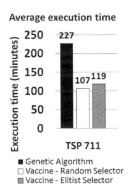

Fig. 10.9 Comparison of the average execution time by the three different algorithms.

10.7 Conclusions

In this chapter, we presented programs that allow implementing the ROE method to achieve the reduction of COP complexity to improve performance. The ROE method is specially suitable for obtaining solutions of very large COPs (Montiel et al. 2013). The idea emphasizes a more optimal utilization of the available computational resources instead of adding more hardware to tackle tougher problems. We have presented the main code of the ROE method in C#.

In order to further support the proposal, we have provided step by step illustrations of the vaccination processed applied to the TSP instance RBX711 from the available public repertoire of TSP called TSPLIB.

Three experiments were conducted in order to demonstrate the usefulness of the proposal with the same TSP instance, which include the benchmarking data for the two main vaccines generating algorithms VRS and VES. The gathered data including the quality of the solutions, convergence values and execution time were recorded. The large amount of cities of the TSP instance allowed the vaccination methodology to produce much better solutions and much faster conversion for the same amount of CPU time, a tendency that will remain as the number of cities grows. Additionally, improved execution times are reported between 47% and 52%, giving the methodology good results in all evaluated parameters.

Acknowledgments

The authors would like to thank the "Instituto Politécnico Nacional (IPN)", "Comisión de Operación y Fomento de Actividades Académicas del IPN (COFAA-IPN)" and the Mexico's entity "Consejo Nacional de Ciencia y Tecnología (CONACYT)" for supporting our research activities.

References

Arora, S. (1998). Artificial immune systems: A new computational intelligence approach. Springer-Verlag, Great Britain. Polynomial time approximation schemes for euclidean traveling salesman and other geometric problems. Journal of the ACM, 45(5): 753–782.

Dantizig, G.B., D.R. Fulkerson and S.M. Johnson (1954). Solution of a large-scale traveling salesman problem. Operation Research, 2: 393–410.

Dasgupta, D. and L.F. Nino (2009). Immunological computation: theory and applications. US: CRC. Press Taylor & Francis Group. Auerbach Publications Boston, MA, USA.

de Castro, L. and J. Timmis (2002). Artificial immune systems: A new computational intelligence approach. Great Britain: Springer.

Montiel, O., F.J. Díaz-Delgadillo and R. Sepúlveda (2013). Combinatorial complexity problem reduction by the use of artificial vaccines. Expert Systems with Applications, 40(5): 1871–1879.

Vazirani, V.V. (2001). Approximation algorithms. US:Springer-Verlag New York, Inc. New York, NY, USA.

Zhao, M.-Y., Ke, T., G, L., Zhou, M. T., Fu, C., Yang, F. and C.-G. Zhang (2008). A novel clonal selection algorithm and its application. ICACIA 2008. Chengdu, China: IEEE, pp. 385–388.

Zhu, Y., T. Zhen, H. Dai and S. Gao (2008). Cooperation Artificial Immune System with Application to Traveling Salesman Problem. An International Journal of Research and Survey, 2(2): 143–148.

11

A Parallel Implementation of the NSGA-II

Oscar Montiel,* Roberto Sepúlveda and
Josué Domínguez

ABSTRACT

Multi-objective optimization methods have to find the best solution
that minimizes or maximizes simultaneously two or more objective
functions; these algorithms can usually find a set of solutions that
fulfill such conditions trying to get the Pareto optimal. Traditional
algorithms have difficulties finding the Pareto optimal since the search
space is complex; for this reason, natural algorithms have found an
important research niche due to its ability to approach to the Pareto
Optimal set at each run. This paper presents a proposal to achieve the
parallel implementation of the NSGA-II algorithm. In the proposal, the
algorithms for the crossover procedure and Pareto rank assignments
were designed using threads for parallel programming. The NSGA-II
was coded in C#, and it was evaluated with five special test functions
using different genetic operators. Experimental results show that these
operators can perform fine for some function, but they do not always
exhibit the best performance for all problems; this is depending on the
problem complexity.

Instituto Politécnico Nacional, CITEDI, Tijuana, México.
  E-mail: oross@ipn.mx
* Corresponding author

11.1 Introduction

A Multi-Objective Optimization Problem (MOOP) has two or more objectives that need to be optimized simultaneously, in contrast to Single-Objective Optimization Problems (SOOPs); in general, MOOPs do not have only one solution, but a set of solutions called the Pareto Optimal Set, which has a good relationship between the objectives.

In the beginning, it was necessary to use mathematical programming techniques, such as the Weighted Sum method (Stadler 1979) and Goal Programing methodology (Jones 2003, Ignizio 1978), to solve MOOPs. These techniques have limitations when the Pareto frontier is concave or discontinuous; moreover, most of these techniques only generate a solution per run for each objective; therefore, for several objectives, to find the Pareto Optimal Set it was necessary to run several times.

According to various bibliographic sources (Coello Coello et al. 2007, Coello 2006, Deb 2001), the first attempt to solve optimization problems with multiple objectives using Evolutionary Algorithms (EAs), was made by David Schaffer in 1984 (Shaffer 1984) who proposed the VEGA (Vector Evaluated English Genetic Algorithm), this is a simple genetic algorithm, where selection, crossing and mutation are independently performed in cycles of classification according to each objective. This algorithm usually trends to find the optimum of each objective function.

In 1989, David Goldberg proposed an implementation of dominance principle in EAs; he suggested a method to sort the non-dominated individuals in a population, and make more copies of these individuals for the next generation. Goldberg recommended a strategy to avoid the convergence to a single point (Goldberg 1989), by maintaining diversity to get as close as possible to the Pareto Optimal Set. He does not provide a practical implementation of the algorithm, but after this publication most EAs used the concept of dominance and various techniques to preserve diversity in the Pareto frontier, based on the idea of dividing the population into niches.

The algorithm Non-Dominated Sorting Genetic Algorithm (NSGA) proposed by Srinivas and Deb (Srinivas and Deb 1994), was the first EA for MOOPs that used the concept of niche dominance and the Goldberg's concept previously explained. The NSGA does not have good performance in concave functions and problems with discontinuity (Coello Coello et al. 2007).

The NSGA can be considered the second generation for solving MOOPs using EAs. It was sought to improve efficiency, with the aim to be as close as possible to the Pareto Optimal Frontier, which was achieved using the mechanism of elitism as the standard practice, Eckart Zitzler was the first to use this mechanism in MOPs successfully (Zitzler 1999). After Zitzler, there were proposed other algorithms with good performance, as it is the case of NSGA-II (Non-Dominated Sorting Genetic Algorithm II) (Deb et al. 2002), and PESA-II (Pareto Envelope-based Selection Algorithm II) (Corne et al. 2001).

11.2 Basic Concepts

The concept of optimization is found in many areas of engineering and economics, primarily in those that involve problems of decision making; optimization refers to obtain the values of decision variables, which correspond to the maximum or minimum of one or more objective functions (Rangaiah 2009). The classical approaches try to find the best solution for a single objective function; however, in most real-world problems it is necessary to find the best solution of two or more objective functions; this concept is called multi-objective optimization. Its goal is to find a set of optimal solutions for different objective functions. In some literature, MOOPs are referred to as multiple criteria decision making.

When solving a MOOP, the aim is to find the Pareto Optimal Set; this set of solutions serves to the humans in the decision-making process to solve problems, knowing some preferential information about the objective functions, from the optimal set, they can choose the best solution for a specific problem.

Mathematically speaking, an optimization problem is defined as follows (Deb 2001):

Minimize/Maximize $f(x)$

subject to:

$$g_i(x) \geq 0, i = \{1, 2, \ldots, m\}$$
$$h_j(x) = 0, j = \{1, 2, \ldots, p\}$$

where x is a vector of n decision variables $(x_1, x_2,..., x_n)$, $g_i(x)$, represents m inequality constraints, while $h_j(x)$ represents p equality constraints in the decision space Ω, see Fig. 11.1. A MOOP can also be written in the form of minimize/maximize $f(x) = [f_1(x), f_2(x), ..., f_k(x)]^T$ for k objective functions $f_i: \mathbb{R}^k \rightarrow Z$ subject to some equality and inequality constraints, where Z is the objective space as it is shown in Fig. 1.11 (Abraham et al. 2005). Given the vector of decision variables $x = [x_1, x_2,..., x_n]^T$, our task is to determine the set $f(x)$ of all vectors which satisfy all the constraints, the particular set of values $x = [x_1^*, x_2^*,..., x_n^*]^T$, and also yields the optimum values for all the objective functions (Konak et al. 2006).

The main purpose of an MOP algorithm is to identify solutions in the Pareto Optimal Set. In general, there are three principal methods of dealing with MOOPs:

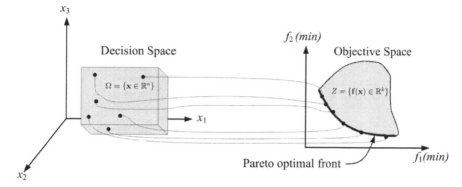

Fig. 11.1 Decision variable and objective space relationship.

1. One method consists in combining all the objective function values to form a single scalar value; here, the weighted sum is used (Stadler 1979), where the obtained scalar value must be optimized. For this approach, this method provides only one solution.
2. Other methods are based on optimizing each objective function separately, assigning hierarchies to the objective values. An algorithm that uses this scheme is the ε-constraint method (Wismer 1971), this technique optimizes only one objective function at time, while the other objective functions are used as constrained functions (Becerra 2006).
3. A third option is to obtain a set of alternative solutions, known as non-dominated set of solutions. Most EAs use this strategy to iteratively find the optimal set, finishing when there is no other non-dominated set in the objective space; this is called the Pareto Optimal Set.

The concept of domination is fundamental to understand multi-objective optimization since in the problem-solving process a multitude of solutions are generated, but only a small subset of the solutions will be of interest, they are those solutions that dominate the others. More formally, the concept of domination can be defined as follows: We can say that a decision vector, $x^{(1)}$ dominates a decision vector $x^{(2)}$, expressed as $x^{(1)} \prec x^{(2)}$, if and only if the next two conditions are fulfilled,

1. The solution $x^{(1)}$ is no worse than $x^{(2)}$ in all objectives, or $f_i(x^{(1)}) \not\rhd f_i(x^{(2)})$ for all $i = 1, 2, \ldots, k$.
2. The solution $x^{(1)}$ is strictly better than $x^{(2)}$ in at least one objective, denoted as $f_i(x^{(1)}) \lhd f_i(x^{(2)})$ for at least one $i \in \{1, 2, \ldots, k\}$.

We have that $x^{(1)}$ does not dominate $x^{(2)}$ if any of the above conditions are not fulfilled (Deb 2001).

The concept of dominance is illustrated in Fig. 11.2.

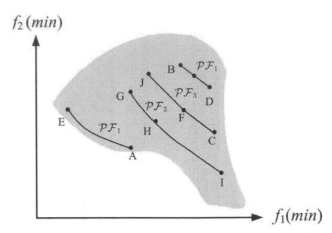

Fig. 11.2 Dominance concept and Pareto front. Points A and E are in the non-dominated set.

The Pareto Optimal Set $\mathcal{P}^*$, for a given MOOP where $\mathbf{F}(\mathbf{x}) = \mathbf{f}(\mathbf{x})^T$, is defined using (11.2.1), where the Pareto optimal solutions are those solutions which dominates the others but do not dominate themselves.

$$\mathcal{P}^* := \{\mathbf{x} \in \Omega | \neg \exists\, \mathbf{x}' \in \Omega, \mathbf{F}(\mathbf{x}') \preceq \mathbf{F}(\mathbf{x})\} \qquad (11.2.1)$$

In the objective space, all the non-dominated vectors are known as the Pareto front, $\mathcal{PF}^*$, and it is defined as (11.2.2), where $\mathbf{u}$ is a vector function.

$$\mathcal{PF}^* := \{\mathbf{u} = \mathbf{F}(\mathbf{x}) | \mathbf{x} \in \mathcal{P}^*\} \qquad (11.2.2)$$

Generating the Pareto Set $\mathcal{P}$ can be computationally expensive, and it is often unfeasible because the complexity of the application prevents exact methods from being applied. For this reason, a number of stochastic search strategies such as EAs, tabu search, Ant Colony Optimization, and others have been used, and they can usually find a good approximation, i.e., a set of solutions whose objective vectors are not too far away from the Pareto Optimal.

The Pareto Optimality concept for minimization problems are as follows: A decision vector $\mathbf{x}^{(1)} \in \Omega$ is a Pareto optimal value $\mathbf{x}^*$, if there is no other decision vector, such as $\mathbf{x}^{(2)} \in \Omega$, where $\mathbf{x}^{(1)} \neq \mathbf{x}^{(2)}$ for which $\mathbf{f}(\mathbf{x}^{(2)}) \trianglelefteq \mathbf{f}(\mathbf{x}^{(1)})$, this can be also expressed as $f_i(\mathbf{x}^{(2)}) \leq f_i(\mathbf{x}^*)\ \forall i = 1, \ldots, l$; so $\mathbf{x}^*$ is known as a strictly Pareto optimal value.

Similarly, if we change the symbols $\trianglelefteq$ by $\triangleleft$, and $\leq$ by $<$ in the above expressions, we are referring to $\mathbf{x}^*$ as the weakly Pareto optimal value (Coello Coello et al. 2007).

11.3 Genetic Algorithms

Holland and his colleagues proposed the concept of Genetic Algorithms (GAs) in the 1960s and 1970s (Holland 1975), they were inspired in the evolution theory that explains the origin of species. In nature, weak and unfit species within their environment are faced with extinction by natural selection. The strong ones have greater opportunity to pass their genes to future generations via reproduction. In the long run, species carrying the correct combination in their genes become dominant in their population. Sometimes, during the slow process of evolution, random changes may occur in genes, if these changes provide additional advantages in the challenge for survival, new species evolve from them; unsuccessful changes are eliminated by natural selection. The evolution process, in the most elementary way, can be algorithmically modeled for computer simulation using the difference equation (11.3.1),

$$\mathbf{x}(t + 1) = s(v(\mathbf{x}(t))) \qquad (11.3.1)$$

where t represents the time, the new population $\mathbf{x}(t + 1)$ is obtained from the actual population $\mathbf{x}(t)$ which is operated by random variation v, and selection s (Fogel et al. 1998).

A solution vector $\mathbf{x} \in \Omega$ is called an individual or a chromosome, which is made up of discrete units called genes. Each gene controls one or more features of the chromosome (Mitchell 1998), genes are assumed to be binary bits. Our population is a collection of N chromosomes that are randomly initialized.

A GA uses two operators to generate new solutions from existing ones: crossover and mutation. In crossover, usually two chromosomes, called parents, are combined to form different chromosomes, called offspring. The parents are selected from among existing chromosomes in the population with preference towards fitness, so the offspring is expected to inherit good genes. By iteratively applying the crossover operator, genes of the selected chromosomes are expected to appear more frequently in the population; eventually, leading to convergence to an overall good solution.

The mutation operator introduces random changes into the chromosomes, it is generally applied at the gene level. In a typical GA implementation, the mutation rate (probability of changing the properties of a gene) is very small and it depends on the length of the chromosome; therefore, the new chromosome produced by mutation will not be very different from the original one. Figure 11.3 illustrates the crossover and mutation operator in the evolutionary process. Reproduction involves selection of chromosomes for the next generation; the fitness of an individual determines the probability of its survival through generations. There are different selection procedures in GA, such as proportional selection, and

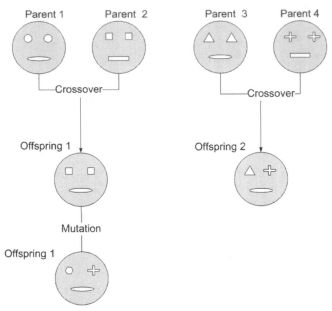

Fig. 11.3 Illustration of crossover and mutation operator process.

ranking and tournament selection, which are the most popular procedures. In the following subsections, we describe the crossover and mutation operators used in this work.

The general procedure for a GA is given in Algorithm 11.1.

Algorithm 11.1 GA general procedure.

Let $t = 0$ be the generation counter;
Create and initialize a population $P(t)$ of size n ;
while Stopping condition != true **do**
    Evaluate the fitness, $f(x_i)$, of each individual, x_i, in the population, $P(t)$;
    Perform cross-over to produce offspring;
    Perform mutation on offspring;
    Using some criteria, select population $P(t+1)$;
    Make $t = t + 1$;
end while

11.3.1 Crossover operators

Linear crossover. One of the first crossover operators implemented for a Real Coded Genetic Algorithm (RCGA) was the linear crossover operator proposed by Wright (Wright 1991). Here two parents are chosen randomly, i.e., $\mathbf{x}^{(1)} = (x_1, x_2, \ldots, x_n)$ and $\mathbf{x}^{(2)} = (x_1, x_2, \ldots, x_n)$ from which three solutions are built, one solution is obtained using (11.3.2) which is in the middle of the two parents; and the other two solutions are displaced to the extremes using (11.3.3) and (11.3.4) (see Fig. 11.4).

$$\mathbf{x}^{(a)} = 0.5(\mathbf{x}^{(1)} + \mathbf{x}^{(2)}) \tag{11.3.2}$$

$$\mathbf{x}^{(b)} = 1.5\mathbf{x}^{(1)} - 0.5\mathbf{x}^{(2)} \tag{11.3.3}$$

$$\mathbf{x}^{(c)} = -0.5\mathbf{x}^{(1)} + 1.5\mathbf{x}^{(2)} \tag{11.3.4}$$

Depending on the separation of the parents, the offspring will also end separated. Wright proposes to eliminate the worst offspring of the two extremes and thus generate only two children (Wright 1991), but also it may be useful to keep the three values. Figure 11.4 shows the lower limits $x_i^{(L)}$ and the upper limit $x_i^{(U)}$ for the variable x, as well as the parents $x_i^{(1)}$ and $x_i^{(2)}$ with their offspring $x_i^{(a)}$, $x_i^{(b)}$ and $x_i^{(c)}$ for one variable (dimension i).

Heuristic Crossover. This operator is also known as HX (Michalewicz et al. 1994). The HX crossover operator generates only one offspring $\mathbf{x}^{(3)}$ from two parents $\mathbf{x}^{(1)}$ and $\mathbf{x}^{(2)}$ using Eq. (11.3.5).

$$\mathbf{x}^{(3)} = \alpha(\mathbf{x}^{(2)} - \mathbf{x}^{(1)}) + \mathbf{x}^{(2)} \tag{11.3.5}$$

where α is a random number uniformly distributed between 0 and 1 (see Fig. 11.5), the parent $\mathbf{x}^{(2)}$ should not be worse than $\mathbf{x}^{(1)}$ in terms of this objective function; this is done to keep the search in the right direction; in other words, for maximization of MOOPs $f(\mathbf{x}^{(2)}) \unrhd f(\mathbf{x}^{(1)})$, or for minimization of MOOPs $f(\mathbf{x}^{(2)}) \unlhd f(\mathbf{x}^{(1)})$. The offspring has to be allocated into Δx_i region, however, in some cases, the operator may not generate a feasible offspring (they are allocated outside the Δx_i region); in this case, another

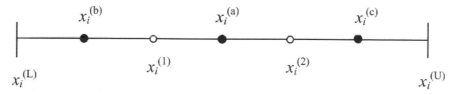

Fig. 11.4 Distribution of the offspring in the linear crossover operator.

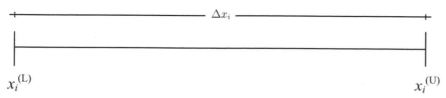

Fig. 11.5 Distribution of the offspring in the heuristic crossover operator.

offspring must be generated using a different α value. If the problem still remains after some intent, then the operator gives up and no offspring is produced. An option is to use the Eq. (11.3.6)

$$\mathbf{x}^{(3)} = \mathbf{x}^{(1)} + \beta \left| \mathbf{x}^{(1)} - \mathbf{x}^{(2)} \right| \tag{11.3.6}$$

where β is a random number generated with the Laplace distribution function.

Blend Crossover. The BLX-α crossover was suggested by Eshelman and Schaffer (Eshelman and Schaffer 1993). From two parents $\mathbf{x}^{(1)}$ and $\mathbf{x}^{(2)}$ (assuming that $\mathbf{x}^{(1)}$ is better than $\mathbf{x}^{(2)}$) the BLX operator randomly generates a solution in the interval $[\mathbf{x}^{(1)} - \vec{\alpha}(\mathbf{x}^{(2)} - \mathbf{x}^{(1)}), \mathbf{x}^{(2)} + \vec{\alpha}(\mathbf{x}^{(2)} - \mathbf{x}^{(1)})]$, this new solution is calculated using Eq. (11.3.7), the offspring are allocated into the $\triangle$ region of the variable x_i (see Fig. 11.6).

$$\mathbf{x}^{(a)} = (1 - \vec{\gamma})\mathbf{x}^{(1)} + \vec{\gamma}\mathbf{x}^{(2)} \tag{11.3.7}$$

where $\vec{\gamma} = (1 + 2\vec{\alpha})\mathbf{u} - \vec{\alpha}$, and $\mathbf{u}$ is a random number between 0 and 1. According to Deb (Deb and Beyer 1999), the best performance of this operator is when $\alpha_i = 0.5$; if the difference between the parents is small, the difference between the child and parents would also be small, this is an adaptive search property. Note that $\vec{\alpha}$ might have a different value for each dimension, hence the variable $\vec{\gamma}$ might also be different.

Simulated Binary Crossover. Deb and his students developed the algorithm SBX (Agrawal and Deb 1994), which creates two offspring from two parents. As its name suggests, it simulates the working principle of the operator single-point crossover, in binary strings. With this operator, from two parents, two offspring in x_i are calculated using Equations (11.3.8), (11.3.9) and (11.3.10).

$$x_i^{(a)} = 0.5[(1 + \beta_{qi})x_i^{(1)} + (1 - \beta_{qi})x_i^{(2)}] \tag{11.3.8}$$

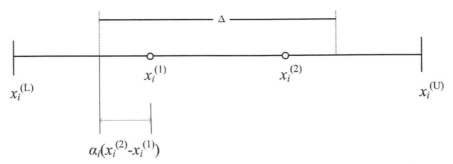

Fig. 11.6 Probability distribution of the offspring in the BLX-α operator for the variable x_i.

$$x_i^{(b)} = 0.5[(1 - \beta_{qi})x_i^{(1)} + (1 + \beta_{qi})x_i^{(2)}] \tag{11.3.9}$$

where,

$$\beta_{qi} = \begin{cases} (2u_i)^{\frac{1}{(\eta_c+1)}}, & \text{if } u_i \le 0.5, \\ \left(\frac{1}{2(1-u_i)}\right)^{\frac{1}{\eta_c+1}}, & \text{otherwise} \end{cases} \tag{11.3.10}$$

u_i is a random number between 0 and 1, η_c is a parameter chosen by the user, the recommended value for η_c is any real number between 0 and 10 and the probability that a child is created close to the father depends on it. The distribution of probability for this operator is shown in Fig. 11.7 (Deb 2001).

Laplace Crossover. This operator was proposed by Deep and Thakur (Deep and Thakur 2007), it is also known as LX operator, which produces two children (offspring $x_i^{(a)}$ and $x_i^{(b)}$) from two parents using Equations (11.3.11), (11.3.12) and (11.3.13).

$$x_i^{(a)} = x_i^{(1)} + \beta |x_i^{(1)} - x_i^{(2)}| \tag{11.3.11}$$

$$x_i^{(b)} = x_i^{(2)} + \beta |x_i^{(1)} - x_i^{(2)}| \tag{11.3.12}$$

$$\beta = \begin{cases} a - b\ln(u), & u \le 0.5, \\ a + b\ln(u), & u > 0.5. \end{cases} \tag{11.3.13}$$

where β is a function between 0 and 1, and u_i is a random number uniformly distributed. β is obtained by inverting the Laplace distribution function. It is

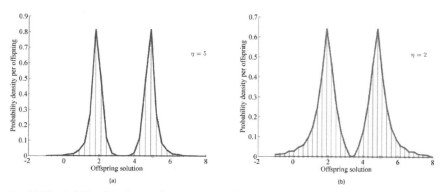

Fig. 11.7 Probability distribution for creating children solutions of continuous variables using the SBX crossover operator.

Color image of this figure appears in the color plate section at the end of the book.

easy that the offspring are placed symmetrically in relation to the position of the parents. We have that for small b values the offspring are likely to be produced near to the parents, and for large b values, the offspring are produced far from the parents. The parameter $a \in \mathbb{R}^+$ is called location parameter, and the parameter $b > 0$ is called the scale parameter. Figure 11.8 illustrates the obtained density function of Laplace distribution for $a = 0, b = 0.5$, and $b = 1$ (Deep and Thakur 2007).

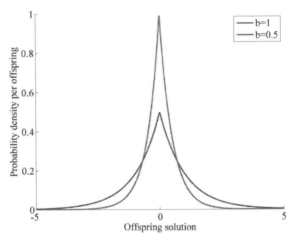

Fig. 11.8 Density function of Laplace distribution ($a = 0$, $b = 0.5$ and $b = 1$).

Color image of this figure appears in the color plate section at the end of the book.

11.3.2 Mutation operators

Uniform mutation (UM) (Michalewicz 1996). It is the simplest mutation operator for real coding given by Eq. (11.3.14); it creates a random solution throughout the search space,

$$x_i^{(a)} = r_i\big(x_i^{(U)} - x_i^{(L)}\big) \tag{11.3.14}$$

where r_i is a random number in $[0, 1]$, $x_i^{(L)}$ is the lower limit of the variable x_i and $x_i^{(U)}$ is the upper limit. This operator is independent of parents and it is equivalent to a random initialization of the population. If the objective is to modify a parent using this operator, it can be performed with Eq. (11.3.15),

$$x_i^{(a)} = x_i^{(1)} + (r_i - 0.5)\Delta_i. \tag{11.3.15}$$

where Δ_i is the maximum perturbation defined by the user, see Fig. 11.9. Special care must be taken about that this disturbance does not produce solutions beyond the limits.

Non uniform mutation (NUM) This operator was proposed by Michalewicz (Michalewicz 1996). Here, the probability of creating a solution near the father is greater than the probability of creating an offspring away. The probability of create a solution close to the parent increases in each generation (t),

$$x_i^{(a)} = x_i^{(1)} + \tau(x_i^{(U)} - x_i^{(L)})(1 - r_i^{(1-t/t_{max})^b})$$
(11.3.16)

In (11.3.16) τ can take the -1 or 1 value, each one with a probability of 0.5. The parameter t_{max} is the allowed maximum number of generations, b is a user defined parameter, in (Michalewicz 1996) the value $b=2$ was used.

Mäkinen, Periaux and Toivanen Mutation (MPTM). This operator is also known as MPTM, it generates using the limits $\mathbf{x}^{(L)}$ and $\mathbf{x}^{(U)}$ of the decision variable $\mathbf{x}$, the offspring $\mathbf{x}^{(a)}$. In (11.3.17) and (11.3.18) the formulas to obtain the offspring for each dimension of decision variable $\mathbf{x} = \{x_1, x_2, \ldots, x_i, \ldots, x_n\}$ are shown, here $r_i \in [0, 1]$ is a random number uniformly distributed (Mäkinen et al. 1999, Deep and Thakur 2007).

$$x_i^{(a)} = (1 - \hat{t})x_i^{(L)} + \hat{t}x_i^{(U)},$$
(11.3.17)

where

$$\hat{t} = \begin{cases} t - t(\frac{t-r}{t})^p & r < t, \\ t & r = t, \\ t + (1-t)(\frac{r-t}{1-t})^p & r > t. \end{cases}$$
(11.3.18)

and

$$t = \frac{x_i^{(1)} - x_i^{(L)}}{x_i^{(U)} - x_i^{(L)}}$$
(11.3.19)

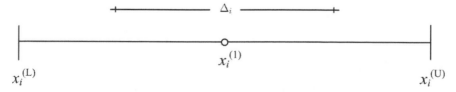

Fig. 11.9 Spread of the new solution with UM operator.

The parameter p defines the distribution of the mutation, it is called the mutation exponent. For uniform mutation, use $p = 1$.

11.4 NSGA-II

The Non-dominated Sorting Genetic Algorithm II (NSGA-II) (Deb 2001) is an improved version of the algorithm NSGA proposed by Srinivas and Deb (Srinivas and Deb 1994). This is a scheme for solving MOOPs using the concept of non-dominance introduced by Goldberg (Goldberg 1989), and a genetic algorithm to produce new solutions, so this algorithm is a multiobjective optimization genetic algorithm (MOGA).

The schematic of this algorithm is shown in Fig. 11.10. First, a started population P_t with N individuals is generated randomly, then a set Q_t of offspring is generated using the crossover and mutation genetic operators. The parents and offspring are combined to form a new set called R_t of size $2N$, in this set a non-dominated sorting is made, then different non-dominated fronts are obtained. In order to create the new parent population, first it is necessary to choose the fronts with the better range of non-dominance, being advisable to reject those members in the least crowded region in that front, and selecting to be part of the new parent population, those members with the best crowding distance. The pseudocode for the NSGA-II is given in Algorithm 11.2 (Deb 2001).

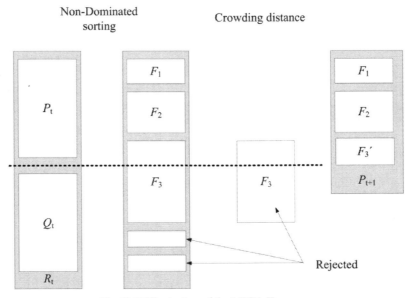

Fig. 11.10 Illustration of the NSGA-II process.

Algorithm 11.2 NSGA-II procedure.

Step 1. Initialize population P of size N.

Step 2. Apply crossover and mutation to P to obtain the population Q_t

Step 3. Combine parents and offspring populations to create $R_t = P_t \bigcup Q_t$ of size $2N$.

Step 4. Perform a nondominated sorting to R_t (see Algorithm 11.3), and identify different fronts $\mathcal{F}_i, \forall i = 1, 2, \ldots$, etc.

Step 5. Initialize a new population $P_{t+1} = 0$, and a counter $i = 0$.
   Until $|P_{t+1}| + ||f_i| < N$, perform $P_{t+1} = P_{t+1} \bigcup \mathcal{F}_i$
   $i = i + 1$

Step 6. Perform the Crowding-sort $(F_i, <_c)$ procedure (see Algorithm 11.4) and include the most widely spread $(N - |P_{t+1}|)$ solutions by using the crowding distance values in the sorted $\mathcal{F}_i$ to P_{t+1}.

Step 7. Create offspring population Q_{t+1} from P_{t+1} by using the crowded tournament selection, crossover and mutation operators.

The fast non-dominated sorting procedure separates the population into different fronts according to their level of dominance, see Algorithm 11.3. The idea of this algorithm is to sort a population of size N according to the level of non-domination, where every solution in the population is compared with each other solution to find whether it is dominated (Deb et al. 2002). Here, for each solution two entities are calculated: 1) the domination count n_p, which is the number of solutions that dominate the solution p, and 2) S_p is a set of solutions that the solution P dominates. It is necessary to identify all those points $n_p = 0$ and put them in a list $\mathcal{F}_1$ called the current front. Next, for each solution in the current front each member j in its set S_p is visited, and reduces its n_j count by one. In this way, if for any member j the count becomes zero, we put it in a separate list Q. When all members of the current front were checked, the list $\mathcal{F}_1$ is the first front. This process continues until using the new front Q as the current front.

Algorithm 11.3 Fast non-dominated sorting procedure.

```
for each p ∈ P
    Sp = 0
    np = 0
    for each q ∈ P
        if (p ≺ q) then                  if p dominates q
            Sp = Sp ∪{q}                  Add q to the set of solutions dominated by p
        else if (q ≺ p) then
            np = np + 1                   Increment the domination counter of p
    If np = 0 then                        p belongs to the first front
    prank = 1
    F1 = F1 ∪{p}
i = 1                                     Initialize the front counter
while Fi ≠ 0
    Q = 0                                 Use to store the members of the next front
    for each p ∈ Fi
        for each q ∈ Sp
            nq = nq − 1
                if nq = 0 then            q belongs to the next front
            qrank = i + 1
            Q = Q ∪{q}
    i = i + 1
    Fi = Q
```

The Crowding distance algorithm described by Algorithm 11.4 is used to maintain diversity within the population in each non-dominated frontier; for each individual, a density estimator named *Crowding distance* is applied. It is an average separation between two contiguous individuals; the extreme points of each front are assigned with an infinite distance. The pseudocode for this procedure is shown in Algorithm 11.4.

Algorithm 11.4 Crowding distance.

crouding-distance-assignment$(\mathcal{F})$
$l = |\mathcal{F}|$
for each $j \in \mathcal{F}$ do
    $\mathcal{F}[j]_{distance} = 0$
    for each objetive m do
        $I = \text{Sort}\ (\mathcal{F}, m)$
        $\mathcal{F}[1]_{distance} = \mathcal{F}[l]_{distance} = \infty$
        for $j = 2$ to $(l-1)$ do
            $\mathcal{F}[j]_{distance} = \mathcal{F}[j]_{distance} + (\mathcal{F}[j+1].m - \mathcal{F}[j-1].m)/(f_m^{max} - f_m^{min})$
        end for
    end for
end for

11.5 Parallel Implementation of NSGA-II (PNSGA-II)

Obtaining results using MOGAs is usually slow to be used in many real-world application that demands high-speed computing; so, the use of parallel programming as well as any technique that helps to speed them up are valuable. In this section, we shall show how to implement MOGAs using parallel programming with the aim to make them suitable to incur into a broader range of applications. Specially, we are going to focus in the NSGA-II algorithm using a real coded GA. First we are going to explain how to parallelize the GA, then how to include it to form part of the NSGA-II.

To implement any GA, one of the first issues that must be addressed is to choose how to encode the individuals. One way is to represent each individual as a string of bits; another one is to consider it as a string of real numbers. Here, we used real encoding, so each individual is represented as a vector of real values formed by *k* independent variables, *m* objective functions and one locality for the crowding distance *d*.

The first step is to initialize the parent population P_t of size N, each individual in the population is an array of real numbers of size $k+m+1$, the domain $(\mathbf{x}^{(L)}, \mathbf{x}^{(U)})$ of the variables and objective function must be established, Fig. 11.11 illustrates the individuals. Using P_t, the population Q_t is generated to obtain the population R_t of size $2N$. The calculation of their objective values is achieved in parallel form using threads. Next, it is

Fig. 11.11 Encoding of individuals in a population.

necessary to identify the different fronts $\mathcal{F}_i$, this process is made using the parallel sorting procedure; we need to use the following procedure where $W = \{w_1, w_2, \ldots w_r\}$ is the set of workers.

a) Divide the population R_t equitatively in the workers w_j, so we have the next subpopulations assigned to each worker $R_t^{(w_1)}, R_t^{(w_2)}, \ldots, R_t^{(w_r)}$.

b) For each parallel worker (w_j), find all the fronts $\mathcal{F}_i^j$ using the fast non-dominated-sorting procedure in each subpopulation $R_t^{(w_j)}$. Here i is the number of the front and j is the number of the worker.

c) Synchronize the workers and join all the fronts $\mathcal{F}_i$.

To compute the crowding distance of each individual j in the $\mathcal{F}_i$, sort the individuals contained in the front, in ascending order of their first objective function, then assign infinite distance to the first and the last individuals. For the other individuals, iteratively compute the average of the normalized values of the objective function of the adjacent individuals using the next sentence,

$$\mathcal{F}[j]_{distance} = \mathcal{F}[j]_{distance} + (\mathcal{F}[j+1].m - \mathcal{F}[j-1].m)/(f_m^{max} - f_m^{min})$$

To achieve the truncation of R_t once it was ordered by fronts, to create a new parent population P_{t+1}, first add the front $\mathcal{F}_1$, then add the next front, and so on. This process must continue until $|P_{t+1}| \geq N|$ in order to avoid overflow in P_{t+1}. The last candidate front to be added needs to be sorted in descending order using the crowding distance of individuals, choosing only the first required individuals to satisfy the condition $|P_{t+1}| = N|$.

Applying the elitism strategy, we have N Parents P with the best Rank and/or crowding distance. Using the crowding comparator, we choose the best solutions and move them to the mating pool to generate new offspring solutions.

The crowding comparator is a binary tournament that compares two solutions randomly selected from P, and the winner has the better Rank. If the Rank is equal, the winners have the biggest crowding distance and move them to the mating pool B, this process continues until B have N solutions.

To create new solutions, the crossover and mutation operators are executed in a parallel form with W workers at the same time. For this process, the mating pool B is a global memory, and each worker generates $N=W$ offspring, and after this, they are synchronized to the offspring population Q.

Each worker chooses two random solutions from B, after generating two random numbers c (crossover probability) and m (mutation probability), if p is bigger than the selected crossover probability, the worker chooses other parents, else apply the crossover operator. If m is bigger than the mutation probability the new solution is added to the offspring population Q, else, the mutation operator is applied, and the new solution is added to the offspring population Q. This process continues until we have N/W offspring. If some offspring is out of the variable boundaries, it is replaced for some parents.

11.6 Results

The PNSGA-II algorithm was tested using five test-bench problems; the first two were proposed by Schaffer and are referred at literature as SCH1, SCH 2 (Shaffer 1984), equations (11.6.1) and (11.6.2). The other three test-bench problems were proposed by Zitzler (Zitzler 1999), they are known as ZDT1, ZDT2 and ZDT3 problems and defined by equations (11.6.3) to (11.6.5).

All these problems have two objective functions to be minimized, and its borders are known as Pareto optimal.

$$SCH1 = \begin{cases} f_1(x) = x^2 \\ f_2(x) = (x-2)^2 \end{cases} \tag{11.6.1}$$

$$SCH2 = \begin{cases} f_1(x) = \begin{cases} -x & \text{si } x \le 1 \\ x-2 & \text{si } 1 < x \le 3 \\ 4-x & \text{si } 3 < x \le 4 \\ x-4 & \text{si } x > 4 \end{cases} \\ f_2(x) = (x-5)^2 \end{cases} \tag{11.6.2}$$

$$ZDT1 = \begin{cases} f_1(\mathbf{x}) = x_1 \\ f_2(\mathbf{x}) = 1 - \sqrt{x_1/g(\mathbf{x})} \\ g(\mathbf{x}) = 1 + \frac{9}{n-1}\left(\sum_{i=2}^{n} x_i\right) \end{cases} \tag{11.6.3}$$

$$ZDT2 = \begin{cases} f_1(\mathbf{x}) = x_1 \\ f_2(\mathbf{x}) = 1 - (x_1/g(\mathbf{x}))^2 \\ g(\mathbf{x}) = 1 + \frac{9}{n-1}\left(\sum_{i=2}^{n} x_i\right) \end{cases} \tag{11.6.4}$$

$$ZDT3 = \begin{cases} f_1(\mathbf{x}) = x_1 \\ f_2(\mathbf{x}) = 1 - \sqrt{x_1/g(\mathbf{x})} - (x_1/g(\mathbf{x}))\sin(10\pi x_1) \\ g(\mathbf{x}) = 1 + \frac{9}{n-1}\left(\sum_{i=2}^{n} x_i\right) \end{cases} \tag{11.6.5}$$

Table 11.1 shows the results obtained with the algorithm PNSGA-II for the SCH1 function using a population of 100 individuals and 10 generations. For all the tests, we used the non-uniform mutation operator combined with different crossover operators (BLX-α, SBX, LX and LinX). Table 11.1 shows the average of all individuals in the decision variable (x_1), the minimum, maximum and average crowding distance. The mean is used just to show that there exists a balance in the population distribution in the front, which can also be confirmed with the average value of the crowding distance.

Table 11.1 Results of the algorithm PNSGA-II for function SCH1.

		x_1			
		BLX	SBX	LX	LINX
NUM	Mean	1.0236	1.0151	0.9951	1.0364
	Min	−0.0007	0.0093	0.0077	0
	Max	2.0083	1.9999	1.9994	1.996
	Crowding	0.1209	0.1263	0.1233	0.1266

For function SCH1, the Pareto optimality is in the interval $x \in [0, 2]$; Table 11.1, shows that with any combination of mutation and crossover operator, all obtained solutions are within this range, the crowding distance shows that all tests have similar distribution in the Pareto front.

Table 11.2 shows results for the SCH2 problem, they were obtained using the BLX-α crossover operation and three different mutation operators, NUM, RM and MPTM. In the experiments, a population of 100 individuals was used, the algorithm runs 10 generations. The expected Pareto optimality value is $x \in [1, 2] \cup [4, 5]$. For all the experiments, the values were obtained within this range, and the separation of the solutions is uniform. Figure 11.12 shows the best set of individuals found after 10 generations, they were obtained using the BLX-α and non-uniform mutation operators.

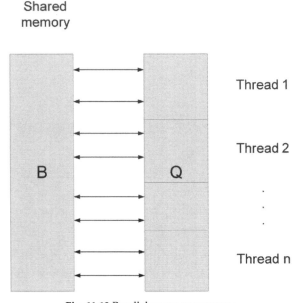

Fig. 11.12 Parallel crossover process.

Table 11.2 Results of the algorithm PNSGA-II for function SCH2 using the BLX-alpha and the three different mutation operators NUM, RM and MPTM.

		x_1		
		NUM	RM	MPTM
BLX-α	Mean	2.0445	2.1544	2.0436
	Min	0.9983	0.9987	1.0016
	Max	4.9824	4.9994	4.9959
	Crowding	0.329	0.3227	0.3276

Table 11.3 shows the results obtained using the PNSGA-II for the ZDT1 problem. It used a population of 100 individuals, because this problem is more complex than the two problems discussed above, it was necessary to run 200 generations. The Pareto optimality for this function occurs when $x_1 \in [1, 2]$ and $x_i = 0$ for i=2,3,...,30. To analyze the results, the variables x_2 to x_{30} are shown in the same column; in each row, for each crossover operator, the average of all the minimum values, maximum and average crowding distance are shown.

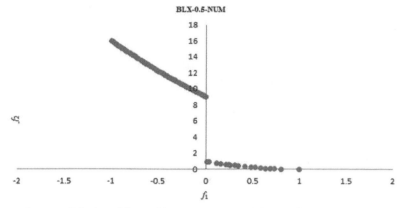

Fig. 11.13 Solution of the problem SCH2 using BLX-0.5 and NUM operators.

For this problem, for the BLX-0.5 crossover operator combined with any of the three mutation operators, the solutions close to the true Pareto optimal were obtained.

The SBX operator, only in combination with the non-uniform mutation operator (NUM), produced good results.

The Laplace crossover operator (LX) only produced good results with the NUM operator.

The linear crossover operator (LinX) produced good results with and all the used mutation operators.

shows that the best result were obtained using the LX and NUM operators.

The worst results were achieved using the LX and RM operators.

The analysis of the results obtained for the ZDT2 problem is shown in Table 11.4, the results using the same procedure that has been used for the ZDT2 function are presented. The Pareto optimal for this not convex function is $x_1 \in [1, 2]$ and $x_i = 0$ for i=2,3,..,30. The algorithm ran for each experiment 200 generations with a population of 100 individuals.

The best result for the problem ZDT2 was achieved using the linear crossover and non-uniform mutation operator. Laplace operator combined with random mutation provided us the worst results, see Fig. 11.14. For this problem, the best results were obtained with the combination of the BLX operator with any of the three mutation operators. With the simulated crossover binary and Laplace mutation operators, the obtained results were far from the true Pareto optimal.

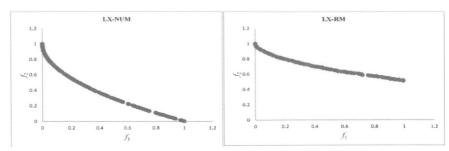

Fig. 11.14 Best and worst solutions of the problem ZDT1 using PNSGA-II algorithm.

Table 11.3 Results of the algorithm PNSGA-II for function ZDT1.

		NUM		RM		MPTM	
		X_1	$X_{2\text{-}30}$	X_1	$X_{2\text{-}30}$	X_1	$X_{2\text{-}30}$
BLX	Mean	0.4339	0.0069	0.4288	0.0157	0.422	0.0073
	Min	7.8E-06	0.0011	0	0.0009	1.7E-05	0.001
	Max	0.9989	0.0209	0.9931	0.4662	0.9996	0.0251
	Crowding	0.0304	0	0.0317	0	0.0309	0
SBX	Mean	0.5013	0.0087	0.4566	0.2304	0.4802	0.336
	Min	0	0	0	0.0054	0	0.0919
	Max	0.9999	0.046	0.9997	0.8453	0.9996	0.6922
	Crowding	0.0369	0	0.0269	0	0.0232	0
LX	Mean	0.4128	0.001	0.468	0.4065	0.4974	0.3853
	Min	0	0	0	0.0273	4.1E-06	0.1451
	Max	0.9996	0.0049	0.9951	0.9249	0.9959	0.7823
	Crowding	0.0335	0	0.0272	0	0.0255	0
LINX	Mean	0.3925	0.001	0.4166	0.0013	0.425	0.0012
	Min	0	0	0	0	0	0
	Max	0.9936	0.0109	0.9999	0.0101	0.9998	0.013
	Crowding	0.0311	0	0.0315	0	0.0317	0

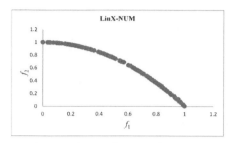
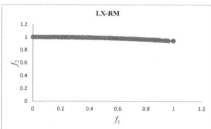

Fig. 11.15 Best and worst solutions of the problem ZDT2 using PNSGA-II algorithm.

Table 11.4 Results of the PNSGA-II algorithm for the function ZDT2.

		NUM		RM		MPTM	
		X_1	$X_{2\text{-}30}$	X_1	$X_{2\text{-}30}$	X_1	$X_{2\text{-}30}$
BLX	Mean	0.5981	0.0039	0.5719	0.0121	0.5515	0.0099
	Min	0.0002	0.0002	0.0000	0.0005	0.0000	0.0006
	Max	0.9994	0.0141	0.9985	0.5131	0.9942	0.3532
	Crowding	0.0298		0.0305	0.0000	0.0300	0.0000
SBX	Mean	0.5496	0.1641	0.5477	0.2547	0.5237	0.2690
	Min	0.0002	0.0871	0.0001	0.0819	0.0000	0.1623
	Max	0.9996	0.2594	0.9991	0.5386	0.9995	0.4357
	Crowding	0.0187	0.0000	0.0185	0.0000	0.0173	0.0000
LX	Mean	0.5153	0.2449	0.4865	0.4077	0.5074	0.2895
	Min	0.0000	0.1598	0.0000	0.0522	0.0001	0.1576
	Max	0.9981	0.6642	0.9988	0.8840	0.9947	0.6993
	Crowding	0.0188	0.0000	0.0203	0.0000	0.0193	0.0000
LINX	Mean	0.5558	0.0008	0.5921	0.0046	0.5712	0.0046
	Min	0.0000	0.0000	0.0000	0.0000	0.0000	0.0000
	Max	0.9991	0.0048	0.9983	0.3912	1.0000	0.2225
	Crowding	0.0311	0.0000	0.0311	0.0000	0.0324	0.0000

For the multi-objective optimization problem ZDT3, the results are shown in Table 11.5; this function has the Pareto optimal $x_i^* = 0$ for $i=2,3,...,30$ and some values in the range $0 \leq x_i^* \leq 0$ which produce a discontinuous Pareto optimal region. This discontinuity prevents convergence to optimal;

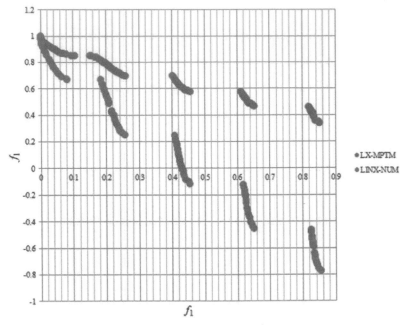

Fig. 11.16 Best and worst solutions for the problem ZDT3 found using the PNSGA-II algorithm.

Color image of this figure appears in the color plate section at the end of the book.

for this reason, each experiment was run for 300 iterations with 100 individuals in the population. The best results were obtained with the linear crossing operator and closer to the Pareto optimal frontier was combined with non-uniform mutation, see Fig. 11.16. With the BLX, we obtained good results using any of the three mutation operators. With the Laplace and simulated binary crossover operators, results were far from the Pareto optimal frontier.

11.7 Conclusions

At present time there are several good algorithms to solve MOOPs being difficult to choose the most appropriated for an application since the selection of one can depend on many aspects, i.e., precision, time to offer good solutions, memory, processor technology, etc. For example, an algorithm can solve a problem very fast offering good solutions using parallel computation; however, if we want to use the same algorithm in a consumer electronic application where usually computational resources are limited, the same algorithm can hardly be implemented.

Table 11.5 Results of the algorithm PNSGA-II for function ZDT3.

		NUM		RM		MPTM	
		X_1	X_{2-30}	X_1	X_{2-30}	X_1	X_{2-30}
BLX	Mean	0.4257	0.0123	0.4036	0.0266	0.3919	0.0181
	Min	0	0.0021	0	0.0069	0	0.0012
	Max	0.8545	0.0415	0.8523	0.6657	0.8532	0.4365
	Crowding	0.0462	0	0.0457	0.0000	0.0463	0
SBX	Mean	0.3532	0.2556	0.3434	0.3321	0.3424	0.2984
	Min	0.0000	0.0339	0.0000	0.0623	0.0001	0.1163
	Max	0.8512	0.6505	0.8498	0.8149	0.8521	0.5206
	Crowding	0.0280	0	0.0268	0	0.0273	0
LX	Mean	0.3481	0.2524	0.3611	0.3574	0.3791	0.3742
	Min	0.0000	0.1979	0.0000	0.2728	0.0000	0.0659
	Max	0.8488	0.6497	0.8532	0.6901	0.8518	0.8109
	Crowding	0.0307	0	0.0284	0	0.0286	0
LINX	Mean	0.3957	0.0017	0.4189	0.0019	0.4156	0.0024
	Min	0	0	0	0	0	0
	Max	0.8528	0.0117	0.8519	0.0157	0.8521	0.0235
	Crowding	0.0482	0	0.0470	0	0.0480	0

The idea of using GA to solve MOOPs is known as MOGA, in general terms practically any evolutionary algorithm can be used instead of a GA; hence, the MOEA acronym is used to mean Evolutionary Algorithm. In this chapter, we used MOGA, and we have mentioned and illustrated the importance of having a good implementation of the GA with good explorative and exploitative characteristics to avoid premature convergence and to find precise solutions. Therefore, choosing the correct combination of genetic operators is crucial, unfortunately, it is difficult to find such appropriated recipe of genetic operators, since there are several combinations of them and their specific values that might produce similar results, due to their stochastic nature; however, testing them many times, some of them offer consistently better results.

We showed ideas to implement the PNSGA-II using threads; particularly we focused on parallelizing a real coded GA that was used jointly with the NSGA-II.

Acknowledgments

The authors would like to thank the "Instituto Politécnico Nacional (IPN)", "Comisión de Operación y Fomento de Actividades Académicas del IPN (COFAA-IPN)" and the Mexico's entity "Consejo Nacional de Ciencia y Tecnología (CONACYT)" for supporting our research activities.

References

Abraham, A., L.C. Jain and R. Goldberg (2005). Evolutionary Multiobjective Optimization: Theoretical Advances and Applications. Springer, London, UK.

Agrawal, R.B. and K. Deb (1994). Simulated Binary Crossover for Continuous Search Space. Complex Systems Journal, Champaing, IL, USA, pp. 115–148.

Becerra, R.L. (2006). Solving hard multiobjective optimization problems using epsilon-constraint with cultured differential evolution. Proceedings of the 9th international conference on Parallel Problem Solving from Nature. Berlin, Heidelberg: Springer-Verlag, pp. 543–552.

Coello Coello, C.A., G.B. Lamong and D.A. Van Veldhuizen (2007). Evolutionary Algorithms for Solving Multi-Objective Problems. New York: Springer.

Coello, C.A. (2006, February). Evolutionary multi-objective optimization: A historical view of the field. IEEE Computational Intelligence, 1(1): 28–35, .

Corne, D.W., N.R. Jerram, J.D. Knowles and M.J. Oates (2001). PESA-II: Region-based Selection in Evolutionary Multiobjective Optimization. Genetic and Evolutionary Computation Conference (GECCO'2001). Morgan Kaufmann Publishers, San Francisco, CA, USA, pp. 283–290.

Deb, K. (2001). Multi-Objective Optimization using Evolutionary Algorithms. Wiley, West Sussex, England.

Deb, K. and H.-g. Beyer (1999). Self-Adaptive Genetic Algorithms with Simulated Binary Crossover. Complex Systems, 9: 431–454.

Deb, K., S. Agrawal, A. Pratap and T. Meyarivan (2002). A Fast Elitist an Elitist Multiobjective Genetic Algorithm: NSGA-II. IEEE Transactions On Evolutionary Computation, 6(2): 182–197.

Deep, K. and M. Thakur (2007). A new crossover operator for real coded genetic algorithms. Applied Mathematics and Computation, 188(1): 895–911.

Eshelman, L.J. and J.D. Schaffer (1993). Real-coded genetic algorithms and interval-schemata. In: D.L. Whitley, Foundation of Genetic Algorithms 2. San Mateo: Morgan Kaufmann, pp. 187–202.

Fogel, D.B., T. Bäck, U. Hammel and H.P. Schwefel (1998). Chapter 1. An Introduction to Evolutionary Computation. In: D.B. Fogel, Evolutionary Computation. The Fossile Record. New York: IEEE Press, p. 1.

Goldberg, D.E. (1989). Genetic Algorithms in Search, Optimization and Machine Learning. Boston: Addison-Wesley Longman Publishing Co., Inc.

Holland, J.H. (1975). Adaptation in Natural and Artificial Systems. Ann Arbor: University of Michigan Press.

Ignizio, J.P. (1978). A Review of Goal Programming: A Tool for Multiobjective Analysis. The Journal of the Operational Research Society, pp. 1109–1119.

Jones, D.a. (2003). Goal Programming in the Period 1990–2000. In M. a. Ehrgott, Multiple Criteria Optimization: State of the Art Annotated Bibliographic Surveys, Springer US, pp. 129–170.

Konak, A., D.W. Coit and A.E. Smith (2006). Multi-objective optimization using genetic algorithms: A tutorial, Reliability Engineering and System Safety 91.

Mäkinen, R.A., J. Periaux and J. Toivanen (1999). Multidisciplinary shape optimization in aerodynamics and electromagnetics using genetic algorithms. International Journal for Numerical Methods in Fluids, 30(2): 149–159.

Michalewicz, Z. (1996). Genetic algorithms + data structures = evolution programs. London, UK, UK: Springer-Verlag.

Michalewicz, Z., T. Logan and S. Swaminathan (1994). Evolutionary operators for continuous convex parameter space. In: A.V. Sebald and L.J. Fogel (Ed.), 3rd Annual Conference on Evolutionary Programming. World Scientific, pp. 84–97.

Mitchell, M. (1998). An Introduction to Genetic Algorithms. (M. Press, Ed.). Cambridge, MA, USA.

Rangaiah, G.P. (2009). Multi-Objective Optimization: Techniques and Applications in Chemical Engineering. World Scientific Publishing CO. Pthe. Ltd.

Shaffer, J.D. (1984). Some experiments in machine learning using vector evaluated genetic algorithms (artificial intelligence, optimization, adaptation, pattern recognition. USA.

Srinivas, N. and K. Deb (1994). Multiobjective Optimization Using Nondominated Sorting in Genetic Algorithms. Evolutionary Computation, 2: 221–248.

Stadler, W. (1979). A Survey of Multicriteria Optimization, or the Vector Maximum Problem. Journal of Optimization Theory and Applications, 29: 1–52.

Wismer, Y.Y. (1971). On a Bicriterion Formulation of the Problems of Integrated System Identification and System Optimization. IEEE Transactions on Systems, Man, and Cybernetics, 296–297.

Wright, A.H. (1991). Genetic Algorithms for Real Parameter Optimization. Foundations of Genetic Algorithms, Morgan Kaufmann, pp. 205–218.

Zitzler, E. (1999). Evolutionary Algorithms for Multiobjective Optimization: Methods and Applications. Zurich, Switzerland.

12

High-performance Navigation System for Mobile Robots

*Ulises Orozco-Rosas, Oscar Montiel** and
Roberto Sepúlveda

ABSTRACT

Robotics is an open issue in its various areas. This chapter presents the
implementation of a high-performance navigation system providing an
alternative solution in the navigation area of mobile robots. It makes
use of parallel computation as a tool to improve the performance
of the navigation system, which is based on the method of artificial
potential field and genetic algorithms. This chapter at the end provides
a comparison between a navigation system based on a sequential
scheme and another based on a parallelized scheme, demonstrating
the advantage of parallel computing implementation in the navigation
system performance.

12.1 Introduction

There is a demand for autonomous mobile robots in various fields of
application, such as material transport, cleaning, monitoring, guiding
people and military applications. These mobile robots must interact with
their environment to accomplish their tasks, this environment may be subject
to change environments and unforeseen. This chapter addresses the problem

Instituto Politécnico Nacional, CITEDI, Tijuana, México.
  E-mail: oross@ipn.mx
* Corresponding author

of navigation of a mobile robot, understanding navigation as the move from point A (start position) to point B (target position) through a defined path. The mobile robot will have to interact with the environment and avoid the obstacles, following the path planned to achieve its established mission. All of these tasks, without the assistance of a human operator. To achieve this, we propose the integration of the method of artificial potential field (APF) with a genetic algorithm and parallel computing to develop a simulation of high-performance navigation system. This works is organized as follows: Section 12.2 shows the fundamentals of the artificial potential field, which is a mathematical method widely used in the path planning for mobile robots, given that it provides an effective control on the movement (Kim et al. 2011). Section 12.3 shows a little reminder of the genetic algorithms. In this work a genetic algorithm is used to mitigate the limitations presented by the artificial potential field method. Section 12.4 presents the implementation of the high-performance navigation system. This section is divided into three phases of development. The first phase shows the implementation of a simple navigation system where only the method of artificial potential field is used. The second phase incorporates a genetic algorithm to achieve the implementation of the complete navigation system, which it operates with complete autonomy. Finally the third phase of the implementation the high-performance navigation system is made. Making use of the high performance parallel computation to computing proportional gains and as a consequence of this, maximizing the performance already achieved faster. In Section 12.5, the conclusions and results of this chapter are presented, the section shows the performance results of the implementations made in phases 2 and 3 of Section 12.4 in terms of performance time and the overall conclusions of this chapter.

12.2 Artificial Potential Field

Depending on the scope, the path planning methods can be classified into two branches: the methods of artificial intelligence (AI) and the method of artificial potential field (Zhang et al. 2006).

We are going to focus on the method of artificial potential field. The artificial potential field is a method of artificial local planning, this method was proposed in 1986 by Khatib. At the beginning, this method was used for obstacle avoidance in path planning for robot manipulators and make obstacle avoidance in real time. The main idea of the method of artificial potential field is to establish an attractive potential field force around the target point, Fig. 12.1a, and establish a repulsive potential field force around obstacles, see Fig. 12.1b. The two potential fields together form a new potential field called artificial potential field. This seeks to address the decline function of the potential to find a collision free path, which is built from the start position to the target position (Zhang et al. 2012).

The approach of the artificial potential field is basically operated by the gradient descent search method, which is directed toward minimizing the total potential function in a position of the robot, in this case a mobile robot. The total potential function can be obtained by the sum of the attractive potential and repulsive potential, see Fig. 12.2. In an attractive potential its energy is usually shaped like a bowl around the target position and this leads the robot to this goal. The repulsive potential energy has a form of hill on the obstacle and this repels the robot (Park and Lee 2003).

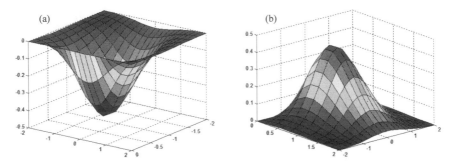

Fig. 12.1 Simulation of (a) an attractive potential surface (b) a repulsive potential surface.

Color image of this figure appears in the color plate section at the end of the book.

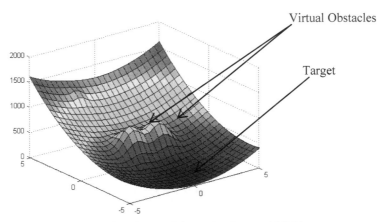

Fig. 12.2 Simulation of the artificial potential field.

Color image of this figure appears in the color plate section at the end of the book.

Using the artificial potential field to achieve the target position of the robot, it means that the robot path is unknown or has not been previously planned. In other words, the robot "chooses" their way independently to achieve the target position.

Particular characteristics of artificial potential field are (Lee 2004):

- Real time use.
- Suitable for redundant robots (with more than six degrees of freedom) and non-redundant.
- Incorporation of the robot dynamics.
- The attractive potential and the repulsive potential are formulated separately and then these are added to get the total artificial potential field.

Some advantages of the artificial potential field are (Lee 2004):

- It does not require complete knowledge about the workspace.
- When the number of obstacles increases in the workspace, the total artificial potential field is modified in the way of superimpose a repulsive potential field by an obstacle, in other words adding the repulsive potential field (obstacle) to the total artificial potential field.

Certain disadvantages are (Zhang et al. 2006, Lee 2004, Weijun et al. 2010):

- *Local minimum exists.* The local minimum problem is found when the mobile robot has reached a point that is not the target, and the mobile robot believes that this point is the target. The following few typical special location relationships would have made the mobile robot unable to reach the target by the artificial potential field method:
  o When the mobile robot, obstacle and target are aligned, the obstacle is between the mobile robot and the target, and the repulsion is greater than or equal to attraction.
  o When the target is within the affected scope of obstacles whose repulsion is rapidly increasing while attraction is reducing.
  o Before the mobile robot is reaching the target, the resultant of repulsion produced by multi-obstacles and the target attraction is zero, this is the best example of local minimum problem.
- It presents oscillation in the presence of obstacles.

12.2.1 Attraction potential

The target position is the point to be achieved, this represents an attractive pole and the obstacles represent repulsive surfaces in a force field where the robot will perform their moves. Therefore, the artificial potential field $U_{tatal}(q)$ comprises two terms, the attractive potential function $U_{att}(q)$ and the repulsive potential function $U_{rep}(q)$. The total artificial potential field $U_{tatal}(q)$ is then the sum of these two potential functions, see Eq. (12.2.1).

$$U_{total}(q) = U_{att}(q) + U_{rep}(q) \qquad (12.2.1)$$

The attractive potential function is described by equation (12.2.2):

$$U_{att}(q) = \frac{1}{2}k_a(q - q_d)^m \qquad (12.2.2)$$

where q represents the robot position vector in a workspace of two dimensions, and in a space of three dimensions $q = [x,y,z]^T$. The vector q_d is representing the point of the target position and k_a is a positive constant scalar that represents the proportional gain of the function. The expression $(q - q_d)$ is the distance between the robot position and the target position, in other words, the length of the line segment connecting them to obtain the Euclidean distance which is a position function, and m is a positive number greater than zero.

For $0 < m < 1$, the attractive potential field has a conical shape and the resultant attractive force has a constant amplitude, except the target position where $U_{att}(q)$ is singular. It is consequently common to use $m \geq 2$, which provides a minimal attractive potential value on the target position. You can use different values for m to avoid local minima in the presence of obstacles; in short, using Eq. (12.2.2) you can change the shape of the attractive potential field by changing the value of m and the effect of the field can be modified by changing the value of the proportional gain k_a (Lee 2004).

By using $m \geq 2$ can easily get the information of the gradient, using (12.2.3):

$$\frac{\partial U_{att}(q)}{\partial q} \qquad (12.2.3)$$

The negative gradient of the total artificial potential field $U_{tatal}(q)$ contains two terms, the attractive force $F_{att} = \nabla U_{att}$ driving the robot to reach the target position, which is located at the target point q_d, and the repulsive force $F_{rep}(q) = \nabla U_{rep}(q)$.

The attractive F_{att} force can be obtained by the negative gradient of the attractive potential function, taking $m = 2$ we can obtain Eq. (12.2.4),

$$F_{att}(q) = -\nabla U_{att} = -k_a(q - q_d) = k_a(q_d - q) \qquad (12.2.4)$$

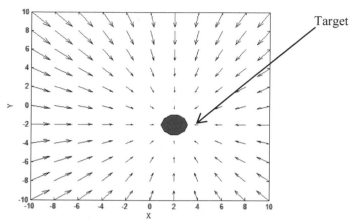

Fig. 12.3 Simulation representing the gradient of an attractive potential field.
Color image of this figure appears in the color plate section at the end of the book.

Considering the quadratic range radio d_a with $m = 2$, Andrews (Park and Lee 2003) proposes the attractive potential function given by (12.2.5),

$$U_{att}(q) = \begin{cases} k_a|q - q_d|^2 & if \ |q - q_d| \le d_a \\ k_a(2d_a|q - q_d| - d_a^2) & if \ |q - q_d| > d_a \end{cases} \tag{12.2.5}$$

where the attractive force F_{att} can be described using (12.2.6),

$$F_{att}(q) = -\nabla U_{att} = \begin{cases} -2k_a(q - q_d) & if \ |q - q_d| \le d_a \\ -2d_a k_a \frac{q-q}{|q-q_d|} & if \ |q - q_d| > d_a \end{cases} \tag{12.2.6}$$

12.2.2 Repulsive potential

The repulsive potential function has a limited range of influence. The movement of the robot is affected by a distant obstacle. The repulsive potential function (12.2.7) is given by Kathib.

$$U_{rep}(q) = \begin{cases} \frac{1}{2}k_r \left(\frac{1}{\rho} - \frac{1}{\rho_0}\right)^2 & if \ \rho \le \rho_0 \\ 0 & if \ \rho > \rho_0 \end{cases} \tag{12.2.7}$$

where ρ_0 represent the limit distance of influence of the potential field and ρ is the shortest distance to the obstacle. The selection of the distance ρ_0 depends of the robot maximum speed and the control period. The constant representing the repulsive proportional gain function is k_r.

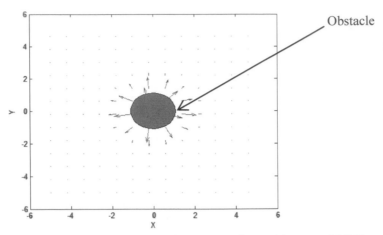

Fig. 12.4 Simulation representing the gradient of a repulsive potential field.

Color image of this figure appears in the color plate section at the end of the book.

The repulsive force $F_{rep}(q)$ keeps the robot away from the obstacles, thus its name: Force Inducing an Artificial Repulsion from the Surface (FIRAS, from the French),

$$F_{rep}(q) = \begin{cases} k_r \left(\frac{1}{\rho} - \frac{1}{\rho_0}\right)^2 \frac{1}{\rho^2} \frac{\partial \rho}{\partial q} & \text{if } \rho \leq \rho_0 \\ 0 & \text{if } \rho > \rho_0 \end{cases} \tag{12.2.8}$$

where $\frac{\partial \rho}{\partial q}$ denotes the partial derivative of the distance vector from the point subject to the potential over an obstacle in a workspace of two dimensions (Kathib 1986):

$$\frac{\partial \rho}{\partial q} = \left(\frac{\partial \rho}{\partial x} \frac{\partial \rho}{\partial y}\right)^T \tag{12.2.9}$$

The generalized force $F(q)$ is obtained by the negative gradient of a generalized potential. The principle of superposition may be used to obtain $F(q)$ because $\nabla(\cdot)$ is a linear operator, see Eq. (12.2.10).

$$F(q) = -\nabla U(q) = -\nabla U_{att}(q) - \nabla U_{rep}(q) = F_{att}(q) + F_{rep}(q) \tag{12.2.10}$$

In an environment with multiple obstacles, the total potential field is the sum of all repulsive fields of the obstacles (He et al. 2011) plus the attractive field. This can be expressed with the function (12.2.11),

$$U_{total}(q) = U_{att}(q) + \Sigma_i U_{rep_i}(q) \tag{12.2.11}$$

and the total force is represented by (12.2.12),

$$F(q) = F_{att}(q) + \Sigma_i F_{rep_i}(q)$$

$$(12.2.12)$$

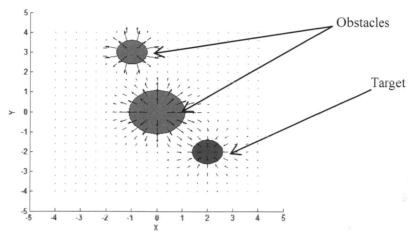

Fig. 12.5 Simulation representing the negative gradient of a total potential field.

Color image of this figure appears in the color plate section at the end of the book.

12.2.3 Limitations of the artificial potential field

Although artificial potential fields are already widely used in the path planning of mobile robots for its simplicity, mathematical elegance and effectiveness with a smooth and safe planning; however, they have limitations in many real-world implementations, where environments are dynamic. In these applications the traditional artificial potential fields are not applicable since they have inefficient path planning (Qixin et al. 2006). Some known problems are:

- **Local minimum problem:** This problem occurs when the robot stops in its movement because it believes that it has reached the target position when, in fact, it has not reached its goal (He et al. 2011).
- **Trajectory prediction problem:** Another disadvantage of the artificial potential field is the difficulty that is presented to predict the current path. In practice, the proportional gains of the potential function should be carefully chosen to ensure the avoidance of obstacles; in some cases, oscillation occurs in the presence of obstacles. In addition to the above, usually generated paths are far from optimal (Zhang et al. 2006, Lee 2004).

- **Goals non-reachable problem:** A problem on the most of the potential field methods was identified by Ge, Cui, Volpe and Khosla, which consists of the goals (target points) non reachable with obstacles nearby (GNRON) (Lee 2004).

For the above reasons the artificial potential field method is complemented by various techniques looking to mitigate or avoid these limitations. This chapter makes use of evolutionary computation, particularly the genetic algorithms. These make more efficient to use the artificial potential field, and potentiates its performance with parallel computing to speed up the processing management in evaluating possible solutions trajectory planning.

12.3 Genetic Algorithms

The objectives of creating artificial intelligence and artificial life go back to the very beginning of the computer age. Early computer scientists, Alan Turing, John von Neumann, Norbert Wiener and others were largely motivated by visions of inducing intelligence on computer programs, with the ability of life to self-replicate and the ability of adaptation to learn and control their environments. These pioneers of computing were so interested in biology, psychology and electronics, and they looked to natural systems as guiding metaphors for how to achieve their visions. Not surprising then, that since the early days computers were applied not only to calculate missile trajectories and deciphering military codes, but also to model the brain, mimic human learning and simulating biological evolution. These biologically motivated computing activities have waxed and waned over the years, but since the early 1980s have all experienced resurgence in the computing community dedicated to research. The first one has grown in the field of neural networks, the second in Machine Learning and the third in what is now called Evolutionary Computation, where the genetic algorithms are the most prominent example (Mitchell 2001).

Genetic algorithms as search technique are powerful, as mentioned above, they are based on the mechanisms of natural selection and natural genetics, they are successfully used to solve problems in various disciplines. The robustness of these algorithms on complex problems has led to an increasing number of applications in the fields of artificial intelligence, numerical and combinatorial optimization, business, management, medicine, computer science, engineering and others (Chande and Sinha 2008).

The method of genetic algorithms was developed by John Holland (1975) during the 1960s and 1970s and finally popularized by one of his students, David Goldberg, who was able to solve a difficult problem involving the control of a pipeline transmission for his dissertation. Holland's original work is summarized in Goldberg's book (Goldberg 1989) and he has probably contributed with energy in their successful applications. He was the first to try to develop a theoretical basis for genetic algorithms through its schema theorem. The work of De Jong (De Jong 1975) proved the utility of genetic algorithms to optimize functions and made the first concerted effort to find the optimized parameters of genetic algorithms. Since then, many versions of evolutionary programming have been tried with varying success (Haupt and Haupt 2004).

Some of the advantages of a genetic algorithm are (Haupt and Haupt 2004):

- It optimizes with continuous or discrete variables.
- It does not require derivative information.
- It simultaneously searches from a wide sampling of the cost surface.
- It deals with a large number of variables.
- It is well suited for parallel computers.
- It optimizes variables with extremely complex cost surfaces (they can jump out of a local minimum).
- It provides a list of optimum variables, not just a single solution.
- It may encode the variables, so that the optimization is done with the encoded variables.
- It works with numerically generated data, experimental data, or analytical functions.

12.3.1 How does a genetic algorithm work?

Genetic algorithms are an abstraction of biological evolution and therefore, a method to go from a population of chromosomes (bit string) to a new population by using a kind of natural selection with the genetic operators of selection, crossover and mutation. Each chromosome is composed of genes (bits) that are instances of "allele" (1 or 0). The functions of genetic algorithms generate a large set of possible solutions to a given problem, then it evaluates each of these solutions and decides a "fitness level" for each solution set, these solutions generate new solutions. The "parents" that are more "fit" are more likely to reproduce, while those that are less "fit" are less likely to do so. In essence, the solutions are evolved over time. In this way, there is evolution of the search space scope to a point where a solution can be found (Chande and Sinha 2008). In Fig. 12.6 we can see the flow chart of the overall operation of genetic algorithms.

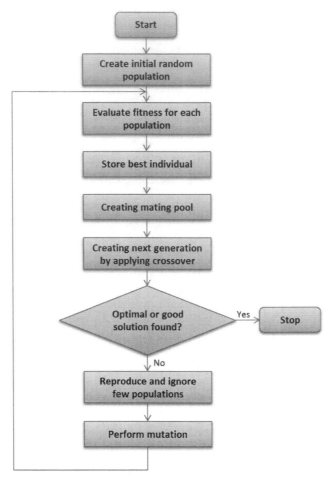

Fig. 12.6 Flowchart of genetic algorithm (Sivanandam and Deepa 2008).

12.3.2 Genetic algorithm operators

Selection

Selection is a method that randomly picks chromosomes out of the population according to their evaluation function. The higher the fitness function is, the more chance an individual has to be selected. The selection pressure is defined as the degree to which the better individuals are favored. The higher the selection pressure is, the more the best individuals are favored. This selection pressure drives the genetic algorithm to improve the population fitness over the successive generations (Sivanandam and Deepa 2008).

Selection has to be balanced with variations from crossover and mutation. Too strong selection means suboptimal highly fit individuals will take over the population, reducing the diversity needed for change and progress; too weak selection will result in too slow evolution (Sivanandam and Deepa 2008). There are various selection methods, but these are not discussed in this chapter. Roulette Wheel Selection, Random Selection, Rank Selection, Tournament Selection, Boltzmann Selection, Stochastic Universal Sampling, are some examples.

Figure 12.7 shows an example of selection, where starting from the upper part of the figure, the descent circles with a dark color are maintained, and the descent circles with a light color are removed and therefore losing the opportunity of reproduction. This process will continue until the bottom of the figure, where only the circles with dark color were kept.

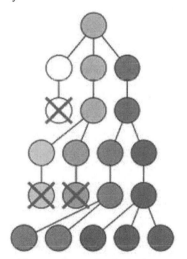

Fig. 12.7 Example of selection.

Crossover

This operator randomly chooses a locus, and exchanges the subsequences before and after that locus between two chromosomes to create two offspring. For example, the strings 1111111 and 0000000 could be crossed over after the third locus in each to produce the two offspring 1110000 and 0001111, see Fig. 12.8. The crossover operator roughly mimics biological recombination between two single–chromosome (haploid) organisms (Mitchell 2001).

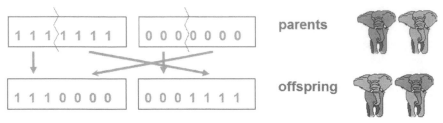

Fig. 12.8 Example of crossover (Rodríguez 2007).

Color image of this figure appears in the color plate section at the end of the book.

Mutation

Random mutations alter a certain percentage of the bits in the list of chromosomes. Mutation is the second way a genetic algorithm explores a cost surface. It can introduce traits not in the original population and keeps the genetic algorithm from converging too fast before sampling the entire cost surface. A single-point mutation changes a 1 to a 0, and vice versa. Mutation points are randomly selected from the $N_{pop} \times N_{bits}$ total number of bits in the population matrix. Increasing the number of mutations increases the algorithm's freedom to search outside the current region of variable space. It also tends to distract the algorithm from converging on a popular solution (Haupt and Haupt 2004).

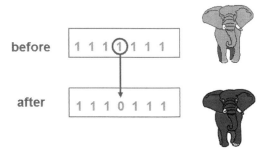

Fig. 12.9 Example of mutation (Rodríguez 2007).

Color image of this figure appears in the color plate section at the end of the book.

12.4 High-performance Implementation

In this section we describe the three phases of development to achieve the high-performance implementation. Starting with the first phase in which it is implemented a simple navigation system with the artificial potential field. To improve the first phase, a genetic algorithm to make a complete navigation system was implemented as the second phase. Finally, the goal of this chapter is to achieve a high-performance navigation system, which was achieved by implementing parallel computing programs in the third phase, and as a result of this, it maximized the performance of what was achieved in the first and second phases.

12.4.1 Phase 1: Simple navigation system with artificial potential field

As already mentioned in Section 12.2 the artificial potential field method has great advantages in the path planning for robots. In this case, the main purpose of the navigation system is to perform the path planning of a mobile robot, where the user sets the starting position and the target position. The target position must be reached by the mobile robot in autonomously form; this should be achieved evading obstacles that lie between set points.

The process followed by the system in the first phase is shown in Fig. 12.10. The first step, requires that the user sets an initial position and a target position. The second step is required to set the proportional gains k_a and k_r which are the gains of attractive and repulsive potential respectively, for more details refer to Section 12.2.1 and 12.2.2 of this chapter. In the choice of the numerical value of the proportional gains, the operator must have enough experience to select the most appropriate, depending on the starting position and the target position set, because on this depends the success of achieving the target position with the simple navigation system. From these parameters established the rest of the process is carried out independently by the simple navigation system.

The next process step consists in calculating the artificial potential field. For this process the system makes use of Eq. (12.2.1) only in case of a single obstacle, or Eq. (12.2.11) in case of multiple obstacles. These equations are the total artificial potential field and they make use of Equations (12.2.2) and (12.2.7) which are the attractive and repulsive potential field respectively. To calculate the gradient of the artificial potential field Eq. (12.2.10) is used in case of a single obstacle or Eq. (12.2.12) for multiple obstacles. These equations in turn make use of the Equations (12.2.4), (12.2.6) and (12.2.8) depending on the case.

In the fourth step of this process, the system calculates the speed and direction of the mobile robot, based on the information obtained in the previous steps. The next step shows a constraint in which it questions whether the robot is trapped in a local minimum. This is detected by showing the same position results or oscillations on the position results. If the answer is in the affirmative, the system proceeds to tell that the mobile robot is trapped in a local minimum and makes a requisition for

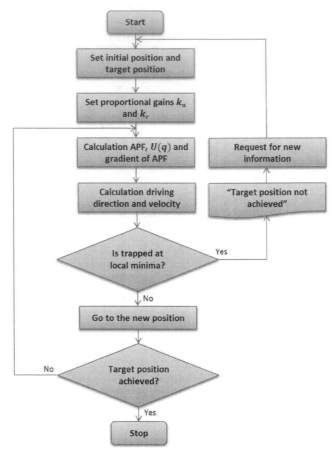

Fig. 12.10 Flowchart of the Simple Navigation System with Artificial Potential Field.

new parameters because the objective was not achieved. Failure to achieve the desired objective may be due to a poor choice of proportional gains or by the limitations of artificial potential fields. To solve this problem the implementation of a genetic algorithm is proposed and this will be discussed in Section 12.4.2.

In case of a negative answer to the condition, the system proceeds to move to the new position calculated. This process is repeated from the step in which it calculates the artificial potential field and the gradient of the artificial potential field until to reach the desired position.

12.4.2 Phase 2: Complete navigation system with artificial potential field and genetic algorithms

In this second phase the navigation system made in the previous phase was improved by integrating a binary genetic algorithm, which is responsible for calculating the proportional gains k_a and k_r automatically, on the basis of the initial position information and the target position set by the user. Based on the above, the whole process is done autonomously; the process is shown in Fig. 12.11.

The process begins by setting the starting position and target position for the mobile robot. In the next step, the system starts with the binary genetic algorithm which is shown in the right part of Fig. 12.11. This genetic

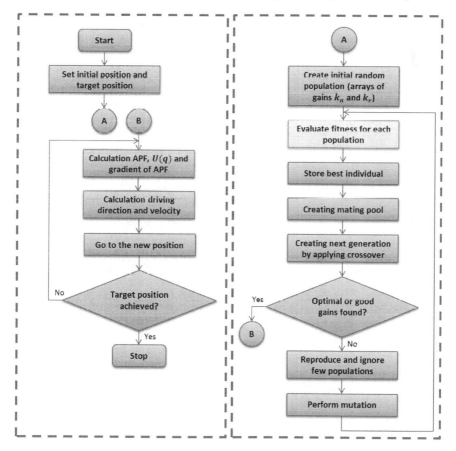

Fig. 12.11 Flowchart of a Complete Navigation System with Artificial Potential Field and Genetic Algorithms.

Color image of this figure appears in the color plate section at the end of the book.

algorithm begins by generating a random population which contains two arrays of numeric values for proportional gains, one for k_a, and other one for the proportional gain k_r. These gains are evaluated in the next step of the process as it is shown in the flowchart of the Fig. 12.11. The proportional gains are evaluated one by one, in pairs consisting of a one gain k_a and another one k_r. This evaluation block called "Evaluate fitness for each population" denoted in green color, comprises an algorithm that describes the flowchart shown in the right part of Fig. 12.12, on it the block called "Evaluate N fitness for the population" denoted in red color, is formed by an algorithm described in the flowchart shown in Fig. 12.10. This means that the algorithm of the flowchart in Fig. 12.10 in this complete navigation system is executed N times, according to the value of N set by the designer.

In the next step of the evaluation the best pairs of the proportional gains k_a and k_r are stored, these are the best combinations of the gains that achieve the target position on the path for the mobile robot. In the next step, the couples are created to form new generations. Then the crossover is applied to obtain new generations of proportional gains.

Now it is achieved the condition where the pair of gains is evaluated, looking for the best or optimal pairs; if they are found, the mobile robot can be achieve the goal. In case of negative response, the next step continues the reproduction of the pairs already formed, and the mutation is conducted to obtain new populations of k_a and k_r values (gains). Thus the process is performed again and again to find the best or optimal pair of values, and when this is achieved, we obtain an affirmative response to the conditioning, and now it will be possible to exit of the cycle.

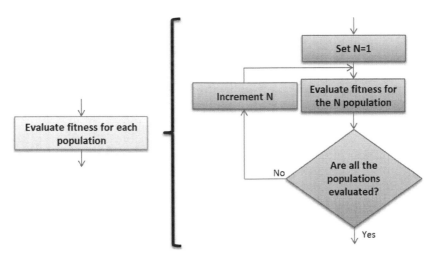

Fig. 12.12 Step by step fitness evaluation process.

Color image of this figure appears in the color plate section at the end of the book.

When the best or optimal pair of gains is found, it may continue with the process established in the flow chart on the left side of Fig. 12.11, which starts at point B and basically it is performed as it is described in section 12.4.1. The advantage that now arises is that the proportional gains k_a and k_r have been calculated automatically and this ensures the correct operation to achieve the target position for the mobile robot.

12.4.3 Phase 3: High-performance navigation system with artificial potential field and parallel genetic algorithms

The final phase to achieve a high-performance navigation system consisted of integrating parallel genetic algorithms with artificial potential field; the first part makes use of parallel computing techniques. In Fig. 12.12 you can see the process called "Evaluate fitness for each population" in the left block in green color, and the right side of the figure shows the steps performed within this block. As it can be seen, it is in sequential form, first a population is evaluated, at the end of this the second population is evaluated, and thus they are evaluated one at a time up to evaluate the population N.

Now, what we want is a high-performance system, and since the evaluation process shown in Fig. 12.12 consumes a lot of time, and each evaluation is independent of the next, there is the opportunity to parallelize the evaluation processes and maximize the performance of the navigation system.

It can be said that parallel processing is a computing technique, which can improve the efficiency of programs by distributing the subtasks of a problem on a set of processors operating in parallel. When it comes to solve a complex problem, usually the number of subtasks that can run in parallel is much greater than the number of physical processors available. In this situation we need to develop an algorithm that allows the processor to multiplex between the physical processes active in each time interval (Rodríguez 2007).

On the right side of Fig. 12.13, you can see the parallelized evaluation process, where the evaluation of each subpopulation is running on a thread and thus for making the evaluation simultaneously or parallel, giving runtime advantages over the sequential process shown in the left of Fig. 12.13 which have been working since the Section 12.4.2.

Up to now we have been specifying the design and the processes of the high-performance navigation system. It is time to talk about the implementation to achieve the simulation of the high-performance navigation system. For this we required two platforms: MATLAB and the INTEL 64 VISUAL STUDIO. The communication of each one is achieved using the tool MEX-file of MATLAB as it is shown in Fig. 12.14.

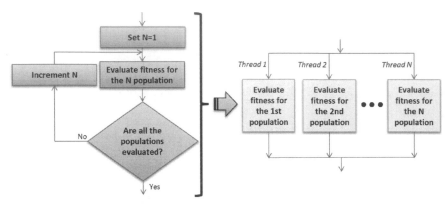

Fig. 12.13 Parallelization of the fitness evaluation process.

Color image of this figure appears in the color plate section at the end of the book.

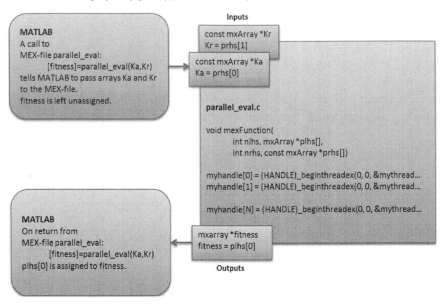

Fig. 12.14 Implementation of the simulation.

MATLAB is a programming language for scientific research. For many applications, the scientific community is exploring new ways to speed up MATLAB functions and programs. One of the options is to use more efficient languages such as C/C++, even C#, with the idea of developing time consuming programs in these languages and then to use the produced code in MATLAB. For doing this, there are two common options, one is to obtain .Net code using an efficient language to integrate it to the MATLAB platform as functions; the second option, is to compile the code using efficient compilers such as the INTEL Parallel Studio Compiler, the Visual

Studio, etc., with the idea to incorporate the developed function through MEX files to the MATLAB.

In some cases implementing a numerical algorithm in C and display graphical results usually involves much more coding and painful debugging. The pain is eased, knowing that it is possible to interface the C implementations back to MATLAB. By creating a MATLAB interface in our code, then compiling the code into a MEX file and the obtuse code can become both fast and easy to use, as a result we can run the experiments of our algorithm written in C and plot the results using MATLAB's plotting tools.

Basically MEX-file stands for MATLAB executable. A MEX file (also written as MEX) provides an interface between MATLAB and subroutines written in C, C++ or Fortran. When compiled, MEX files are dynamically loaded and allow non- MATLAB code to be invoked from within MATLAB, as if it was a built-in function. To support the development of MEX files, MATLAB offers external interface functions that facilitate the transfer of data between MEX files and MATLAB.

Now, placing the ideas, the high-performance navigation system was achieved through the implementation of the algorithm described by the flowchart in Fig. 12.11. The high-performance navigation system code was written in MATLAB and just the evaluation process code was written in C language using parallel computing techniques on the INTEL 64 VISUAL STUDIO. To achieve the communication between both programs the tool MEX-file of MATLAB was used, as it is shown in Fig. 12.14.

In this manner, it was obtained a high-performance navigation system in which the operator only had to set the starting position and the target position, so the navigation system will autonomously generate the path for the mobile robot. The robot must follow this path to meet the goal of reaching the desired position, all avoiding obstacles in its path.

12.5 Results and Conclusions

In Fig. 12.15a, it can be observed the surface of the artificial potential field is generated with the conditions given for the simulation in the high-performance navigation system. These conditions are: start position of the mobile robot in the coordinates (5, 9), target position of the mobile robot in the coordinates (7, 1). There are three obstacles present. The first obstacle is located in the coordinates (5, 6), the second obstacle is in the coordinates (7, 7), and the last obstacle is in the coordinates (7, 3). The simulation of the surface shows that the obstacles are represented as hills, and the target position as the bottom of a kind of bowl (valley) of the entire surface, which is represented by the lowest point in a deep blue color.

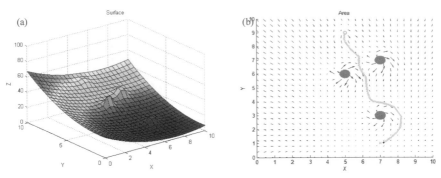

Fig. 12.15 (a) Surface generated by artificial potential field. (b) Path generated by the high-performance navigation system.

Color image of this figure appears in the color plate section at the end of the book.

The navigation path for mobile robot obtained by the simulation of the high-performance navigation system is shown in Fig. 12.15b, where the path is shown in green color. As you can see, the start position is shown with an 'o', and the target position is shown with an 'x', the obstacles are shown with red circles, which have coordinates described in the previous paragraph. Through this simulation, we showed that the mobile robot reached its goal evading obstacles in their way.

The simulation was developed on a computer with Intel Core i5-2410M, which is a fast dual core processor at the time of introduction in the year of 2011. It is based on the Sandy Bridge architecture and offers Hyperthreading to handle 4 threads at once (for a better usage of the pipeline). Sandy Bridge is the evolutionary successor of the Arrandale architecture. The most notable improvements are the improved Turbo 2.0 and the integration of the graphics card into the 32nm CPU core. The base clock speed is 2.3 GHz but due to Turbo Boost, it can reach 2.6 GHz (2 cores active) and 2.9 GHz (1 core active).

The high-performance navigation system simulation was performed using four threads of execution, this condition was dictated by the architecture used in this system. In these four threads there were tested a total population of size N = 16, which means that we had two proportional gains arrangements in size of 16x1 each one, one arrangement is for gain k_a

and a second one for gain k_r as can be seen in the left part of Fig. 12.16. On the right side of this figure we can see the distribution of work by thread. Every thread works four times to evaluate a pair of gains by turn.

As a comparison between the full navigation system (evaluation process in sequential form) described in Section 12.4.2, and the high-performance navigation system (evaluation process in parallel form) described in Section 12.4.3, we found that the average execution time is 0.934 seconds and 0.442 seconds respectively, which talks about the effectiveness of the high-performance navigation system. To find these results there were carried out 45 test executions to measure the time for each navigation system. The time results can be seen in Fig. 12.17.

To conclude this chapter, we have considered three phases of development to reach a high-performance navigation system. This system is made and based on the method of artificial potential field, genetic algorithms and parallel computing. The artificial potential field is an effective method in the path planning for mobile robots, but it has some limitations, and these limitations were attacked by the implementation of a parallel genetic algorithm. This chapter shows a system that can be improved in future work, but for now it presents great opportunities to be implemented in real-world applications. Robotics is a fascinating field which is presented as an open problem where there are many challenges to address and solve yet. In terms of navigation, this chapter presents an alternative solution in time and effectiveness to the problem of mobile robot navigation.

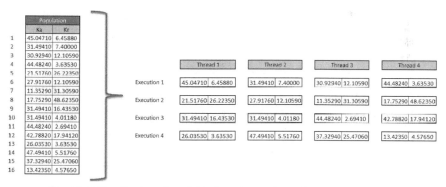

Fig. 12.16 Distribution of work by thread to evaluate the total population.

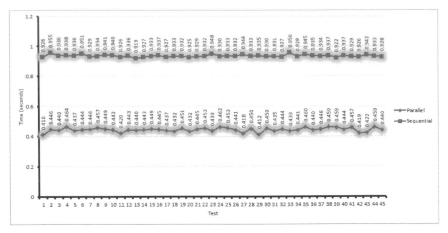

Fig. 12.17 Time of performance for the navigation system working in sequential form (red line) and parallel form (blue line). The vertical axis represents time in seconds and the horizontal axis represents the number of tests.

Color image of this figure appears in the color plate section at the end of the book.

Acknowledgments

The authors would like to thank the "Instituto Politécnico Nacional (IPN)", "Comisión de Operación y Fomento de Actividades Académicas del IPN (COFAA-IPN)" and the Mexico's entity "Consejo Nacional de Ciencia y Tecnología (CONACYT)" for supporting our research activities.

References

Chande, S. and M. Sinha (2008). Genetic Algorithms: A Versatile Optimization Tool. BVICAM'S International Journal of Information Technology (BIJIT), New Delhi, pp. 7–13.

De Jong, K. (1975). An Analysis of the Behavior of a Class of Genetic Adaptive Systems. Ann Arbor: University of Michigan.

Goldberg, D. (1989). Genetic Algorithms in Search, Optimization and Machine Learning. Boston, MA, USA: Addison-Wesley.

Haupt, R. and S. Haupt (2004). Practical Genetic Algorithms. Hoboken, New Jersey, USA: Wiley.

He, L., W. Gao and L. Nan (2011). A Route Planning Method Based on Improved Artificial Potential Field Algorithm. IEEE, China, pp. 550–554.

Kathib, O. (1985). Real-Time Obstacle Avoidance for Manipulators and Mobile Robots. IEEE-International Conference on Robotics and Automation, Stanford, CA, pp. 500–505.

Kim, M., J. Heo, Y. Wei and M. Lee (2011). A Path Planning Algorithm using Artificial Potential Field Based on Probability Map. IEEE - 8th International Conference on Ubiquitous Robots and Ambient Intelligence, Korea, pp. 41–43.

Lee, L. (2004). Decentralized Motion Planning within an Artificial Potential Framework (APF) for Cooperative Payload Transport by Multi-Robot Collectives. USA: State University of New York.

Mitchell, M. (2001). An Introduction to Genetic Algorithms. Cambridge, Massachusetts: Bradford.

Park, M. and M. Lee (2003). Artificial Potential Field Based Path Planning for Mobile Robots Using a Virtual Obstacle Concept. IEEE/ASME - International Conference on Advanced Intelligent Mechatronics, Korea, pp. 735–740.

Qixin, C., H. Yanwen and Z. Jingliang (2006). An Evolutionary Artificial Potential Field Algorithm for Dynamic Path Planning of Mobile Robot. IEEE/RSJ, China, pp. 3331–3336.

Rodríguez, M.C. (2007). iScOp—Intelligent Scheduling an Optimization Research Group University of Oviedo. Retrieved from http://www.aic.uniovi.es/tc.

Sivanandam, S. and S. Deepa (2008). Introduction to Genetic Algorithms. Berlin: Springer-Verlag.

Weijun, S., M. Rui and Y. Chongchong (2010). A Study on Soccer Robot Path Planning with Fuzzy Artificial Potential Field. IEEE - International Conference on Computing, Control and Industrial Engineering, Wuhan, China, pp. 386–390.

Zhang, B., W. Chen and M. Fei (2006). An Optimized Method for Path Planning Based on Artificial Potential Field. IEEE - Sixth International Conference on Intelligent Systems Design and Applications. China.

Zhang, Q., D. Chen and T. Chen (2012). An Obstacle Avoidance Method of Soccer Robot Based on Evolutionary Artificial Potential Field. Elsevier B.V.—International Conference on Future Energy, Environment, and Materials, China, pp. 1792–1798.

13

A Method Using a Combination of Ant Colony Optimization Variants with Ant Set Partitioning

Evelia Lizarraga, Oscar Castillo* and José Soria

ABSTRACT

In this paper, we propose an ant's partition method for Ant Colony Optimization (ACO), a meta-heuristic that is inspired in ant's behavior and how they collect their food. The proposed method equivalently divides the total number of ants in three different subsets, and each one is evaluated separately by the corresponding variation of ACO (AS, EAS, MMAS) to solve different instances of the Traveling Salesman Problem (TSP). This method is based in the common saying "divide and conquer" to be applied in the repartition of the work, as the ants are evaluated in different ways in the same iteration. This method also includes a stagnation mechanism that stops a certain variation if it is not working properly after several iterations. This allows us to save time performing tests and have less overhead in comparison with the conventional method, which uses just one variation of ACO in all iterations.

Tijuana Institute of Technology, México.
  E-mail: ocastillo@tectijuana.mx
* Corresponding author

13.1 Introduction

Nowadays, there are open NP-complete problems where the complexity to construct the right configuration of a solution lies in finding a possible solution within a great number of possible combinations. In the past it was assumed that this kind of problems were impossible to solve since the feasibility to find the right answer was attached to the idea that an exhaustive search was needed resulting in the impossibility to find an answer in a reasonable frame of time. In many cases even when a super computer was used, the amount of time needed to find a solution was monumental causing overhead and great computational complexity. Therefore, NP-hard problems are approached finding the solution in a subset of the decision problem; this means finding an optimal solution using approximate algorithms that can give us good solutions efficiently; sacrificing finding very good solutions in a polynomial time. One of the most known examples of this kind of problems is TSP (Traveling Salesman Problem), which is represented figuratively as a salesman who wants to travel from city to city starting in an original point (city) without passing a city twice before returning home (original city). This is considering the length between cities as the cost that needs to be minimized. This problem can be represented as an undirected weight graph where the cities are the vertices of the graph, paths are graph's edges and the distance of the path is the length of the edge. There are many approaches proposed to solve this problem, one of the most used are meta-heuristics inspired in biological behavior such as Ant Colony Optimization (ACO) (Dorigo and Stützle 2004, Dorigo et al. 2006), Genetic Algorithms (GA), Particle Swarm Optimization (PSO) (Eberhart and Kennedy 1995, Engelbrecht 2005), among others. These techniques are known to be efficient in the search for a space solution providing optimal values. Given that ACO can be represented as graph, this meta-heuristic is one of the most used to represent and solve TSP. Dorigo and Stützle proposed the solution of TSP in (Dorigo and Stützle 2004, Stützle and Hoos 1996, 1999, Stützle and Linke 2002) using ACO according to the biological information that an ant provides when a good path is found adding pheromone to the nodes of that specific path. ACO uses the amount of pheromone and heuristic information to find the probability of the next node, where the heuristic information in this case (TSP) is the distance between cities. Owing to the fact, that ACO has many variations in how the pheromone is applied, the way to solve each problem is narrowed to how a designer uses these variations and how good is a specific variation in a specific problem. In this paper, we propose a strategy to compare these variations, such as Elitist Ant System (EAS), Ant System (AS), and Max Min Ant System (MMAS) within the same iteration; hence minimizing the number of tests needed. For this paper, there were performed 30 experiments using the proposed method (P-ACO) compared to 30 experiments for each ACO variation (AS, EAS, MMAS) resulting in a good performance and reduction of time. The main contribution in this work is the proposed P-ACO approach that combines the use of EAS, AS and MAS in a single method that is more efficient in finding solutions for complex problems, such as the TSP.

13.1.1 Ant Colony Optimization (ACO)

ACO is a meta-heuristic introduced by Dorigo and Stützle that is based in ant's behavior and how they collect their food following biological and heuristic information. Artificially speaking each ant represents a possible solution in which each one of them cooperates to find an optimal global solution. ACO works creating a construction graphGc(C=L) identical to the problem graph (Fig. 13.1). Each node connects to a corresponding set of arcs (i.e., L=A) and those arcs have a weight that corresponds to the distance d_{ij} between nodes i to j.

After the graph is constructed, a set of constraints are defined, which depend on the problem. In this case, the only constraint is, that a city is visited only once within the feasible neighborhood N_i^k where k is the ant and i is the node to be visited choosing the next node probabilistically. The ants "memorize" the visited nodes as well as the cost of that node (heuristic information). The evaporation of the pheromone, helps to forget bad decisions and follow the best path; also delimitating the maximum amount of pheromone deposited by the ants. The following step defines how the pheromone τ_{ij} is updated and the evaporation rate ρ that is needed, also the heuristic information η_{ij} that is going to be used; the distance between nodes gives this information. Finally, we construct a solution setting the ants randomly in nodes to start searching paths. The algorithm ends when all the cities have been visited.

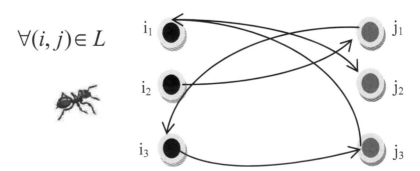

$$\forall(i, j) \in L$$

Fig. 13.1 Diagram that shows the nodes and the corresponding arcs.

13.1.2 ACO variations

According to literature (Dorigo and Stützle 2004) there exist many variations of ACO, the slightly difference from each other is in how they update the pheromone and the use of some parameters to perform such update. Some variations are the best for certain problems, including the TSP like MMAS, EAS and RankAS. In the same way these variations have different functionality in specific problems (Dorigo and Stützle 2004, Stützle and Hoos 2000, Dorigo et al. 2006, Stützle and Linke 2002).

The main justification for using these variations is to proof the feasibility of the proposed method P-ACO using the variations that have different performance in TSP like AS that perform badly in comparison with the others, EAS which has proved that can provide good solutions and finally MMAS which is one of the most used variations in TSP because it gives excellent results. This will show us if the algorithm is working when it compares the different results of the different variations per iteration, expecting the best one (MMAS) as the winner, this means the variation that throws the best global in each test.

13.2 Proposed Method

This method provides a way to divide the total number of ants in different subsets. This means given a set of m ants the method equivalently divides the total number of ants in three different subsets and each one is evaluated separately by the corresponding variation of ACO (AS, EAS and MMAS) to solve different instances of the Traveling Salesman Problem (TSP). The method is illustrated in Fig. 13.2.

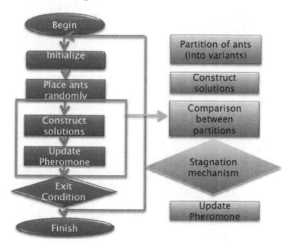

Fig. 13.2 Proposed variation of the ACO algorithm (P-ACO).

The stagnation mechanism works evaluating the best result per iteration and counting which variation has poor results. Making several experiments we concluded that the feasible number of allowed poor results is five; this means if the variation best result is not the global best in that iteration for five consecutively times then the variation it's stopped. This could cause that none of the variations are stopped if there isn't five consecutively bad results in any of them, also could cause that one by one the variations are stopped leaving only one "winner" variation at the end of the maximum iterations which will make the algorithm faster because it will perform the best variation(s), not wasting time in evaluating the ants in the variations with poor results.

13.2.1 Methodology

Firstly, we choose the desired parameters. For practicality purposes, we made an interface to choose our parameters and TSP instances, which is illustrated in Fig. 13.3.

Subsequently the best ant of each partition is compared with each other; obtaining the best global ant i (global best i), which is shown in Fig. 13.4.

The evaluation of the partitions is made sequentially in the same iteration i. Therefore in each iteration the best global ant i is compared with the best global ant i-1 following the conventional ACO algorithm. This allows us to compare in one iteration three different variations, saving time performing tests and having less overhead in comparison with the conventional method, which uses just one variation of ACO in all iterations. This means n (iterations) x 3 (variations) total tests in comparison with the

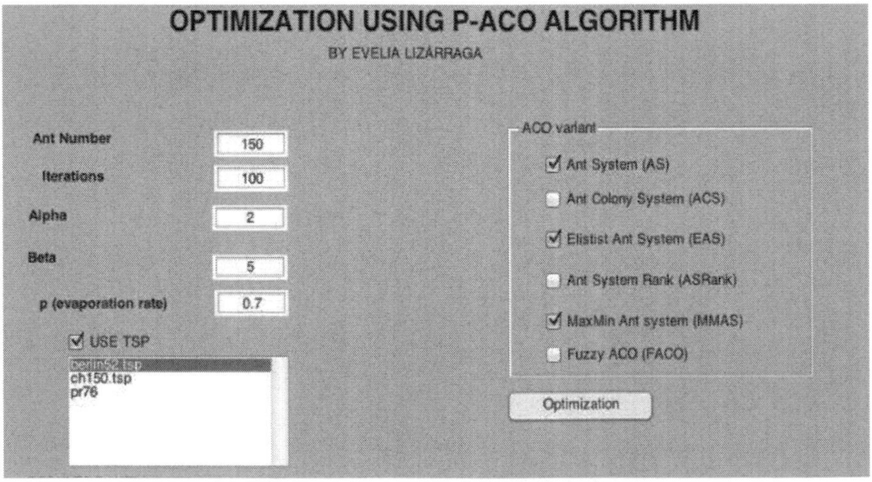

Fig. 13.3 Proposed user interface.

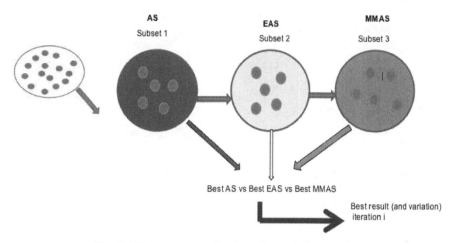

Fig. 13.4 Picture representing how the partition is made.

Color image of this figure appears in the color plate section at the end of the book.

proposed method, which only needs n tests. Because the proposed method selects one ant per iteration that represents also a variant of ACO; the end result provides the most used variant hence the one with best performance in a particular instance of TSP.

13.3 Experiments

We made tests with different instances of TSP to test the algorithm using different parameters for two types of experiments.

- TSP Instances us, ed: Berlin52 (52 cities) and bicr127 (127 cities)
- 30 experiments with P-ACO
- 30 experiments with AS
- 30 experiments with EAS
- 30 experiments with MMAS
- We used alpha=2, beta=5 and rho=0.7 in every experiment.
- Hardware information:
- Mac OS X 2.3 Ghz Intel Core i5 4GB 1333 Mhz DDR3.

13.4 Simulation Results

Variables used: It(Iterations), G.B. (Global Best), Avg. (Average), T(Time),BKSF(Best Result Known so far) according to literature (Stützle and Linke 2002), E (Error), N.E. (Normalized Error). The average is calculated by the global best of each experiment dived in the number of experiments (30). The error is calculated as the following:

Error=Best Result so Far—Average
Normalized Error=1—(Best Result so Far/Average)

13.4.1 Berlin 52 cities

Tables 13.1, 13.2, 13.3 and 13.4 show the simulation results for the benchmark TSP problem of Berlin with 52 cities.

Table 13.1 Experiments of type 1 using proposed algorithm P-ACO.

Ants	It	G.B.	Avg.	V	T(min)	BKSF	E	N.E
120	150	7549.29	7549.29	P-ACO (MMAS)	14.2	7542	7.29	0.00096

Table 13.2 Experiments of type 1 using individual algorithms.

Ants	It	G.B.	Avg.	V	T(min)	BKSF	E	N.E
120	150	7713.03	7961.82	AS	37.95	7542	171.03	0.052729
120	150	7544.37	7628.35	EAS	37.53	7542	2.366	0.011321
120	150	7549.29	7944.86	MMAS	839.09	7542	7.29	0.050707
Total time used					914.57 min(15.24 hours)	Average Error		0.038252

Table 13.3 Experiments of type 2 using proposed algorithm P-ACO.

Ants	It	G.B.	Avg.	V	T(min)	BKSF	E	N.E
150	100	7544.37	7544.37	P-ACO (MMAS)	15.2	7542	2.37	0.000314

Table 13.4 Experiments of type 2 using individual algorithms.

Ants	It	G.B.	Avg.	V	T(min)	BKSF	E	N.E
150	100	7857.17	7857.17	AS	23.62	7542	315.16	0.040112
150	100	7544.2	7688.23	EAS	31.44	7542	2.37	0.019020
150	100	7658.96	7941.22	MMAS	451.55	7542	116.96	0.050272
Total time used					506.61 min(8.44 hours)	Average Error		0.036468

13.4.2 Bier 127 (127 cities)

Tables 13.5, 13.6, 13.7 and 13.8 show the simulation results for the benchmark TSP problem of Bier with 127 cities.

Table 13.5 Experiments type 1 using proposed algorithm P-ACO.

Ants	It	G.B.	Avg.	V	T(min)	BKSF	E	N.E
120	150	126551	126551.72	P-ACO (EAS)	69.85 (1.16 hours)	118282	8269	0.065

Table 13.6 Experiments type 1 using individual algorithms.

Ants	It	G.B.	Avg.	V	T(min)	BKSF	E	N.E
120	150	133400.8	136754.79	AS	149.97	118282	15118.8	0.135
120	150	125788.79	128061.79	EAS	136.20	118282	7506.79	0.076
120	150	126554.00	126781.43	MMAS	3071.89	118282.	8272	0.067
			Total time used		3357.06 min (55.95hours)	Average Error		0.092

Table 13.7 Experiments type 2 using proposed algorithm P-ACO.

Ants	It	G.B.	Avg.	V	T(min)	BKSF	E	N.E
150	100	123903	123912.33	P-ACO (EAS)	50.46	118282	5621	0.045

Table 13.8 Experiments type 2 using individual algorithms.

Ants	It	G.B.	Avg.	V	T(min)	BKSF	E	N.E
150	100	134660.87	137786.55	AS	127.21	118282	16378.87	0.142
150	100	123903.00	123915.46	EAS	101.3	118282	5621.00	0.045
150	100	123903.00	123921.21	MMAS	1463.34	118282.	5621.00	0.045
		Total time used			1691.85 min (28.19 hours)	Average Error		0.077

From the simulation results and previous Tables, it can be clearly appreciated that the proposed P-ACO approach outperforms the individual variants of ACO.

13.5 Conclusions

We proposed a method for ant set partitioning in ACO. The proposed method is to divide the total number of ants into different partitions for a corresponding variant of ACO. Simulation results show that the proposed approach is working, in some cases 64 times faster than the conventional methodology and also providing better results. Performing two types of experiments varying the number of ants and iterations, we made a comparison between the proposed algorithm P-ACO and the conventional methodology. In experiment type 1 we used 120 ants and 150 iterations and type 2 we used 150 ants and 100 iterations for each instance of TSP. Using 52 cities (Berlin52) with type 1 experiment we compared Table 13.1 with Table 132, in which the P-ACO reached a better result in global best 64 faster, with a much less average normalized error of 0.00096. In experiment type 2 the comparison between Tables 13.3 and 13.4 shows that P-ACO reached a better result 33 times faster, with a normalized error of 0.000314. Using 127 cities (bier127) with type 1 experiment we compared.

Tables 13.5 and 13.6, where P-ACO showed a better global best 48 times faster with an average normalized error of 0.065. In experiment type 2 we compared Tables 13.7 and 13.8 obtaining a better global best of 0.045, 33.5 times faster than the conventional method. These results clearly demonstrate the feasibility of dividing the work between the ants for the evaluation, also providing a stagnation mechanism that controls the waste of computational calculations reducing only the work in the variations that shows good results.

References

Engelbrecht, A. (2005). Fundamentals of Computational Swarm Intelligence, Wiley, London, UK.

Dorigo, M. and T. Stützle (2004). Ant Colony Optimization, Massachusetts Institute of Technology, Cambridge, USA.

Dorigo, M., M. Birattari and T. Stützle (2006). Ant Colony Optimization, IEEE Computational Intelligence Magazine, November, pp. 28–39.

Eberhart, R.C. and J. Kennedy (1995). A new optimizer using particle swarm theory, Proceedings of the Sixth International Symposium on Micro Machine and Human Science, Nagoya, Japan, pp. 39–43.

Stützle, T. and H.H. Hoos (1996). Improving the Ant System: A detailed report on the MAX-MIN Ant System. Technical report AIDA-96-12, FG Intellektik, FB Informatik, TU Darmstadt, Germany.

Stützle, T. and H.H. Hoos (1999). MAX-MIN Ant System and local search for combinatorial optimization problems. In: S. Voss, S. Martello, I. Osman and C. Roucairol (Eds.), Meta-Heuristics: Advances and Trends in Local Search Paradigms for Optimization. Dordrecht, Netherlands, Kluwer Academic Publishers, pp. 137–154.

Stüttzle, T. and H.H. Hoos (2000). MAX-MIN Ant System, Future Generation Computer Systems 16(8): 889–914.

Stützle, T. and S. Linke (2002). Experiments with variants of ant algorithms, Mathware and Soft Computing 9(2–3): 193–207.

14

Variants of Ant Colony Optimization: A Metaheuristic for Solving the Traveling Salesman Problem

*Iván Chaparro, Fevrier Valdez and Patricia Melin**

ABSTRACT

Ant Colony Optimization (ACO) has been used to solve several optimization problems. However, in this paper, the variants of ACO have been applied to solve the Traveling Salesman Problem (TSP), which is used to evaluate the ACO variants as Benchmark problems. Also, we developed a graphic interface to allow the user to input parameters and having as objective to reduce processing time through a parallel implementation. We are using ACO because for TSP is easily applied and understandable. In this paper, we used the following variants of ACO: Max-Min Ant System (MMAS), Ant Colony System (ACS), Elitist Ant System (EAS) and Rank Based Ant System (ASrank).

14.1 Introduction

There are different algorithms based on the simulation of natural processes and genetics such as genetic algorithms and Ant Colony Optimization

Tijuana Institute of Technology, Tijuana, México.
  E-mail: pmelin@tectijuana.mx
* Corresponding author

(ACO), based on heuristic problem solving (Bianchi et al. 2002). Nowadays, ACO is used to solve more complex problems, which require a lot of processing time for achieving results (Barán and Sosa 2000). Therefore, we can work with highly complex problems getting results with less processing time with a parallel implementation. In this paper, we describe several variants of Ant Colony Optimization (ACO) to solve the Traveling Salesman Problem (TSP) allowing the user to input parameters using a graphical interface and performing parallel processing.

14.2 ACO Variants

The ACO metaheuristic is inspired by observing the behavior of real ant colonies, which presented as an interesting feature how to find the shortest paths between the nest and food, on its way the ants deposit a substance called pheromone (Favareto et al. 2004). This trail allows the ants back to their nest from the food; it uses the evaporation of pheromone to avoid an unlimited increase of pheromone trails and allows to forget the bad decisions, thus avoiding the persistence of the pheromone trails and therefore, the stagnation in local optima (Dorigo and Stützle 2004, Favaretto et al. 2004).

14.2.1 Traveling Salesman Problem (TSP)

This problem is defined as to visit "n" cities, starting and ending with the same city, visiting each city once and making the tour with the lowest cost, this cost can be expressed in terms of time or distance, i.e., travel a minimum of kilometers or perform a tour in the shortest time possible (Website Ant Col. 2012). More formally, the TSP can be represented by a complete weighted graph $G= (N, A)$ with N being the set of nodes representing the cities, and A being the set of arcs. Each arc $(i, j) \in A$ is assigned a value (length) d_{ij}, which is the distance between cities i and j, with $i,j \in N$. In the general case of the asymmetric TSP, the distance between a pair of nodes i,j is dependent on the direction of traversing the arc, that is, there is at least one arc (i, j) for $d_{ij} \neq d_{ji}$. In the symmetric TSP, $d_{ij} = d_{ji}$, holds for all the arcs in A. The goal in TSP is to find a minimum length Hamiltonian circuit of the graph, where a Hamiltonian circuit is a closed path visiting each of $n = |N|$ nodes of G exactly once (Dorigo and Gambardella 1997a, b). Thus, an optimal solution to the TSP is a permutation π of the node indices $\{1,2,...,n\}$ such that the length $f(\pi)$ is minimal, where $f(\pi)$ is given by:

$$f(\pi) = \sum_{i=1}^{n-1} d_{\pi(i)\pi(i+1)} + d_{\pi(n)\pi(1)} \qquad (14.2.1)$$

14.2.2 Elitist ant system

This algorithm was introduced by Marco Dorigo in 1992, the idea is to provide strong additional reinforcement to the arcs belonging to the best tour found since the start of the algorithm; this tour is denoted as T^{bs} (best-so-far tour) (Almiron et al. 1999).

14.2.3 Rank based ant system

This algorithm was introduced by Bullnheirmer in 1998, in ASrank each ant deposits an amount of pheromone that decreases with its rank. Additionally, as in EAS, the best-so-far ant always deposits the largest amount of pheromone in each iteration (Almiron et al. 1999).

14.2.4 Max-Min ant system

This algorithm introduces four main modifications with respect to the Ant System (De la Cruz et al. 2003, Cordon et al. 2002):

- It strongly exploits the best tours found.
- It limits the possible range of pheromone trail values to the interval $[\tau_{min}, \tau_{max}]$.
- The pheromone trails are initialized to the upper pheromone trail limit, which, together a small pheromone evaporation rate, increases the exploration of tours at the start of the search.
- Pheromone trails are reinitialized each time the system approaches stagnation or when no improved tour has been generated for a certain numbers of consecutive iterations.

14.2.5 Ant Colony System (ACS)

This algorithm was proposed by Dorigo and Gambardella in the year 1997, and differs from Ant System in three main points (Dorigo and Stützle 2004):

- It exploits the search experience accumulated by the ants more strongly, than Ant System does, through the use of a more aggressive action choice rule.
- Pheromone evaporation and pheromone deposit takes place only on the arcs belonging to the best-so-far tour.
- Each time an ant uses an arc (i,j) to move from city i to city j, it removes some pheromone from the arc to increase the exploration of alternative paths (Stützle 1998).

14.3 Graphical Interface in Matlab

In this section, we show an interface graphical allowing the user to select the mode in which he wants to work, either sequentially or in parallel form, as shown in Fig. 14.1 (Website of Interface Desig 2012, Website of Matlab 2012).

To go out of the system press the exit button, showing the confirmation message, as illustrated in Figs. 14.1 and 14.2:

Fig. 14.1 Main Interface.

Color image of this figure appears in the color plate section at the end of the book.

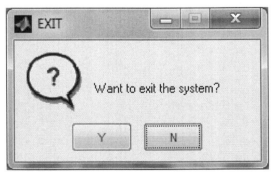

Fig. 14.2 Interface confirmation message.

14.4 Sequential Processing

In Fig. 14.3 is shown an example of how the user can select the type of variant ACO and the option Run to execute the algorithm, and Fig. 14.4 shows in which different parameters are introduced, or press the back button to go to the main interface.

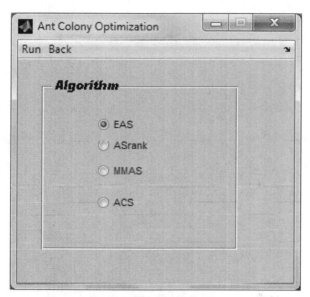

Fig. 14.3 Menu of options.

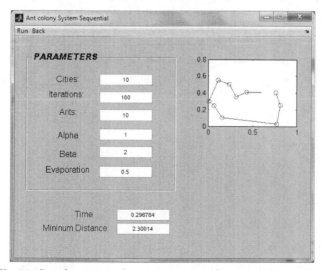

Fig. 14.4 Interface to introduce parameters with sequential processing.

In Fig. 14.4, the user can introduce the parameters to run the algorithm, the execution time and minimum distance that is the cost of taking the tour of cities in addition to plot the location of cities of any algorithm that has been selected for execution are presented.

14.5 Parallel Processing

The user can select the number of labs to work in parallel form. In Fig. 14.5 is shown an example with a lab, the parallel mode is initialized with the number of Labs selected.

In Fig. 14.6, the user can select the algorithm and introduce the parameters. This figure shows the runtime and the minimum distance of the tour, and plotting of the cities.

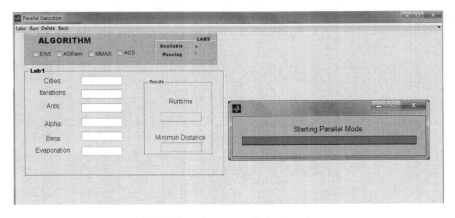

Fig. 14.5 Interface to parallel processing.

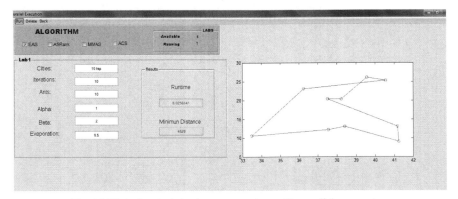

Fig. 14.6 Interface to introduce parameters with parallel processing.

Table 14.1 Sequential Variant for EAS.

EXPERIMENT	CITIES	ANTS	ITERATIONS	ALPHA	BETA	RHO	MINIMUN DISTANCE	RUNTIME
1	10	100	100	1	2	0.3	4371	0.0844685
2	10	100	100	1	2	0.3	4411	0.0638465
3	10	100	100	1	2	0.3	4411	0.0512251
4	10	100	100	1	2	0.3	4455	0.0502664
5	10	100	100	1	2	0.3	4455	0.0718149
6	10	100	100	1	2	0.3	4453	0.0663716
7	10	100	100	1	2	0.3	4463	0.0573872
8	10	100	100	1	2	0.3	4371	0.0568037
9	10	100	100	1	2	0.3	4549	0.0577863
10	10	100	100	1	2	0.3	4613	0.0560225
11	10	100	100	2	3	0.5	4530	0.0539174
12	10	100	100	2	3	0.5	4515	0.0530887
13	10	100	100	2	3	0.5	4389	0.0652427
14	10	100	100	2	3	0.5	4453	0.0523123
15	10	100	100	2	3	0.5	4371	0.0531191
16	10	100	100	2	3	0.5	4509	0.0693982
17	10	100	100	2	3	0.5	4463	0.0551354
18	10	100	100	2	3	0.5	4453	0.0531819
19	10	100	100	2	3	0.5	4389	0.0543939
20	10	100	10	2	3	0.5	4411	0.0565103
21	10	100	10	2	3	0.5	4371	0.0641863
22	10	100	10	2	3	0.5	4518	0.0537091
23	10	10	10	2	3	0.5	4714	0.0235802
24	10	10	9	2	3	0.5	4636	0.0267826
25	10	10	8	2	3	0.5	4453	0.0218175
26	10	10	7	2	3	0.5	4463	0.0226498
27	10	10	6	2	3	0.5	4549	0.0242553
28	10	10	5	2	3	0.5	4544	0.0241957
29	10	10	4	2	3	0.5	4723	0.0289262
30	10	10	3	2	3	0.5	4533	0.0224091
31	22	100	100	2	3	0.5	7124	5.265
Average runtime of 30 experiments performed with 10 cities							0.049826813	Seconds
Runtime with 22 cities.							5.265	Seconds

14.6 Simulation Results

This section shows the results obtained from experiments with the different variants of ACO with sequential and parallel processing.

It was found a minimum distance of 4371 in only 10 iterations; if the number of iterations decreases we observed the minimum distance increases. We obtained an average runtime of 0.049826813 seconds of 30 experiments with 10 cities. We also performed an experiment with 22 cities showing the results in the experiment 31 (see Table 14.1).

High Performance Programming for Soft Computing

Table 14.2 Parallel Variant for EAS.

EXPERIMENT	CITIES	ANTS	ITERATIONS	ALPHA	BETA	RHO	MINIMUN DISTANCE				RUNTIME			
							Lab1	Lab 2	Lab3	Lab 4	lab1	lab2	lab3	lab4
1	10	100	100	1	2	0.3	4530	**4389**	4453	4515	0.043017	**0.080566**	0.222274	0.075892
2	10	100	100	1	2	0.3	**4453**	4528	4455	4455	**0.042892**	0.068253	0.070235	0.072571
3	10	100	100	1	2	0.3	4371	4453	4453	4533	**0.037814**	0.070587	0.110031	0.118336
4	10	100	100	1	2	0.3	4371	4515	4411	4533	**0.050405**	0.070338	0.068501	0.084073
5	10	100	100	1	2	0.3	4371	4528	4528	4411	**0.058342**	0.10221	0.138822	0.037909
6	10	100	100	1	2	0.3	**4389**	4491	4389	4530	**0.048681**	0.061963	0.114626	0.106219
7	10	100	100	1	2	0.3	4453	**4389**	4463	4389	0.070888	**0.037568**	0.07354	0.098251
8	10	100	100	1	2	0.3	4411	4453	4371	4389	0.049216	0.103535	**0.057889**	0.121885
9	10	100	100	1	2	0.3	4453	4491	4371	**4371**	0.040456	0.111546	0.110886	**0.091063**
10	10	100	100	1	2	0.3	4371	4453	4549	4389	**0.046605**	0.095228	0.072813	0.105032
11	10	100	100	2	3	0.5	4371	4371	4520	4371	**0.049047**	0.066378	0.073185	0.094319
12	10	100	100	2	3	0.5	4389	4411	4371	4371	0.071819	0.071145	0.070115	**0.038032**
13	10	100	100	2	3	0.5	4530	4411	4371	4389	0.057607	0.113547	**0.067831**	0.071338
14	10	100	100	2	3	0.5	**4389**	4491	4389	4509	**0.037438**	0.057938	0.070763	0.085679
15	10	100	100	2	3	0.5	4389	4411	4453	4371	0.038878	0.103606	0.09811	**0.068744**
16	10	100	100	2	3	0.5	4371	4453	4371	4411	**0.049174**	0.107479	0.069143	0.116024
17	10	100	100	2	3	0.5	4463	4509	4453	**4453**	0.06947	0.037432	0.131665	**0.104728**
18	10	100	100	2	3	0.5	4371	4389	4411	4411	**0.049558**	0.067379	0.083924	0.11382
19	10	100	100	2	3	0.5	4491	4411	**4411**	4509	0.073657	0.072778	**0.045065**	0.134579
20	10	100	100	2	3	0.5	4389	4389	4389	4371	0.06711	0.124622	0.049275	**0.069494**
21	10	100	100	2	3	0.5	4371	4371	4371	4455	0.054952	0.073171	**0.053148**	0.061512
22	10	100	100	2	3	0.5	4371	4389	4463	4371	0.173185	0.069074	0.038107	**0.073517**
23	10	10	10	2	3	0.5	**4515**	4695	4562	4675	**0.015138**	0.012603	0.012828	0.014746
24	10	10	9	2	3	0.5	4704	4784	**4638**	4695	0.008227	0.010756	**0.013944**	0.007778
25	10	10	8	2	3	0.5	4710	**4411**	4453	5055	0.013592	**0.013418**	0.016794	0.013086
26	10	10	7	2	3	0.5	4714	4740	**4533**	4678	0.016751	0.014296	**0.014109**	0.013085
27	10	10	6	2	3	0.5	4613	4669	4509	**4491**	0.009598	0.013065	0.013546	**0.013961**
28	10	10	5	2	3	0.5	4515	4455	4741	**4453**	0.01441	0.0123	0.012987	**0.015576**
29	10	10	4	2	3	0.5	4590	4648	4849	**4453**	0.011439	0.01315	0.014472	**0.013587**
30	10	10	3	2	3	0.5	4678	4741	**4453**	4675	0.009111	0.012995	**0.014797**	0.013643
31	22	100	100	2	3	0.5	7013	7058	7013	7089	2.854154	10.912829	10.501515	11.172711
Average runtime of 30 experiments performed with 10 cities													0.0457377 Seconds	
Runtime with 22 cities.													2.854154 Seconds	

It was found the same minimum distance of 4371, but with a smaller runtime. In the parallel processing the same algorithm is executed in four labs at the same time. An experiment was performed with 22 cities achieving the minimum distance in the lab1 with a runtime of 2.854154, see Table 14.2.

With the algorithm ASrank in only two experiments the minimum distance of 4371 was found with a runtime similar to EAS algorithm. 30 experiments were performed with 10 cities and obtained an average by execution of 0.047427413 and also performed an experiment with 22 cities achieving a runtime of 0.609622 (see Table 14.3).

The minimal distance is obtained with less than 10 iterations, and Experiment 31 with 22 cities has a runtime of 2.05922 seconds, as show in Table 14.4.

Table 14.3 Sequential Variant for ASrank.

EXPERIMENT	CITIES	ANTS	ITERATIONS	ALPHA	BETA	RHO	MINIMUN DISTANCE	RUNTIME
1	10	100	100	1	2	0.3	4389	0.0621181
2	10	100	100	1	2	0.3	4549	0.0619001
3	10	100	100	1	2	0.3	4549	0.0561063
4	10	100	100	1	2	0.3	4371	0.0561858
5	10	100	100	1	2	0.3	4389	0.0680707
6	10	100	100	1	2	0.3	4518	0.0605986
7	10	100	100	1	2	0.3	4389	0.0575612
8	10	100	100	1	2	0.3	4389	0.0585765
9	10	100	100	1	2	0.3	4515	0.0364701
10	10	100	100	1	2	0.3	4389	0.0573131
11	10	100	100	2	3	0.5	4463	0.0578759
12	10	100	100	2	3	0.5	4389	0.0540786
13	10	100	100	2	3	0.5	4389	0.0554142
14	10	100	100	2	3	0.5	4530	0.0443401
15	10	100	100	2	3	0.5	4371	0.0585429
16	10	100	100	2	3	0.5	4411	0.0588792
17	10	100	100	2	3	0.5	4463	0.0530441
18	10	100	100	2	3	0.5	4411	0.0643091
19	10	100	100	2	3	0.5	4389	0.0467455
20	10	100	10	2	3	0.5	4389	0.0573013
21	10	100	10	2	3	0.5	4455	0.0686291
22	10	100	10	2	3	0.5	4463	0.0437769
23	10	10	10	2	3	0.5	4549	0.0241601
24	10	10	9	2	3	0.5	4530	0.0231446
25	10	10	8	2	3	0.5	4741	0.0241621
26	10	10	7	2	3	0.5	4573	0.0229088
27	10	10	6	2	3	0.5	4575	0.0246832
28	10	10	5	2	3	0.5	4664	0.0235117
29	10	10	4	2	3	0.5	4520	0.0226166
30	10	10	3	2	3	0.5	4626	0.0197979
31	22	100	10	2	3	0.5	7113	0.609622
Average runtime of 30 experiments performed with 10 cities							0.047427413	Seconds
Runtime with 22 cities.							0.609622	Seconds

The algorithm MMAS obtained the minimum distance of 46.5519 in all experiments. Average obtained by execution of all experiments is 5.81 seconds and performed an experiment with a 22 cities with a time of 20.329047 seconds (see Table 14.5).

In parallel form the same distance was obtained only reducing the processing time. The results are show in Table 14.6.

30 experiments were performed with different values for alpha, beta, rho factor and the number of iterations using 10 cities (Website TSPLIB95

Table 14.4 Parallel Form for ASrank.

EXPERIMENT	CITIES	ANTS	ITERATIONS	ALPHA	BETA	RHO	MINIMUN DISTANCE				RUNTIME			
							Lab1	Lab2	Lab3	Lab4	lab1	lab2	lab3	lab4
1	10	100	100	1	2	0,3	4463	4371	4491	4515	0,04652	0,126119	0,101735	0,071915
2	10	100	100	1	2	0,3	4389	4389	4509	4463	0,040875	0,076842	0,162788	0,06672
3	10	100	100	1	2	0,3	4371	4389	4411	4509	0,042861	0,107839	0,050918	0,080174
4	10	100	100	1	2	0,3	4463	4371	4518	4371	0,066909	0,071897	0,035517	0,123025
5	10	100	100	1	2	0,3	4411	4371	4371	4491	0,046248	0,06397	0,10232	0,104193
6	10	100	100	1	2	0,3	4371	4463	4593	4520	0,045326	0,062343	0,053035	0,035717
7	10	100	100	1	2	0,3	4389	4371	4520	4453	0,055982	0,036379	0,064768	0,063569
8	10	100	100	1	2	0,3	4515	4389	4520	4518	0,051951	0,041698	0,0362	0,064881
9	10	100	100	1	2	0,3	4491	4608	4411	4463	0,050026	0,140128	0,069062	0,114939
10	10	100	100	1	2	0,3	4453	4411	4463	4371	0,049439	0,115964	0,064522	0,15331
11	10	100	100	2	3	0,5	4463	4371	4463	4411	0,040401	0,112488	0,062922	0,040672
12	10	100	100	2	3	0,5	4371	4463	4371	4530	0,04201	0,06508	0,066247	0,066217
13	10	100	100	2	3	0,5	4371	4411	4389	4389	0,044015	0,11249	0,0725	0,116122
14	10	100	100	2	3	0,5	4371	4491	4411	4463	0,044923	0,07038	0,066934	0,08869
15	10	100	100	2	3	0,5	4389	4463	4371	4389	0,052802	0,051153	0,044031	0,097829
16	10	100	100	2	3	0,5	4411	4371	4411	4389	0,043143	0,064863	0,117396	0,082392
17	10	100	100	2	3	0,5	4411	4463	4389	4371	0,052055	0,040745	0,064351	0,09744
18	10	100	100	2	3	0,5	4509	4509	4598	4371	0,035271	0,138232	0,068309	0,146618
19	10	100	100	2	3	0,5	4544	4453	4371	4371	0,087299	0,066493	0,035956	0,069683
20	10	100	100	2	3	0,5	4530	4411	4371	4463	0,046814	0,063827	0,064743	0,069777
21	10	100	100	2	3	0,5	4530	4411	4411	4389	0,129253	0,066622	0,035473	0,11328
22	10	100	100	2	3	0,5	4453	4371	4411	4371	0,137742	0,05089	0,035669	0,065665
23	10	10	10	2	3	0,5	4528	4607	4549	4455	0,013484	0,013401	0,014246	0,016877
24	10	10	9	2	3	0,5	4608	4647	4530	4411	0,009992	0,012655	0,015725	0,012781
25	10	10	8	2	3	0,5	4411	4455	4520	4548	0,016226	0,022883	0,015989	0,007896
26	10	10	7	2	3	0,5	4544	4593	4598	4635	0,014057	0,013884	0,014716	0,01273
27	10	10	6	2	3	0,5	4371	4693	4695	4509	0,012845	0,017888	0,013506	0,013861
28	10	10	5	2	3	0,5	4548	4371	4637	4732	0,013803	0,014044	0,022557	0,013252
29	10	10	4	2	3	0,5	4411	4635	4695	4371	0,014171	0,007645	0,008237	0,013389
30	10	10	3	2	3	0,5	4664	4689	4757	4711	0,00817	0,01326	0,015733	0,012655
31	22	100	10	2	3	0,5	7289	7287	7348	7747	1,768417	2,05972	2,865184	1,87398
Average runtime of 30 experiments performed with 10 cities													0,054126567 Seconds	
Runtime with 22 cities.													2,05972 Seconds	

2012), the results are shown in Table 14.7. We obtained a minimum distance of 2.3001 with 100 and seven iterations of the algorithm; however, for fewer generations, the processing time is lower reducing the runtime to 0.113762 seconds in experiment number 26. Average was obtained by execution of 1.48 seconds. For more complexity, an experiment was performed only with 1000 cities and it was obtained an execution time of 16 seconds.

The same number of experiments were performed using the same parameters for parallel execution, finding the minimum distance of 2.3001 also with 7 iterations but with a lesser processing time of 0.079525 second in experiment number 26 (see Table 14.8). In the parallel mode, four processes are executed simultaneously, therefore, the average of execution for one process is 0.01988125 seconds and it was obtained an average of 1.05 seconds for all experiments. We obtained 8.7125918 seconds for the experiment done with 1000.

Table 14.5 Sequential Variant for MMAS.

EXPERIMENT	CITIES	ANTS	ITERATIONS	ALPHA	BETA	RHO	MINIMUN DISTANCE	TIME
1	10	100	100	1	2	0,3	46,5519	8,513724
2	10	100	100	1	2	0,3	46,5519	8,540868
3	10	100	100	1	2	0,3	46,5519	8,561641
4	10	100	100	1	2	0,3	46,5519	8,786021
5	10	100	100	1	2	0,3	46,5519	8,463686
6	10	100	100	1	2	0,3	46,5519	8,513592
7	10	100	100	1	2	0,3	46,5519	8,530144
8	10	100	100	1	2	0,3	46,5519	8,531728
9	10	100	100	1	2	0,3	46,5519	8,537607
10	10	100	100	1	2	0,3	46,5519	8,532742
11	10	100	100	2	3	0,5	46,5519	8,757501
12	10	100	100	2	3	0,5	46,5519	8,574803
13	10	100	100	2	3	0,5	46,5519	8,584994
14	10	100	100	2	3	0,5	46,5519	8,647263
15	10	100	100	2	3	0,5	46,5519	8,658588
16	10	100	100	2	3	0,5	46,5519	8,748089
17	10	100	100	2	3	0,5	46,5519	8,594535
18	10	100	100	2	3	0,5	46,5519	8,743204
19	10	100	100	2	3	0,5	46,5519	8,732453
20	10	100	10	2	3	0,5	46,5519	0,996702
21	10	100	10	2	3	0,5	46,5519	0,985875
22	10	100	10	2	3	0,5	46,5519	0,980589
23	10	10	10	2	3	0,5	46,5519	0,991487
24	10	10	9	2	3	0,5	46,5519	0,985393
25	10	10	8	2	3	0,5	46,5519	0,988249
26	10	10	7	2	3	0,5	46,5519	0,990082
27	10	10	6	2	3	0,5	46,5519	0,989426
28	10	10	5	2	3	0,5	46,5519	0,98293
29	10	10	4	2	3	0,5	46,5519	0,979736
30	10	10	3	2	3	0,5	46,5519	0,992027
31	22	100	100	2	3	0,5	75,3097	20,329047
Average runtime of 30 experiments performed with 10 cities							5,813855967	Seconds
Runtime with 22 cities.							20,329047	Seconds

14.6.1 Speedup

In the parallel form, the speedup measures how much a parallel algorithm is faster than a corresponding sequential algorithm. Speedup is defined by the following formula: $S_p = \dfrac{T_s}{T_p}$ where: Ts, is the runtime of the sequential algorithm and Tp is the runtime of the parallel algorithm.

Table 14.6 Parallel Form for MMAS.

EXPERIMENT	CITIES	ANTS	ITERATIONS	ALPHA	BETA	RHO	MINIMUN DISTANCE Lab1, Lab2, Lab3, Lab4	lab1	lab2	lab3	lab4
1	10	100	100	1	2	0.3	46.5519	15.616404	15.656199	15.725126	**15.825714**
2	10	100	100	1	2	0.3	46.5519	15.678724	15.587746	15.675984	**15.923952**
3	10	100	100	1	2	0.3	46.5519	15.774443	15.751579	15.747124	**15.884351**
4	10	100	100	1	2	0.3	46.5519	**15.771519**	15.598393	15.7615	15.716631
5	10	100	100	1	2	0.3	46.5519	15.910909	15.829529	15.946475	**16.128292**
6	10	100	100	1	2	0.3	46.5519	15.742674	15.663364	15.72186	**15.988445**
7	10	100	100	1	2	0.3	46.5519	15.708747	15.692472	15.700473	**15.800296**
8	10	100	100	1	2	0.3	46.5519	15.767879	15.576825	15.869953	**15.948438**
9	10	100	100	1	2	0.3	46.5519	15.919635	15.658901	**15.972495**	15.800514
10	10	100	100	1	2	0.3	46.5519	15.724937	15.706604	15.862844	**15.87672**
11	10	100	100	2	3	0.5	46.5519	15.861694	15.737939	15.916608	**16.02257**
12	10	100	100	2	3	0.5	46.5519	**15.872514**	15.749501	15.821333	15.8214
13	10	100	100	2	3	0.5	46.5519	**15.975199**	15.763913	15.827095	15.844473
14	10	100	100	2	3	0.5	46.5519	15.827191	15.798001	15.831821	**16.165865**
15	10	100	100	2	3	0.5	46.5519	15.827191	15.798001	15.831821	**16.165865**
16	10	100	100	2	3	0.5	46.5519	15.891914	15.836529	16.037666	**16.1051**
17	10	100	100	2	3	0.5	46.5519	15.874714	15.852839	**16.935693**	15.865242
18	10	100	100	2	3	0.5	46.5519	15.87535	15.805074	**15.996528**	15.843767
19	10	100	100	2	3	0.5	46.5519	15.866498	15.82524	15.784151	**16.050466**
20	10	100	100	2	3	0.5	46.5519	15.850193	15.708793	15.816849	**16.069687**
21	10	100	100	2	3	0.5	46.5519	15.823103	15.77423	**16.148439**	15.962383
22	10	100	100	2	3	0.5	46.5519	15.872622	15.737472	**16.052308**	15.989298
23	10	10	10	2	3	0.5	46.5519	1.765158	**1.765994**	1.761965	1.729142
24	10	10	9	2	3	0.5	46.5519	1.737158	**1.790317**	1.788179	1.740829
25	10	10	8	2	3	0.5	46.5519	1.750755	1.71053	1.783875	**1.794154**
26	10	10	7	2	3	0.5	46.5519	1.778715	1.790239	1.770882	**1.825998**
27	10	10	6	2	3	0.5	46.5519	1.757422	1.717733	1.760698	**1.844163**
28	10	10	5	2	3	0.5	46.5519	**1.786446**	1.754178	1.772288	1.771232
29	10	10	4	2	3	0.5	46.5519	1.741479	1.724236	**1.784747**	1.783529
30	10	10	3	2	3	0.5	46.5519	**1.80604**	1.687742	1.78379	1.770516
31	22	100	100	2	3	0.5	75.3097	38.742169	38.24625	37.974074	38.057135
Average runtime of 30 experiments performed with 10 cities										12.23594383	Seconds
Runtime with 22 cities.										38.742169	Seconds

In Table 14.9, is shown the runtime as well the speedup of the best five experiments with the MMAS algorithm with sequential and parallel processing; the efficiency of executing processes in parallel is shown.

Figure 14.7 shows in the graph that the runtime is lower by processing in parallel.

14.7 Conclusions

Several experiments were performed with different algorithms (EAS, ASrank, MMAS, ACS) and a graphical interface was used to change the following parameters: number of cities, number of ants, number of iterations and also the alpha, beta and rho, which modify the increase of pheromone and the evaporation of pheromone.

A minimum distance of 43.72 was obtained in the EAS, ASrank algorithms, with a runtime smaller than one second.

In all experiments that were performed using the MMAS algorithm, we obtained a minimum distance greater and therefore, more runtime.

Table 14.7 Sequential Variant for ACS.

EXPERIMENT	CITIES	ANTS	ITERATIONS	ALPHA	BETA	RHO	MINIMUN DISTANCE	RUNTIME
1	10	100	100	1	2	0.3	2.4064	2.297667
2	10	100	100	1	2	0.3	2.4064	2.204522
3	10	100	100	1	2	0.3	2.4064	2.151696
4	10	100	100	1	2	0.3	2.4064	2.191324
5	10	100	100	1	2	0.3	2.4064	2.1652
6	10	100	100	1	2	0.3	2.4064	2.185889
7	10	100	100	1	2	0.3	2.4064	2.171482
8	10	100	100	1	2	0.3	2.4064	2.172033
9	10	100	100	1	2	0.3	2.4064	2.175278
10	10	100	100	1	2	0.3	2.4064	2.379699
11	10	100	100	2	3	0.5	2.3001	2.293828
12	10	100	100	2	3	0.5	2.3001	2.343802
13	10	100	100	2	3	0.5	2.3001	2.364691
14	10	100	100	2	3	0.5	2.3001	2.3829
15	10	100	100	2	3	0.5	2.3001	2.330182
16	10	100	100	2	3	0.5	2.3001	2.277928
17	10	100	100	2	3	0.5	2.3001	2.261657
18	10	100	100	2	3	0.5	2.3001	2.288932
19	10	100	100	2	3	0.5	2.3001	2.321291
20	10	100	10	2	3	0.5	2.3001	0.301492
21	10	100	10	2	3	0.5	2.3001	0.266089
22	10	100	10	2	3	0.5	2.3001	0.269683
23	10	10	10	2	3	0.5	2.3001	0.072738
24	10	10	9	2	3	0.5	2.3001	0.125904
25	10	10	8	2	3	0.5	2.3001	0.078677
26	10	10	7	2	3	0.5	2.3001	0.113762
27	10	10	6	2	3	0.5	2.5144	0.075725
28	10	10	5	2	3	0.5	2.5319	0.081266
29	10	10	4	2	3	0.5	2.4064	0.07106
30	10	10	3	2	3	0.5	2.53.19	0.078017
31	1000	100	10	2	3	0.5	40.4873	16.62779848
Average runtime of 30 experiments performed with 10 cities							1.483147133	Seconds
Runtime with 1000 cities.							16.62779848	Minutes

Finally, in the ACS algorithm we obtained a minimum distance of 2.3001, where the cities were randomly while in the other algorithms are indicated the number and location of each city through file tsp.

In conclusion, the EAS and ASrank algorithms obtained the best results in terms of minimum distance and runtime.

We also achieved decrease runtime by parallel processing.

Parallel computing should be used when you have problems with high complexity and not enough run sequentially processes.

Table 14.8 Parallel Form for ACS.

EXPERIMENT	CITIES	ANTS	ITERATIONS	ALPHA	BETA	RHO	MINIMUN DISTANCE Lab1, Lab2, Lab3, Lab4.	RUNTIME lab1	lab2	lab3	lab4
1	10	100	100	1	2	0,3	2,3001	4,985979	4,670103	**5,803356**	4,450128
2	10	100	100	1	2	0,3	2,3001	4,717038	**5,766149**	4,855537	4,638474
3	10	100	100	1	2	0,3	2,3001	**5,663317**	4,626035	4,773888	4,315661
4	10	100	100	1	2	0,3	2,3001	4,76076	4,650837	**5,480135**	4,432192
5	10	100	100	1	2	0,3	2,3001	4,562846	4,682623	4,794606	**5,590443**
6	10	100	100	1	2	0,3	2,3001	4,793052	4,706704	**5,581461**	4,903088
7	10	100	100	1	2	0,3	2,3001	4,675758	4,627855	**5,557725**	4,282445
8	10	100	100	1	2	0,3	2,3001	5,676394	4,586213	5,067412	4,835595
9	10	100	100	1	2	0,3	2,3001	4,690962	**5,544794**	4,667791	4,765641
10	10	100	100	1	2	0,3	2,3001	4,640137	4,616093	**5,856676**	4,217882
11	10	100	100	2	3	0,5	2,3001	4,740937	4,697812	4,502633	**5,72732**
12	10	100	100	2	3	0,5	2,3001	4,509803	4,65397	4,752962	**5,862888**
13	10	100	100	2	3	0,5	2,3001	5,163928	4,935564	4,534519	**5,81736**
14	10	100	100	2	3	0,5	2,3001	5,184917	**5,842456**	4,668559	4,347392
15	10	100	100	2	3	0,5	2,3001	4,7837	4,693819	4,962278	**6,054081**
16	10	100	100	2	3	0,5	2,3001	4,615381	**5,656933**	4,580497	4,195778
17	10	100	100	2	3	0,5	2,3001	**5,58897**	4,735533	5,434506	4,359453
18	10	100	100	2	3	0,5	2,3001	5,051927	4,583253	**5,678237**	4,344494
19	10	100	100	2	3	0,5	2,3001	4,30736	4,675053	4,67073	**5,894579**
20	10	100	100	2	3	0,5	2,3001	4,667901	4,375274	4,707069	**5,797504**
21	10	100	100	2	3	0,5	2,3001	**5,730341**	4,426355	4,604745	4,909657
22	10	100	100	2	3	0,5	2,3001	4,583651	4,664926	**5,881932**	4,22536
23	10	10	10	2	3	0,5	2,3001	0,070359	**0,077827**	0,070337	0,05964
24	10	10	9	2	3	0,5	2,3001	**0,108555**	0,072157	0,072857	0,05581
25	10	10	8	2	3	0,5	2,3001	**0,096793**	0,065903	0,045364	0,072012
26	10	10	7	2	3	0,5	2,3001	0,062876	0,058779	**0,079525**	0,045117
27	10	10	6	2	3	0,5	2,5144	0,052096	0,04201	**0,106847**	0,042266
28	10	10	5	2	3	0,5	2,5319	0,040768	**0,061523**	0,03778	0,052522
29	10	10	4	2	3	0,5	2,4064	**0,132485**	0,060001	0,047999	0,046622
30	10	10	3	2	3	0,5	2,5319	**0,059991**	0,044205	0,050919	0,032897
31	1000	100	10	2	3	0,5	38,7935	1656,8	1671,1	**1672,5**	1660,9
Average runtime of 30 experiments performed with 10 cities										4,205587167	Segundos
Runtime with 1000 cities.										27,875	Minutos

Table 14.9 Speedup for MMAS algorithm.

Sequential Processing	Parallel Processing	Speedup
8.786021	1.770882	4.961381391
8.757501	1.754178	4.992367365
8.748089	1.724236	5.073603033
8.743204	1.717733	5.089966834
8.732453	1.687742	5.174044967

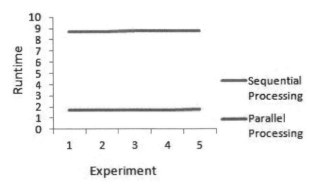

Fig. 14.7 Runtime sequential and parallel.
Color image of this figure appears in the color plate section at the end of the book.

References

Almirón, M., B. Barán and E. Chaparro (1999). Ant Distributed System for Solving the Traveling Salesman Problem. XXV Informatic Latinoamerican Conf.-CLEI, Paraguay, pp. 779–789.

Barán, B. and R. Sosa (2000). A New approach for AntNet routing. IEEE Ninth International Conference on Computer Communications and Networks. Las Vegas, NV, USA.

Bianchi, L., L.M. Gambardella and M. Dorigo (2002). Solving the homogeneous probabilistic traveling salesman problem by the ACO metaheuristic. In: M. Dorigo, G. Di Caro and M. Sampels (Eds.), Proceedings of ANTS 2002—From Ant Colonies to Artificial Ants: Third International Workshop on Ant Algorithms, vol. 2463 of Lecture Notes in Computer Science, Springer Berlin Heidelberg, pp. 176–187.

Cordón, O., F. Moya and C. Zarco (2002). A new evolutionary algorithm combining simulated annealing and genetic programming for relevance feedback in fuzzy information retrieval systems, Soft Computing, 6(5): 308–319.

De la Cruz, J., A. Mendoza, A. Del Castillo and C. Paternina (2003). Comparative Analysis of heuristic Approaches Ant Q, Simulated Annealing and Tabu Search in Solving the Traveling Salesman, Departamento de Ciencias de la Computación e Inteligencia Artificial, E.T.S. Ingeniería Informática, Granada, España.

Dorigo, M. and L.M. Gambardella (1997a). Ant colonies for the traveling salesman problem. BioSystems, 43(2): 73–81.

Dorigo, M. and L.M. Gambardella (1997b). Ant Colony System: A cooperative learning approach to the traveling salesman problem. IEEE Transactions on Evolutionary Computation, 1(1): 53–66.

Dorigo, M. and T. Stützle (2004). Ant Colony Optimization, Bradford, Cambridge, Massachusetts Institute of Technology, MIT Press.

Favaretto, D., E. Moretti and P. Pellegrini. (2006). An ant colony system approach for variants of the traveling salesman problem with time windows, Journal of Information and Optimization Sciences, Vol. 27, Issue 1, pp. 35–54.

Stützle, T. (1998). Parallelization Strategies for Ant Colony Optimization, Proc. of Parallel Problem Solving from Nature – PPSN-V, Springer Verlag, 1498: 722–731.

Website of Ant Colony Optimization Algorithms, www.aco-metaheuristic.org, official, Accessed 5 May 2012.

Website of interface design, www.matpic.com, MC. Diego O. Barragán Guerrero, Universidad Estatal de Campinas (www.unicamp.br), Brasil, Accessed May 2012.

Website of Matlab, www.mathworks.com, Accessed May 2012.

Website http://comopt.ifi.uni-heidelberg.de/software/TSPLIB95/, Accessed September 2012.

15

Quantum Computing

Oscar Montiel

ABSTRACT

We present concise information of the main topics that are needed to understand QC. In the introduction the great potential of this field, in special for soft computing, is presented. We depart from classical computing explaining the limitations of classical Turing machines with the aim to introduce the Quantum Turing Machine to simulate Quantum algorithms; we use digital logic to introduce the necessary concepts about the circuit model and reversible computing, with the purpose of extending these concepts to quantum gates and algorithms. Brief overviews of the mathematics used in QC and the basic principles of quantum mechanics also are included. Basic concepts such as the Bloch sphere, quantum bits, registers, gates, circuits and measurement are explained.

15.1 Introduction

In the twenty-first century, many significant developments in science and engineering have been achieved through interdisciplinary research. Quantum computing (QC) is about computing with quantum systems, called quantum computers. A quantum computer is a computational device that makes use of the quantum mechanics (QM) phenomena such as entanglement, superposition and interference to perform operations on data. Quantum computers are different from classical binary digital

Instituto Politécnico Nacional, CITEDI, Tijuana, México.
E-mail: oross@ipn.mx

computers since they work with quantum bits. The theoretical model is the quantum turing machine (QTM) which is a normal turing machine (TM) with quantum parallelism. Richard Feynman introduced this field for the first time in the 80's (Feynman 1982) since he observed that certain quantum-mechanical effects cannot be simulated in an efficient way using a classical computer. Such observation let to speculate that using quantum effects may become more efficient computing. The interest grew up in 1994 when Peter Shor published the paper entitled "Algorithms for Quantum Computation: Discrete Logarithms and Factoring" (Shor 1994) where he described polynomial time quantum algorithms for finding discrete logarithms and factoring integers.

Quantum computing is a computational model that may provide exponential advantages over classical computational models such as probabilistic and deterministic turing machines (Arora and Barak 2010). It combines QM, information theory and aspects of computer sciences; QC differs from classical computing based on traditional logic in many respects, for example, instead of operating on bits and Boolean logic, QC is based on quantum logic and operates on qubits. The idea is to exploit quantum effects to compute in an efficient and faster way than using conventional computers. QC uses a specific implementation to gain a computation advantage over conventional computers. Although the field is relatively new, it promises a dramatic computing speed increase using properties such as superposition and entanglement; it is expected to obtain, in some cases, an exponential amount of parallelism.

There are several recent works focussing on the application of QC in the soft computing area, for example: In (Visintin et al. 2012) there is a proposal to associate the states of a quantum register with membership functions of fuzzy subsets, and the rules in the fuzzyfication process, are performed using unitary quantum transformations; also it is presented a proposal of t-norms and t-conorms based on unitary and controlled quantum gates. With respect to the use of QC with neural networks, in (Nayak et al. 2011) it is presented the basic concepts of quantum neural computing as well as a mathematical model of a quantum neuron to discuss the training algorithm. It is discussed that Quantum Neural Network is a new paradigm based on the combination of classical neural networks and quantum computation, and it is presented an introductory representation of quantum artificial neural network modeled on the basis of double-slit experiment. In (Finch 2012) the author proposes two algorithms for training neural nets on a quantum computer, through a simulation it is shown that they can be used to train neural nets much faster than using the same methods on a classical computer. In the robotics field we can find other applications of quantum computing like an improved hierarchical Q-learning algorithm in conjunction with quantum parallelization computation for a mobile-robot global navigation, where the idea is to speed up the learning process, which consists of two learning levels for local navigation and topological navigation (Chen and Dong 2012). In (Rybalov et al. 2010) Fuzzy Model of Control for Quantum-Controlled Mobile Robots, it is considered the navigation of quantum-controlled mobile robot using fuzzy methods; the idea is to obtained a model of the robot behavior through the transformation of the robot's states, which appear like qubits, into the pairs of membership functions.

This work does not pretend to provide a deep understanding of quantum mechanics nor quantum computing. Instead, we will give a brief mathematical formalisms review to understand the basis of quantum mechanics and extend them for quantum computing.

15.2 Classic Computation

In general, circuits are networks composed of gates, registers, and wires. Gates perform logical operations on inputs; registers are made up of cells that contain bit values, and the wires carry signals between gates and registers. In general, for any binary logic operation we must provide n-bits to recover an m-bit output; i.e., we must compute a logical function of the form,

$$f : \{0, 1\}^n \rightarrow \{0, 1\}^m \tag{15.2.1}$$

Digital computers represent numbers in the binary form; the **Turing machines** (TM), the **Universal TM** (UTM), the **Circuit Model** (CM), and other computer science models of computation make use of formal language to represent inputs and outputs. A TM is composed by four elements:

1. A tape which is made up of cells containing 0, 1 or blank.
2. A read/write head that in each step can read and overwrite the current symbol, or move one cell to right or to the left.
3. A controller that controls the elements of the machine doing the operation of step 2, change state and halt.
4. A finite state automata (FSA) that describes the controller's behavior, i.e., the way the TM switches states.

The UTM is a TM that can read from a tape a program describing the behavior of another TM. A **probabilistic TM** is one with the capacity of making a random binary choice at each step. The architecture of digital computers is based in digital logic circuitry implemented with gates and registers. The (Classical) **Strong Church-Thesis** states that a probabilistic TM can efficiently simulate any realistic model of computation. However, based on the observations of Feyman about the necessity to design a computer to exploit the quantum physics, we can conclude that we are unable to simulate such a computer using a UTM. The fundamental problem with the classical Strong Church-Thesis is that apparently we cannot efficiently simulate quantum physics using classical physics because it is not enough powerful, what we need is a computer model capable to simulate realistic physical devices including quantum devices. Benioff (Benioff 1982) and Deutsch (Deutsch 1985) independently, proposed the first models of a **Quantum TM (QTM)** as well as the CM. The Quantum Strong Church-Thesis states that QTM can efficiently simulate any realistic model of computation, understanding by "realistic model of computation" a model that is consistent with the laws of physics and in which all the physical resources used by the model can explicitly be taken into account (Kaye et al. 2007).

The basic gates in digital logic are the NOT that inverts the logic input, the OR which returns "1" if either of the inputs is "1" and the AND that only returns a "1" if both inputs are "1". Derived gates from the basic ones are the NOR which is an inverted OR (the OR gate has at its output a NOT gate); the NAND is an inverted AND (AND and NOT gate combined), and the XOR that returns a logic "1" if only one of its inputs is "1" (for the inputs A and B, the output is given by $out = A\overline{B} + \overline{A}B$. The NAND gate is considered as a **universal gate** because any circuit can be implemented using only a combination of NAND gates regardless of how complicated it can be. In general, a **set of gates is universal** for classical computation if a circuit can be constructed for computing (15.2.1) for m and n integers, using only gates from that set. Relevant characteristics of the gates are the **fanout** and **crossover**. The fanout (also known as **copy**) gate takes one bit into two bits, in other words the next operation is achieved: COPY: $a \rightarrow (a, a)$. The crossover (or **swap**) interchanges the value of two bits, i.e., it performs the next operation: CROSSOVER: $(a, a) \rightarrow (b, a)$ (Benenti et al. 2004).

Figure 15.1 illustrates the concept of using classical gates and circuits vs. using the reversible version to achieve the same function. In general, a non-reversible circuit with n input variables $(x_1, \cdots, x_n)$, and j bits

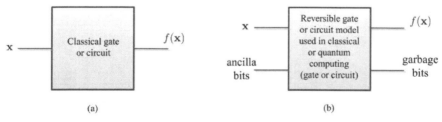

(a) (b)

Fig. 15.1 Black box model of a classical gate or circuits vs. a reversible gates or circuit used in the CM. a) The classical circuit has a vector x of binary imputs, the output *f(x)* is obtained after permorm the binary logic operations. b) A reversible circuit is shown, the vector x and the ancilla bits are shown at the imput, whereas *f(x)* and the garbage bits are shown at the ouput.

$(j = 1, \cdots, n)$ at the output given by the function $f_j(x_1, \cdots, x_n)$ can be transformed into a reversible circuit by the use of an **ancilla** such as it was illustrated in Figs. 15.3 and 15.4. The extra bits that are needed to make invertible the circuit are known as **garbage**. In classical computation, it is not common to use the extra bits to make circuits reversible; however, in QC discarding the redundant information can drastically change the outcomes.

The CM is based on a uniform family of reversible gates and circuits; all the circuits are considered **acyclic** which means that the bits move through the circuit in a linear fashion, and the wires never feedback to a prior position (location) in the circuit. In Fig. 15.2, a black-box circuit diagram of the computational CM is shown; there, the input bits are written onto the wires i_1 to i_4, the information contained in the input bits is traveling from left to right of the circuit to reach the outputs o_1 to o_4. For each time, step t (time slice), the input bits advance at most one gate. The total number of wires is called the width or space (S) of the circuit. The deep of the circuit (T) is given by the total number of time-slices of the circuit (Kay et al. 2007)

The only **reversible** classical one-bit gate is the NOT gate because no matter what the output value is, it is possible to know the input bit value always; i.e., it is reversible in a natural way because it is not necessary to add extra circuitry to infer the input value from the output value. As a rule, an n-bit gate has 2^n possible input combinations as well as 2^n possible outputs. The **Controlled NOT** (CNOT) gate is a 2-bit input 2-bit output reversible NOT gate, its equivalent schematic using the CM is shown in Fig. 15.3a and the equivalent classical schematic in Fig. 15.3b. The first one is a typical diagram of a gate used to construct quantum circuit diagrams. In Fig. 15.3a and b, for a given output, the input can be found using the copy of the input x and the output $x \oplus y$; i.e., using $x \wedge (x \oplus y)$. In circuit diagrams using the CM style, the symbol • is used to denote the control bit, and the symbol $\oplus$ is denotes a conditional negation.

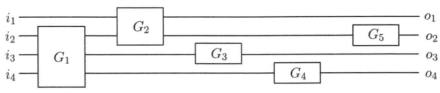

Fig. 15.2 Black-box circuit diagram to illustrate the CM of computation. The four basic components of this model are the wires, gates (G_1 to G_4), inputs (i_1 to i_4), and outputs (i_1 to i_4). Here, the width is $S=4$, and deep is $T=5$.

A generalization to three bits is the **Controlled-Controlled-Not** (CCNOT) known as the **Toffoli gate** that can emulate any classical circuit. In Fig. 15.4a the typical circuitry used in quantum circuits is illustrated and in Fig. 15.4b the classical logic circuit.

15.3 Basic Mathematics Used in Quantum Computing

In this chapter, we used $\mathbb{N}$, $\mathbb{Z}$, $\mathbb{Q}$, $\mathbb{R}$, and $\mathbb{C}$ to denote the natural, integer, rational, real and complex numbers respectively; therefore, we have that $\mathbb{N} \subset \mathbb{Z} \subset \mathbb{Q} \subset \mathbb{R} \subset \mathbb{C}$. Most of the formulations in QC are achieved using the sets $\mathbb{R}$ and $\mathbb{C}$, since QM is founded in complex vectors. A **vector space** V is a set on which two operations: the addition (vector addition), and scalar multiplications are defined.

The **Dirac notation** proposed by Paul Dirac is used in QM, whereas in mathematics and computer science to distinguish a vector from a scalar, it is common to use a boldface symbol, for example "**u**", or to write an arrow over the symbol "$\vec{u}$". In Dirac notation, the symbol for identifying a vector is written inside a "ket", and it looks like $|u\rangle$. The **dual vector** for "**u**" is writing with a *bra* as $\langle u|$. Using this notation, the inner product of "**u**" and "**v**" is written in *bra-kets* as $\langle u|v\rangle$.

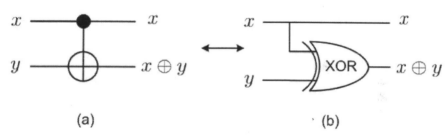

(a)   (b)

Fig. 15.3 The CNOT gate is reversible in a natural way; i.e., it is not necessary to add extra circuitry to infer the input value from the output value. Quantum computing uses the CM model of computation to construct quantum diagrams, for example, the NOT gate is schematized as it is shown in (a), instead the classical way shown in (b). In these two diagrams, the inputs are binary bits.

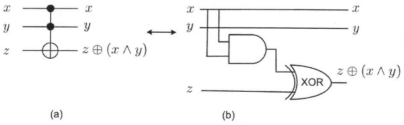

(a)   (b)

Fig. 15.4 The CCNOT (Toffoli) gate is universal for computation. For example, if we set $x=1$, and $y=1$, we have the CNOT gate.

Real vectors in $\mathbb{R}$ are written using bold letters, for example $\mathbf{u} = (u_x, u_y, u_z)$ and $\mathbf{v} = (v_x, v_y, v_z)$ are two vectors in $\mathbb{R}^3$. The dot (.) product (inner product) is given by $\mathbf{u} \cdot \mathbf{v} = (u_x v_x + u_y v_y, u_z)$. Complex vectors in $\mathbb{C}$ will be written using the *bra-ket* notation. Let $V = \mathbb{C}^n$ a complex vector space with n dimensions, the set containing all column vectors with n complex numbers laid out vertically, i.e., the column notation is used. For example in $\mathbb{C}^2$ the quantum notation to represent a *ket* vector is

$$|x\rangle = \begin{pmatrix} x_1 \\ x_2 \end{pmatrix} = \begin{pmatrix} a_1 + b_1 \\ a_2 + b_2 \end{pmatrix} = (x_1, x_2)^T \tag{15.3.1}$$

note that this notation also fulfills the complex vector space properties such as the zero vector, scalar multiplication and addition, vector addition, commutative and associative properties for the addition. Vectors for QC are in the complex vector space $\mathbb{C}^n$ and are finite-dimensional, such vector spaces are members of a class of vector spaces named Hilbert Spaces $\mathcal{H}$; however, in QC literature, a finite-dimensional complex vector space is named $\mathcal{H}$ (Kaye et al. 2007, Perry 2012).

In general, for $|x\rangle = (x_1, x_2, \ldots, x_n)^T$, $|y\rangle = (y_1, y_2, \ldots, y_n)^T \in \mathbb{C}^n$ and $a \in \mathbb{C}$, vector addition and scalar multiplication are defined by (15.3.2) and (15.3.3), respectively

$$|x\rangle + |y\rangle = \begin{pmatrix} x_1 + y_1 \\ x_2 + y_2 \\ \ldots \\ x_n + y_n \end{pmatrix} \tag{15.3.2}$$

and

$$a|x\rangle = \begin{pmatrix} ax_1 \\ ax_2 \\ \ldots \\ ax_n \end{pmatrix}, \tag{15.3.3}$$

it is important to note that any linear combination $c_1|x\rangle + c_2|y\rangle$ where $c_1, c_2 \in \mathbb{C}$ are also elements of $\mathbb{C}^n$.

A set of k vectors $\{|x_1\rangle, \ldots, |x_k\rangle\}$ in $V = \mathbb{C}^n$ is **linearly dependent** if the equation

$$\sum_{i=1}^{k} c_i |x_i\rangle = |0\rangle \tag{15.3.4}$$

i.e., any one of the vectors $\{|x_i\rangle\}$ is expressed as a linear combination of the other vectors; therefore, any set containing the zero-vector $|0\rangle$ is linearly dependent; this is always true if $k > n$. On the other hand, for a set of k

vectors in $\mathbb{C}^n$ where $k \leq n$, the vectors might be **linearly independent** (Nakahara and Ohmi 2008).

Let us consider having n linearly independent vectors $\{|v_i\rangle\}$ in $\mathbb{C}^n$, the set is called the basis of $\mathbb{C}^n$, and the vectors are called **basis vectors**. We can express any $|x\rangle \in \mathbb{C}^n$ as a linear combination of the n vectors,

$$|x\rangle = \sum_{i=1}^{n} c_i |v_i\rangle, \quad c_i \in \mathbb{C}, \tag{15.3.5}$$

the vector space spanned by a basis $\{|v_i\rangle\}$ can be denoted by $\mathrm{Span}(\{|v_i\rangle\})$.

A function $f : \mathbb{C}^n \to \mathbb{C}$; where, $f : |x\rangle \mapsto f(|x\rangle) \in \mathbb{C}$ is called a **linear function** if it satisfies the linearity condition given by,

$$f(c_1|x\rangle + c_2|y\rangle) = c_1 f(|x\rangle) + c_2 f(|y\rangle), \quad \forall |x\rangle, |y\rangle \in \mathbb{C}^n, \forall c_1, c_2 \in \mathbb{C} \tag{15.3.6}$$

A row vector $\langle \alpha |$ is called a bra vector, or simply *bra*; it is given by (15.3.7),

$$\langle \alpha | = (\alpha_2, \ldots, \alpha_2), \ \alpha_i \in \mathbb{C} \tag{15.3.7}$$

The inner product using the *bra-ket* notation is given by (15.3.8)

$$\langle \alpha | x \rangle = \sum_{i=1}^{n} \alpha_i x_i \tag{15.3.8}$$

The inner product of a bra vector induces the linear function $\langle \alpha |(|x\rangle) = \langle \alpha | x \rangle$, which can be seen using (15.3.5) and the linearity condition given by (15.3.6), so we have,

$$\langle \alpha |(c_1|x\rangle + c_2|y\rangle) = \sum_i \alpha_i (c_1 x_i + c_2 y_i)$$

$$= c_1 \sum_i \alpha_i x_i + c_2 \sum_i \alpha_i y_i \tag{15.3.9}$$

$$= c_1 \langle \alpha | x \rangle + c_2 \langle \alpha | y \rangle$$

contrarily, any linear function can be written as a linear function induced by a bra vector.

The dual vector space denoted by V^* is the vector space of linear functions on a vector space V, in this case, the symbol * denotes the dual instead of complex conjugation.

Given a ket vector $|x\rangle = (x_1, \ldots, x_n)^T \in \mathbb{C}^n$ we can obtain a bra vector using the association (15.3.10), where each component is the complex-conjugated.

$$|x\rangle \mapsto \langle x | = (x_1^*, \ldots, x_n^*) \in \mathbb{C}^n \tag{15.3.10}$$

The norm of the *ket vector* $|x\rangle$ is defined by (15.3.11)

$$\| |x\rangle \| = \sqrt{\langle x|x\rangle} = \left[\sum_{i=1} x_i^* x_i\right]^{1/2} = \left[\sum_{i=1} |x_i|^2\right]^{1/2} \geq 0 \tag{15.3.11}$$

The inner product for the vectors $|x\rangle$, $|y\rangle \in \mathbb{C}^n$ using the physicists' convention is given by,

$$\langle x|y\rangle = \sum_{i=1}^{n} x_i^* y_i \tag{15.3.12}$$

From equation (15.3.5) that express an arbitrary vector $|x\rangle \in \mathbb{C}^n$ as a linear combination of the basis vectors, the n complex numbers c_i are called the **components** of $|x\rangle$ with respect to the set of linearly independent vectors $\{|v_1\rangle, \ldots, |v_n\rangle\}$ in $\mathbb{C}^n$ (Nakahara and Ohmi 2008).

A basis $\{|e_i\rangle\}$ that satisfies (15.3.13) is called an orthonormal basis

$$\langle e_i|e_j\rangle = \delta_{ij} \tag{15.3.13}$$

Using in Eq. (15.3.5) the basis $\{|e_i\rangle\}$ instead of $\{|v_i\rangle\}$, we obtain the inner product of $|x\rangle$ and $\langle e_j$. we have the next expression,

$$\langle e_j|x\rangle = \sum_{i=1}^{n} c_i \langle e_j|e_i\rangle, \tag{15.3.14}$$

using (15.3.13) in (15.3.14) yields,

$$\langle e_j|x\rangle = \sum_{i=1}^{n} c_i \delta_{ji} = c_j \tag{15.3.15}$$

Substituting (15.3.15) in the expansion of $|x\rangle$ given by $|x\rangle = \sum_{i=1}^{n} c_i e_i\rangle$ yields,

$$|x\rangle = \sum_{i=1}^{n} \underbrace{\langle e_i|x\rangle}_{c_i} |e_i\rangle = \sum_{i=1}^{n} |e_i\rangle\langle e_i|x\rangle = \underbrace{\left(\sum_{i=1}^{n} |e_i\rangle\langle e_i|\right)}_{I} |x\rangle \tag{15.3.16}$$

From (15.3.16), since $|x\rangle$ is an arbitrary ket vector, the **completeness relation** is given by,

$$\sum_{i=1}^{n} |e_i\rangle\langle e_i| = I \tag{15.3.17}$$

Let us consider that a vector $|v\rangle$ is projected by the action of the matrix P_k given by (15.3.18), the direction of such projection is defined by the unit vector $|e_k\rangle$, resulting in that $|v\rangle - P_k|v\rangle$ is orthogonal to $|e_k\rangle$, see Fig. 15.5 (Nakahara and Ohmi 2008).

$$P_k \equiv |e_k\rangle\langle e_k| \tag{15.3.18}$$

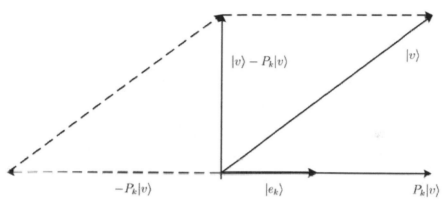

Fig. 15.5 The orthonormal vector $|v\rangle - P_k|v\rangle$ is obtained using the parallelogram rule. Matrix P_k induces a projection operation; therefore, it is called a projection operator.

The matrix P_k satisfies the next three conditions:

1. $P_k^2 = P_k$, $\tag{15.3.19}$

2. $P_k P_j = 0 \quad \forall k \neq j$ $\tag{15.3.20}$

3. $\sum_k P_k = I$ $\tag{15.3.21}$

The Gram-Schmidt orthonormalization process is necessary in QM. It is the process that allows us to construct a set of k orthonormal linearly independent vectors $\{|v_i\rangle\}$ in $\mathbb{C}^n$ for $k \leq n$. This process can be described in three steps:

1. Define a clearly normalized vector,
$$|e_1\rangle = \frac{|v_1\rangle}{\||v_1\rangle\|} \tag{15.3.22}$$

2. Define a vector which is clearly orthogonal to $|e_1\rangle$; for example, $\langle e_1|f_2\rangle = \langle e_1|v_2\rangle - \langle e_1|e_1\rangle\langle e_1|v_2\rangle = 0$, Remember that the component of a vector $|u\rangle$ along $|e_k\rangle$ is given by $\langle e_k|u\rangle$; so, we have,
$$|f_s\rangle = |v_2\rangle - |e_1\rangle\langle e_1|v_2\rangle \tag{15.3.23}$$

3. The orthogonal vector should be normalized,

$$|e_2\rangle = \frac{|f_2\rangle}{|||f_2\rangle||}$$

(15.3.24)

4. In the j_{th} step, the vector (15.3.25) is found,

$$|e_j\rangle = \frac{|v_j\rangle - \sum_{i=1}^{j-1}\langle e_i|v_j\rangle|e_j\rangle}{\left|\left||v_j\rangle - \sum_{i=1}^{j-1}\langle e_i|v_j\rangle|e_j\rangle\right|\right|}$$

(15.3.25)

Therefore, the Gram-Schmidt process allows us to construct the orthonormal set $\{e_1\rangle, \ldots, |e_k\rangle\}$ which spans a k-dimensional subspace in $\mathbb{C}^n$.

For arbitrary $|x\rangle, |y\rangle \in \mathbb{C}^n$ and $c_k \in \mathbb{C}$, the **linear operator** A provides a map $\mathbb{C}^n \rightarrow \mathbb{C}$ if (15.3.26) is satisfied.

$$A(c_1|x\rangle + c_2|\rangle) = c_1 A|x\rangle + c_2 A|y\rangle,$$

(15.3.26)

For the orthonormal basis $\{|e_k\rangle\}$ and the arbitrary vector $|v\rangle$ in $\mathbb{C}^n$, where $|v\rangle = \sum_{k=1}^{n} v_k|e_k\rangle$, using the linear operator implies that,

$$A|v\rangle = \sum_k v_k A|e_k\rangle$$

(15.3.27)

The **Hermitian conjugate** $A^\dagger$ of the linear operator $A : \mathbb{C}^n \rightarrow \mathbb{C}^n$ for the arbitrary vectors $|u\rangle, |v\rangle \in \mathbb{C}^n$ is defined as,

$$\langle u|A|v\rangle \equiv \langle A^\dagger u|v\rangle = \langle v|A^\dagger|u\rangle^*,$$

(15.3.28)

By applying the Hermitian conjugate to a ket vector, we obtain a bra vector,

$$|x\rangle^\dagger = (x_1^*, \ldots, x_n^*) = \langle x|$$

(15.3.29)

The **Hermitian matrix** $A^\dagger$ is defined as a matrix that satisfies $A^\dagger = A$ for a $A : \mathbb{C}^n \rightarrow \mathbb{C}^n$. The **Unitary matrix** U is defined as a matrix that satisfies $U^\dagger = U^{-1}$ for $U : \mathbb{C}^n \rightarrow \mathbb{C}^n$. When $U = 1$, U is known as **special unitary matrix**. U(n) denotes a group of unitary matrix, and SU(n) a group of **special unitary group**.

Let A be a matrix, and $|v\rangle \in \mathbb{C}$ is a vector where $|v\rangle \neq 0$, $\lambda \in \mathbb{C}$ is an eigenvalue of A if,

$$A|v\rangle = \lambda|v\rangle$$

(15.3.30)

in the above equation, $|v\rangle$ is called the eigenvector, the symbol $|\lambda\rangle$ indicates the eigenvector corresponding to the eigenvalue λ. To obtain the eigenvalues of (15.3.30) the **characteristic equation** given by (15.3.31) is used,

$$D(\lambda) = \det(A - \lambda I) = 0. \tag{15.3.31}$$

For a Hermitian matrix, all the eigenvalues are real numbers and two eigenvectors for two different eigenvalues are orthogonal.

The Pauli matrices are a set of three 2×2 Hermitian complex and unitary matrices since $(\sigma_i = \sigma_i^*)$ and $(\sigma_i^* = \sigma_i^{-1})$. The complete set is

$$\sigma_x = \begin{pmatrix} 0 & 1 \\ 1 & 0 \end{pmatrix} \quad \sigma_y = \begin{pmatrix} 0 & -i \\ i & 0 \end{pmatrix} \quad \sigma_z = \begin{pmatrix} 1 & 0 \\ 0 & 1 \end{pmatrix} \tag{15.3.32}$$

To find the eigenvalues of σ_y the characteristic equation given by (15.3.31) is used. Hence, we have $\lambda^2 - 1 = 0$, then the eigenvalues are $\lambda_1 = 1$ and $\lambda_2 = -1$, and the corresponding eigenvectors $|\lambda_1\rangle = (x, y)^T$ is obtained using (15.3.30); i.e., $\sigma_y|\lambda_1\rangle = |\lambda_1\rangle$, from which the normalized eigenvectors are,

$$|\lambda_1\rangle = \frac{1}{\sqrt{2}} \begin{pmatrix} 1 \\ i \end{pmatrix} \quad |\lambda_2\rangle = \frac{1}{\sqrt{2}} \begin{pmatrix} i \\ 1 \end{pmatrix} \tag{15.3.33}$$

The Spectral decomposition of a normal matrix A with $\{|\lambda_i\rangle\}$ eigenvectors and $\{\lambda_i\}$ eigenvalues that are assumed to be orthonormal is given by

$$A = \sum_i \lambda_i |\lambda_i\rangle\langle\lambda_i| \tag{15.3.34}$$

The tensor product also known as the **Kronecker product** is given by (15.3.35), which result is a matrix of size $(mp) \times (nq)$.

$$A \otimes B = \begin{pmatrix} a_{11}B, a_{12}B, \dots, a_{1n}B \\ a_{21}B, a_{22}B, \dots, a_{2n}B \\ \vdots \\ a_{m1}B, a_{12}B, \dots, a_{mn}B \end{pmatrix} \tag{15.3.35}$$

where A and B are matrices of sizes $m \times n$ and $p \times q$, respectively.

15.4 Quantum Mechanic: basic principles

The standard model of a fundamental particle is a mathematical point because they have no spatial extent; the electron is the most common example of the standard model. Atoms are made of three types of particles: electrons, protons and neutrons. The last two made up of the nuclei. Protons and neutrons may be treated as particles in QM, however, they are no fundamental particles. Fundamental particles do not have any size, they can have a position and a momentum. Main properties of a particle are mass, electric charge, and angular momentum. Experiments have shown that indeed electrons behave as whether they have angular momentum, and because they have exactly the same electric charge as well as mass, all electrons are spinning at the same rate. We refer to the angular momentum in a conceptual way as the spin of the particle. Considering a beam of electrons with randomly oriented angular momentum, if we measure the z component of angular momentum, only two different values will get along z-axes, it is either $+5.27 \times 10^{-35} \text{kg m}^2 \text{ s}^{-1}$ or $-5.27 \times 10^{-35} \text{kg m}^2 \text{ s}^{-1}$. The $z-$component of the angular momentum of the spin is quantized, that is, when measured, it takes on one of a finite number of possible values. QM uses the Plank's constant $\hbar = 1.054 \times 10^{-43} \text{kg m}^2 \text{ s}^{-2}$ as a convenient way to express the $z-$spin of an electron, it comes out to either $+\hbar/2$ or $\hbar/2$, so we refer to the electron as spin-1/2 particle which means that its angular momentum is $\hbar/2$. An eigenstate is a state that corresponds to some observable, and it has a definite value.

The Pauli matrices defined in (15.3.32), known as Pauli Spin Matrices are important in QM because they allow to represent the eigenstates for angular momentum of a spin-1/2 particle along each of the three spatial axes (x, y and z) with column vectors. Therefore, we have for σ_x, $|+x\rangle = (1/\sqrt{2}, 1/\sqrt{2})^T$ and $|-x\rangle = (1/\sqrt{2}, -1/\sqrt{2})^T$. For σ_y, $|+y\rangle = (1/\sqrt{2}, i/\sqrt{2})^T$ and $|-y\rangle = (i/\sqrt{2}, 1/\sqrt{2})^T$. For σ_z, $|+z\rangle = (1, 0)^T$ and $|-z\rangle = (0, 1)^T$. Here, $|+z\rangle$ and $|-z\rangle$ are eigenstates of the z component of angular moment. In fact, QM do not have to be in an eigenstate; but, the act of measurement causes a quantum state to **collapse** to an eigenstate of the observable that was measured.

Considering that a quantum state is an eigenstate of a given observable, then the state vector that represents that state is an eigenvector. For example, if $\hat{S}_z$ is the spin $z-$operator, then $|+z\rangle$ is the eigenvector of $\hat{S}_z$, then we have,

$$\hat{S}_z|+z\rangle = \frac{\hbar}{2}|+z\rangle \tag{15.4.1}$$

$$\hat{S}_z|-z\rangle = \frac{\hbar}{2}|-z\rangle \tag{15.4.2}$$

or, using a single expression,

$$\hat{S}_z = \frac{\hbar}{2}\sigma_z = \frac{\hbar}{2}\begin{pmatrix} 1 & 0 \\ 0 & 1 \end{pmatrix} \tag{15.4.3}$$

Therefore, when the $z-$component of an angular momentum $\hat{S}_z$ is diagonalized, it can take two quantum states: the spin-up and the spin-down which are shown in (15.4.4). Similarly, if $\hat{S}_x$ is diagonalized, the quantum states can be either spin-right or spin-left.

$$| \uparrow \rangle = \begin{pmatrix} 1 \\ 0 \end{pmatrix}, \quad | \downarrow \rangle = \begin{pmatrix} 0 \\ 1 \end{pmatrix} \tag{15.4.4}$$

In quantum information, it is common to use the equivalent notations $|0\rangle = | \uparrow \rangle$ and $|1\rangle = | \downarrow \rangle$. Moreover, they are not necessarily associated with spins. The states $|0\rangle$ and $|1\rangle$ may represent two mutually orthogonal states, for example, horizontally and vertically polarized photons.

Energy is observable for a quantum system or for a quantum particle. The corresponding operator to the observable energy is called the Hamiltonian denoted usually with the symbol $\hat{H}$. For an energy state $|\psi\rangle$ which is an energy eigenstate, the usual eigenvalue equation is given by,

$$\hat{H}|\psi\rangle = E|\psi\rangle \tag{15.4.5}$$

where E is the eigenvalue that corresponds to the eigenvector $|\psi\rangle$. This equation only works if the state has definite energy, and E is the value of that energy, in other words, $\hat{H}$ is the operator that corresponds to energy as an observable; so, $|\psi\rangle$ is an eigenvector for the Hamiltonian, this technically is called the Time-Independent Schrödinger equation which is given by (15.4.6).

QM is a theory in the mathematical sense founded on several postulates that can be justified only through empirical facts. The Copenhagen interpretation developed in 1920s by Niels Bohr and Werner Heisenberg, is an earlier and accepted attempt to summarize QM into a coherent mathematical formulation. Other accepted interpretations of QM are the "Many-worlds interpretation", and "alternative interpretations".

The Copenhagen interpretation has the next three basic axioms (Nakahara and Ohmi 2008):

1. In QM, a pure state is represented in terms of a normalized vector $|\psi\rangle$ in a Hilbert space $\mathcal{H}$.
2. Particles have properties called observables, which can be measured. For any observable quantity (physical quantity) α, there exist a corresponding Hermitian operator A acting on $\mathcal{H}$.
3. The Schrödinger equation (15.4.6) dictates the time dependence of a state, in the equation H is the **Hermitian operator**.

$$i\hbar\frac{\partial|\psi\rangle}{\partial t} = H|\psi\rangle \tag{15.4.6}$$

Three important effects for QC are: Superposition and interference, uncertainty, and entanglement. **Superposition** means that a system can be in two or more states at the same time, to explain this, consider a single particle traveling along two different paths at once, which implies that the particle has wave-like properties that can interfere with each other, therefore, **interference** causes that the particle acts in an unpredictable way. The Heisenberg **uncertainty** principle relates the position and momentum of a particle. It states that the more accurate is our knowledge about the position of the particle, the more inaccurate is our knowledge of its momentum, and vice versa; so, there is no way of making an accurate prediction of all the properties of a particle. The ability for pairs of particles to interact over any distance instantaneously is called **entanglement**.

15.5 Elements of Quantum Computing

In classical information theory, the building blocks are Boolean bits, a bit can assume two distinct values, 0 and 1. In quantum information theory, a quantum bit or a qubit is a unit of information describing a two dimensional Hilbert space (Yansofsky and Mannucci 2008). In other words, a qubit is a vector in the vector space $\mathbb{C}^2$, with the next basis vectors denoted as,

$$|0\rangle = \begin{pmatrix} 1 \\ 0 \end{pmatrix}, \quad |1\rangle = \begin{pmatrix} 0 \\ 1 \end{pmatrix} \tag{15.5.1}$$

Since this notation can represent two mutually orthogonal states, $|0\rangle$ may stand for a photon $|\updownarrow\rangle$ which is vertically polarized, while $|1\rangle$ represents a horizontally polarized photon $|\leftrightarrow\rangle$. Other arrow representation might correspond to photon polarized in different directions. In other cases, $|0\rangle$ and $|1\rangle$ represent the spin state of an electron $|\uparrow\rangle$ and $|\downarrow\rangle$ respectively. Two truncated states from many levels also can be represented using a qubit, $|0\rangle$ and $|1\rangle$ may represent ground level and the first excited level, respectively. In mathematical terms, the state of system is described by a unit vector in a Hilbert space $\mathcal{H}$

For convenience's sake, we assume that the vector $|0\rangle$ corresponds to the classical value 0, while $|1\rangle$ to 1 in QC. Usually, a qubit's state during computational phase is represented by a combination of states; so, it is possible for a qubit to be in a superposition state.

$$\psi = \alpha|0\rangle + \beta|1\rangle = a \begin{pmatrix} 1 \\ 0 \end{pmatrix} + b \begin{pmatrix} 0 \\ 1 \end{pmatrix} = \begin{pmatrix} \alpha \\ \beta \end{pmatrix} \quad a, b \in \mathbb{C}, \ |a|^2 + |b|^2 = 1 \tag{15.5.2}$$

Figure 15.6 shows the representation in value and relative phase in two dimensions (2D) of a single qubit, this representation is similar to the way of representing polar coordinates for complex numbers. To carry out a quantum information processing it is necessary to fix a set of basis vectors.

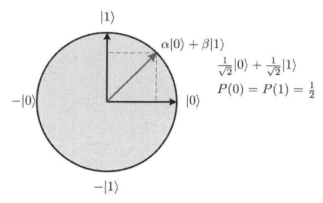

Fig. 15.6 2D representation of the state of a qubit. The sign in the middle of the two values changes the internal evolution of the qubit, but not the outcome of the measurement. Therefore, $\frac{1}{\sqrt{2}}|0\rangle + \frac{1}{\sqrt{2}}|1\rangle$ and $\frac{1}{\sqrt{2}}|0\rangle - \frac{1}{\sqrt{2}}|1\rangle$ have the same output values and probabilities, but different computational phase.

Color image of this figure appears in the color plate section at the end of the book.

Information from qubits can be extracted only through measurements. A qubit has infinitely different states, however, we can extract the same amount of information from a qubit than from a boolean bit. If we make a measurement on the qubit described by (15.5.2) to see if it is $|0\rangle$ or $|1\rangle$, the result will be 0 or 1 with probability $|a|^2$ or $|b|^2$, and the state immediately after the measurement will be $|0\rangle$ or $|1\rangle$. This is because, with the measurement, the sate vector collapses to the corresponding eigenvalue observed.

15.5.1 The Bloch sphere

The Bloch sphere shown in Fig. 15.7 is a 3D geometric representation of the state space of a two levels quantic system. It is important since many operations on single qubits that are commonly used in quantum information processing can be neatly described within the Bloch sphere picture. It is known that the qubit (15.5.2) can be written in a general way as follows:

$$|\psi\rangle = e^{i\gamma}(\cos\frac{\theta}{2}|0\rangle + e^{i\phi}\sin\frac{\theta}{2}|1\rangle) \qquad (15.5.3)$$

where θ, ϕ and γ are real numbers. θ and ϕ define a point in the Bloch sphere of unit radius, $0 \le \theta \le \pi$ and $0 \le \phi \le 2\pi$. Factor $e^{i\gamma}$ is called the **global phase**, qubit states with arbitrary values of γ are all represented by the same point on the Bloch sphere because the factor $e^{i\gamma}$ has no observable effects; i.e., the global phase does not influence the measurement statistics. For this

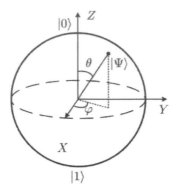

Fig. 15.7 Bloch Sphere.

reason, the global phase is often omitted during the analysis of quantum algorithms and circuits (Imre and Baláz 2005); therefore, the state of a single qubit can be written as follows:

$$|\psi\rangle = \cos\frac{\theta}{2}|0\rangle + e^{i\phi}\sin\frac{\theta}{2}|1\rangle \tag{15.5.4}$$

To obtain (15.5.4), we used the complex number $z = x + iy$, where $x, y \in \mathbb{R}$, and $z \in \mathbb{C}$, we let be $|z|^2 = 1$. Hence, $|z|^2 = z * z = (x - iy)(x + iy) = x^2 + y^2$, which is the equation of a unit circle and center at the origin. With the aim of using polar coordinates, for arbitrary $z = x + iy$ we can write $x = r\cos\theta$ and $y = r\sin\theta$, so $z = r(\cos\theta + i\sin\theta)$. The Euler's identity states that $e^{i\theta} = \cos\theta + i\sin\theta$, and considering that $r = 1$ (unit circle), therefore, $z = re^{i\theta} = e^{i\theta}$ (Benenti et al. 2004). To write (15.5.2) using polar coordinates, the normalization constrain $\langle\psi|\psi\rangle = 1$ requires that $|\alpha|^2 + |\beta|^2 = 1$, so we can express the state using polar coordinates as follows,

$$\psi\rangle = r_\alpha e^{i\phi_\alpha}|0\rangle + r_\beta e^{i\theta_\beta}|1\rangle \tag{15.5.5}$$

where r_α, ϕ_α, r_β and ϕ_β are real numbers. As it was stated before, the only measurable quantities are the probabilities $|\alpha|^2$ and $|\beta|^2$, hence multiplying the state by an arbitrary factor $e^{i\gamma}$ has no observable consequences, because:

$$|e^{i\gamma}\alpha|^2 = (e^{i\gamma}\alpha) * (e^{i\gamma}\alpha) = (e^{-i\gamma}\alpha^*)(e^{i\gamma}\alpha) = \alpha^*\alpha = |\alpha|^2 \tag{15.5.6}$$

and similarly for. Therefore, we are free to multiply the state (15.5.5) by $e^{-i\phi_\alpha}$, and using $\phi = \phi_\beta - \phi_\alpha$, we have,

$$|\psi'\rangle = r_\alpha|0\rangle + r_\beta e^{i(\phi_\beta - \phi_\alpha)}|1\rangle$$
$$= r_\alpha|0\rangle + r_\beta e^{i\phi}|1\rangle \tag{15.5.7}$$

where, we have now only three real parameters, r_α, r_β and ϕ. Now, considering the Cartesian representation of the coefficient $|1\rangle$,

$$|\psi'\rangle = r_\alpha|0\rangle + r_\beta e^{i\phi}|1\rangle = r_\alpha|0\rangle + (x + iy)|1\rangle \tag{15.5.8}$$

And using the normalization constraint,

$$|r_\alpha|^2 + |x + iy|^2 = r_\alpha^2 + (x + iy) * (x + iy)$$
$$= r_\alpha^2 + x^2 + y^2 = 1 \tag{15.5.9}$$

the result of (15.5.9) is the equation of a unit sphere in real 3D space with Cartesian coordinates (x, y, r_α).

The Polar coordinates and Cartesian coordinates are related as follows,

$$x = r\sin\theta\cos\phi$$
$$y = r\sin\theta\sin\phi \tag{15.5.10}$$
$$z = r\cos\theta$$

since $r = 1$ and renaming r_α to z, we can write

$$|\psi'\rangle - z|0\rangle + (x + iy)|1\rangle$$
$$= \cos\theta|0\rangle + \sin\theta(cos\phi + i\sin\phi)|1\rangle \tag{15.5.11}$$
$$= \cos\theta|0\rangle + e^{i\phi}\sin\theta|1\rangle$$

now, in (15.5.11) we have just two parameters defining points on a unit sphere. To obtain (15.5.4), note that if $\theta' = 0 \implies \psi\rangle = |0\rangle$ and $\theta' = \frac{\pi}{2} \implies |\psi\rangle = e^{i\phi}|1\rangle$ which suggest that $0 \le \theta' \le \frac{\pi}{2}$ may generate all points on the Bloch sphere. Now, considering that state $|\psi'\rangle$ corresponds to the opposite point on the Bloch sphere, which has polar coordinates $(1, \pi - \theta', \phi + \pi)$,

$$|\psi'\rangle = \cos(\pi - \theta')|0\rangle + e^{i(\phi+\pi)}\sin(\pi - \theta)'|1\rangle$$
$$= -\cos(\theta')|0\rangle + e^{i\phi}e^{i\pi}\sin(\theta')|1\rangle$$
$$= -\cos(\theta')|0\rangle - e^{i\phi}\sin(\theta')|1\rangle \tag{15.5.12}$$
$$|\psi'\rangle = -|\psi\rangle$$

Considering only the upper hemisphere $0 \le \theta' \le \frac{\pi}{2}$, as opposite point in the lower hemisphere differs only by a phase factor of -1 and so are equivalent in the Bloch sphere representation. Hence, we can map points on the upper hemisphere onto points on a sphere using,

$$\theta = s\theta' \implies \theta' = \frac{\theta}{2} \tag{15.5.13}$$

so, we have

$$|\psi\rangle = \cos\frac{\theta}{2}|0\rangle + e^{i\phi}\sin\frac{\theta}{2}|1\rangle$$

$$(15.5.14)$$

here, $0 \leq \theta \leq \pi, 0 \leq \phi \leq 2\pi$ are the coordinates of point on the Bloch sphere.

15.5.2 Quantum registers

A quantum system that admits three different states is called **qutrit**, while when it admits d different states is called a **qudit**, whereas a **qubyte** consists of eight qubits. A system with n qubits (**multi-qubit**) is called a **quantum register**, and it behaves different from a classical one. In a multi-qubit system, each of the qubits is described using a two-dimensional complex vector; i.e., using the tensor product state given by,

$$(\alpha_1|0\rangle + \beta_1|1\rangle) \otimes (\alpha_2|0\rangle + \beta_2|1\rangle) \otimes \cdots \otimes (\alpha_n|0\rangle + \beta_n|1\rangle) \quad (15.5.15)$$

so, we need $2n$ complex numbers of the form $\{a_i, b_i\}_{1 \leq i \leq n}$ to specify a state. In general, the state vector can be represented by,

$$|\Psi\rangle = \sum_{i_k=0,1} a_{i_1 i_2}|i_1\rangle \otimes \cdots \otimes |i_n \quad (15.5.16)$$

Let us consider a quantum register with two qubits, then we will have 2^2 computational basis states. They are:

$$|00\rangle, |01\rangle, |10\rangle, \text{ and } |11\rangle \quad (15.5.17)$$

Here $|10\rangle$ means that the first qubit is in state $|1\rangle$ and the second qubit is in state $|0\rangle$, this expressed using the tensorial product is $|10\rangle = |1\rangle \otimes |0\rangle$. For example, a two-qubit system can be decomposed into a single qubit as follows:

$$\frac{1}{2}(|00\rangle + |01\rangle + |10\rangle + |11\rangle) = \frac{1}{\sqrt{2}}(|0\rangle + |1\rangle) \otimes \frac{1}{\sqrt{2}}(|0\rangle + |1\rangle) \quad (15.5.18)$$

A quantum register can have the four states shown in (15.5.17) in superposition, so, we have,

$$|\Psi\rangle = \alpha_0|00\rangle + \alpha_1|01\rangle + \alpha_2|10\rangle + \alpha_3|11\rangle \quad (15.5.19)$$

and as it was expected, the sum of all of the probabilities must be 1, for the general case of n qubits. The sum is given by,

$$\sum_{i=0}^{2^n-1} \left(|\alpha_i|^2 + |\beta_i|^2\right) = 1 \quad (15.5.20)$$

15.5.3 Quantum measurements

QM does not allow us to determine the values α and β of the quantum bit (15.5.2); however, we can read out limited information about these two values by measurement the computational basis, so, the $P(0) = |\alpha|^2$ and $P(1) = |\beta|^2$. Unavoidably, taking a measurement disturbs the system, leaving it in a state $|0\rangle$ or $|1\rangle$ which is determined by the outcome.

In a more formal way, quantum measurements are described by a set $\{M_m\}$ of linear operators for $1 \leq m \leq n$, where n is the number of outcomes, they act on the state vector of the measured system. If the state of the system before measurements is $|\psi\rangle$, the outcome m occurs with probability p_m,

$$p_m = \langle\psi|M_m^\dagger M_m|\psi\rangle \tag{15.5.21}$$

and the new state (immediately after the measurement) is,

$$|m\rangle = \frac{M_m|\psi\rangle}{\sqrt{p_m}} \tag{15.5.22}$$

Similarly as in (15.5.20), the probabilities of all outcomes sum up to one,

$$\sum_{i=1}^{m} p_m = \sum_{i=1}^{m} \langle\psi|M_m^\dagger M_m|\psi\rangle = 1 \tag{15.5.23}$$

so that,

$$\sum_{m} M_m^\dagger M_m = 1 \tag{15.5.24}$$

A ket can evolve in two ways:

1. Using an unitary operator: $|\psi\rangle \xrightarrow{U} |m\rangle$ (i.e., $|m\rangle = U|\psi\rangle$), where U is a unitary operator.
2. By the application of a measurement operator: $|\psi\rangle \xrightarrow{\{M_i\}_k} |m\rangle$.

For example, if we have the measurement operators M_0 and M_1 given by (15.5.25) and (15.5.26) respectively, and we have $\psi = \alpha|0\rangle + \beta|1\rangle$, in order to measure a qubit to be $|0\rangle$ or $|1\rangle$ we need to calculate the respective probabilities p_0 and p_1, calculated with (15.5.27) and (15.5.28),

$$M_0 = |0\rangle\langle 0| = \begin{pmatrix} 1 & 0 \\ 0 & 0 \end{pmatrix} \tag{15.5.25}$$

and

$$M_1 = |1\rangle\langle 1| = \begin{pmatrix} 0 & 0 \\ 0 & 1 \end{pmatrix} \tag{15.5.26}$$

The state vector (15.5.2) has the probability given by (15.5.27) of yielding $|0\rangle$, and the probability to obtain $|1\rangle$ is calculated with (15.5.28),

$$p_0 = \langle\psi|M_0^\dagger M_0|\Psi\rangle = \langle\psi|M_0|\psi\rangle = |\alpha|^2 \tag{15.5.27}$$

$$p_1 = \langle\psi|M_1^\dagger M_1|\Psi\rangle = \langle\psi|M_1|\psi\rangle = |\beta|^2 \tag{15.5.28}$$

and the new states are

$$\frac{M_0|\psi\rangle}{|a|} = \frac{a}{|a|}|0\rangle \tag{15.5.29}$$

and

$$\frac{M_1|\psi\rangle}{|b|} = \frac{b}{|b|}|1\rangle \tag{15.5.30}$$

respectively.

15.5.4 Quantum Gates

In Section 15.2 we have explained some concepts that are very important in classical computation; moreover, we introduce notions about reversible computing and CM. In this section, we are going to extend such knowledge to the field of QC. We will start defining a quantum gate as follows (Yansofsky and Mannucci 2008):

"A **quantum gate** is simply an operator that acts on qubits, and they are represented by unitary matrices."

The most general quantum logic gate is a $2^2 \times 2^n$ unitary matrix operation in the $\mathcal{H}^{\otimes n}$. There are several quantum gates, for example, the identity operator I, the NOT gate, the CNOT gate, the Hadamard gate, the Toffoli gate, etc. To illustrate how to calculate the output of a quantum gate, we will use Pauli X gate to perform the quantum NOT operation. In Fig. 15.8 the Pauli X gate is shown, the matrix σ_X acts as a NOT quantum operator on a input qubit; hence, for the qubits $|0\rangle = [1 \ 0]^T$ and $|1\rangle = [0 \ 1]^T$ the operations shown in (15.5.31) and (15.5.32) are performed (Perry 2012).

$$X|0\rangle = \begin{bmatrix} 0 & 1 \\ 1 & 0 \end{bmatrix} \begin{bmatrix} 0 \\ 1 \end{bmatrix} = |1\rangle \tag{15.5.31}$$

$$X|1\rangle = \begin{bmatrix} 0 & 1 \\ 1 & 0 \end{bmatrix} \begin{bmatrix} 1 \\ 0 \end{bmatrix} = |0\rangle \tag{15.5.32}$$

In Fig. 15.9 the Hadamard and CNOT gates are illustrated; in the Hadamard gate, the input computational base $\{|0\rangle, |1\rangle\}$ is transformed in the new base $\{|+\rangle, |-\rangle\}$ according to (15.5.33) and (15.5.34), respectively. In the CNOT gate, the first qubit acts as the control, and the second qubit like the target; i.e., the gate changes the quantum state of the qubit target if the control qubit is in the quantum state $|1\rangle$.

$$H|0\rangle = \frac{1}{\sqrt{2}}(|0\rangle + |1\rangle) \equiv |+\rangle \tag{15.5.33}$$

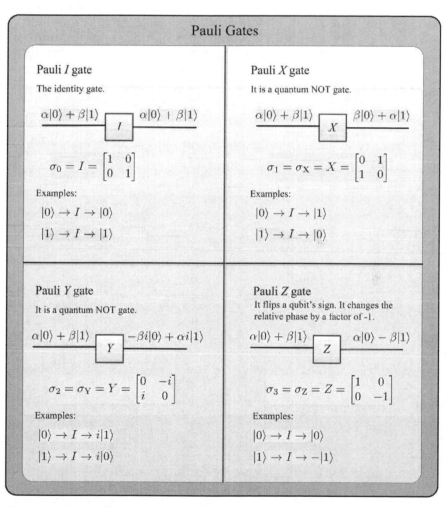

Fig. 15.8 Pauli Gates. For each gate, the name, action, circuit diagram and operation examples are shown.

$$H|1\rangle = \frac{1}{\sqrt{2}}(|0\rangle - |1\rangle \equiv |-\rangle \tag{15.5.34}$$

15.5.5 Quantum circuits

The actions of quantum gates can be represented by drawing a circuit diagram consisting of a sequence of individual diagrams of quantum gates, or unitary operators such as measurement elements (circuits). Figures 15.3 and 15.4, illustrate the CM for classical (Boolean) circuits. These representation are straightforward valid to represent quantum circuit, the only modifications we have to do are to change Boolean variables by quantum variables. Therefore, quantum gates such as the one illustrated in Figs. 15.8 and 15.9 can be used to construct complex quantum algorithms using CM. Figure 15.10 shows a quantum circuit with a three-qubit state at the input; from the top, the first qubit is the input to a Pauli Y gate which performs the NOT quantum operation, hence the operation $|0\rangle \xrightarrow{Y} |1\rangle$ is achieved, the output value $|1\rangle$ enters to the unitary matrix U_2. Following with the next bit, the qubit $|1\rangle$ is the input to the unitary matrix U_1. At the last input, the qubit $|0\rangle$ is processed by the Hadamard gate, so we perform the operation $|0\rangle \xrightarrow{H} |+\rangle$ [(see Eq. (15.5.33)], the output of this gate is also input to the unitary matrix U_1.

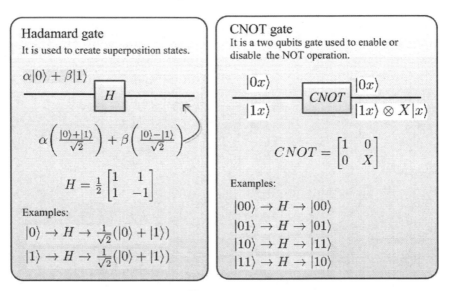

Fig. 15.9 Hadamard and CNOT Gates.

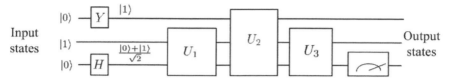

Fig. 15.10 A generic quantum circuit using the CM. There are three unitary matrices (U_1, U_2, U_3) and a meter.

15.6 Concluding Remarks

The main concepts involved in quantum computing have been presented in a very concise way with the aim of giving to the reader a bigger picture of this promissory field that is still in its infancy; undoubtedly, there is a long way to go in the different disciplines that are intrinsically related with QC, such as computer architecture, development of new quantum algorithm, study of quantum complexity, quantum error correction, applications of quantum algorithms to solve science and engineering problems, etc. At the present time, the literature that allows going deep in the field has grown very fast and new and good test platforms have been developed. Finally, with this brief introduction to QC, we hope that new people have been motivated to be interested in this field.

Acknowledgments

The author would like to thank the "Instituto Politécnico Nacional (IPN)", "Comisión de Operación y Fomento de Actividades Académicas del IPN (COFAA-IPN)" and the Mexico's entity "Consejo Nacional de Ciencia y Tecnología (CONACYT)" for supporting my research activities.

References

Arora, S. and B. Barak (2010). Computational Complexity: A Modern Approach. USA: Cambridge University Press.

Benenti, G., G. Casati and G. Strini (2004). Principles of Quantum Computation and Information. Volume I: Basic Concepts. Singapore: World Scientific.

Benioff, P. (1982). Quantum mechanical models of Turing Machines that dissipate no energy. Physical Review Letters, American Physical Society 48: 1581–1585.

Chen, C. and D. Dong (2012). Quantum parallelization of hierarchical Q-learning for global navigation of mobile robots. 2012 9th IEEE International Conference on Networking, Sensing and Control (ICNSC), China, pp. 163–168.

Deutsch, D. (1985). Quantum theory, the Church-turing principle and the universal quantum computer. Proceedings of the Royal Society of London A 400, London, pp. 97–117.

Feynman, R. (1982). Simulating physics with computers. International Journal of Theoretical Physics, 21(6 & 7): 467–488.

Finch, G. (2012). Training Neural Networks on a Quantum Computer. IV Workshop-school on Quantum Computation and Information (WECIQ 2012), Brazil, pp. 162–167.

Imre, S. and F. Baláz (2005). Quantum Computing and Communications. An Engineering Approach. Germany: Wiley & Son.

Kaye, P., R. Laflamme and M. Mosca (2007). An Introduction to Quantum Computing. New York, US.: Oxford University Press.

Nakahara, M. and T. Ohmi (2008). Quantum Computing From Linear Algebra to Physical Realizations. Boca Raton, Fl: CRC Press.

Nayak Shaktikanta, Nayak Sitakanta and J.P. Singh (2011). An Introduction to Quantum Neural Computing. Journal of Global Research in Computer Science, Technical Note, pp. 50–55.

Perry, R.T. (2012). Quantum Computing for the Ground Up. Singapore: World Scientific.

Rybalov, A., E. Kagan, Y. Manor and I. Ben-Gal (2010). Fuzzy Model of Control for Quantum-Controlled Mobile Robots. 2010 IEEE 26-th Convention of Electrical and Electronics Engineers, Israel: IEEE, pp. 19–23.

Shor, P.W. (1994). Algorithms for quantum computation: discrete logarithms and factoring proceedings of the 35th Annual Symposium on Foundations of Computer Science, 1994, Santa Fe, USA, pp. 124–134.

Visintin, L., A. Maron and R.H. Reiser (2012). Interpreting Fuzzy Set Operations from Quantum Computing. IV Workshop-school on Quantum Computation and Information (WECIQ 2012), Fortaleza, Brasil, pp. 142–151.

Yansofsky, N.S. and M.A. Mannucci (2008). Quantum Computing for Computer Scientist. Cambridge, New York: Cambridge.

Index

.NET Framework 185, 216

A

Absolute programming 178
ACO 148, 150, 152, 153, 155–157, 160, 164,
    166, 168, 182, 183, 185–195, 282–292,
    295, 297
Acyclic 20, 21, 310
ADALINE 16, 19, 24
adaptability 198
Adaptation and diversity 201, 212
Adaptive Neuro-Fuzzy Inference System
    (ANFIS) 1, 27–30, 75
Addresses 35, 176
affinity 205, 208
affinity maturation 204, 206
Agent based systems 202
aiNet 207–209
algorithm 14, 16, 23–27, 29, 30, 61, 72, 74,
    76, 79, 83, 84, 86, 90, 95, 96, 100, 108,
    110–113, 116, 119, 120, 122, 127, 128,
    131, 149, 152, 156, 158, 160–169, 184,
    187–192, 194, 196, 200, 202, 203, 205,
    206–210, 211, 216–218, 222–226, 229,
    230, 232–235, 238, 240, 244–246, 248–
    255, 259, 267–270, 272–275, 277, 279,
    284–289, 291, 293, 295, 296, 298–304,
    307, 329
Allen Cahn 110, 111, 124, 125
Amdahl's law 63, 64, 66, 67, 192
Ancilla 309, 310
ANFIS 1, 27–30, 75
Ant Colony Optimization 74, 95, 148–150,
    152, 153, 155, 156, 166, 168, 172,
    182–186, 190, 236, 282–284, 291, 292
Ant Colony System 152, 160, 165, 291, 293
Ant System 148, 152–154, 160, 163, 165, 166,
    182, 189, 283, 290, 291, 293
antibody 199, 200, 206, 207
antigens 200–202, 204, 207

Ant-Q 148, 158, 160–168
ANTS 148–156, 161–163, 165, 182, 183,
    188–192, 282, 284–286, 288–290, 292,
    293, 297–304
applications 3, 4, 15, 27, 34, 35, 38, 39, 48,
    55, 58, 61, 67, 68, 73, 75, 94–97, 99,
    104, 108, 144, 150, 169, 195, 197–200,
    202, 206, 207, 211, 212, 215, 246, 258,
    265–267, 276, 279, 307, 329
Approximate Nondeterministic Tree Search
    163
Artificial Immune Systems (AIS) 197–202,
    204, 206, 210, 211, 215, 216
Artificial Immune System (AIS) paradigm
    216
Artificial intelligence (AI) 259
Artificial life 202, 266
Artificial Neural Network 23, 307
Artificial potential field (APF) 259
Artificial Vaccination 211, 217, 219, 229
Artificial Vaccination methodology 211, 219
artificial vaccines 216, 217
Attractive potential 259–263
AutoCAD™ 184, 185
autonomous control 198, 202

B

Back-propagation 16, 25, 26, 30
Back-propagation algorithm 16, 25, 26
Basis vector 313, 314, 320
B-cells 204, 207
better quality solutions 210
biconjugate gradient 111
bioinformatics 198
Blend Crossover 240
Bloch sphere 306, 321–324
Block 16, 17, 47, 48, 54, 64, 88, 97–99, 101–
    105, 108, 173–176, 199, 219, 274, 275
bra 311–313, 316

Bra-ket 312, 313
BWAS 152, 163–165

C

C# .NET Framework 185
CAD 173, 177, 184, 195
CADL 177
CAM 173, 177, 184, 195
Canned cycles 176, 179–181
Cell processor 39, 40
cells 55, 134, 135, 149, 198, 200–204, 207,
     209, 210, 308
Church-Thesis 308, 309
Circuit Model 306, 308
Classification 23, 35, 52, 56, 73, 74, 149, 150,
     198, 201–203, 212, 215, 233
clonal expansion 204, 206
Clonal Selection Algorithm (CSA) 199, 205,
     206, 211
clonal selection immune theory 204
CLONALG 205–207
CNOT 310, 311, 326–328
Combinatorial Optimization 150, 206, 215,
     266
Combinatorial Optimization Problem
     (COP) 215, 216, 230
commercial applications 198, 200
Commercial software 172, 177, 193–195
Communication overhead 167, 188, 189
Computational complexity 68, 161, 167,
     189, 283
computational paradigms 216
Computer architectures 15, 35, 52, 73
Computer Numerical Control (CNC) 172–
     175, 177, 184, 185, 188, 192, 193, 195
Computer performance 35, 46, 59
computer science 197, 198, 201, 202, 207,
     266, 308, 311
Conjugate Gradient 25, 75, 78, 79, 83, 90,
     110, 111, 120, 122, 131
Conjugate Gradient Method 78, 79, 110,
     111, 120, 122
Controlled NOT 310, 311
Controlled-Controlled-Not (CCNOT) 311
Convergence 29, 90, 115, 116, 119, 128, 139,
     141, 166, 187, 230, 231, 233, 237, 253,
     255
Coolant 173, 175, 177
CPU 3, 34, 36, 42–46, 48–53, 60, 61, 72, 79,
     94–96, 99–102, 106, 108, 127, 128, 131,
     215, 231, 278

Crossover 72, 84, 211, 219, 229, 232,
     237–241, 244, 245, 248–251, 254, 256,
     267, 269, 270, 274, 309
Crowding distance 244–250
Crowding distance algorithm 246
CUBLAS 110, 111, 127, 128, 130, 132
CUDA 70, 72, 86–88, 90, 94–104, 108, 110,
     111, 127, 128, 131, 133, 134, 146, 168
CUSPARSE 110, 111, 127, 128, 130, 132
Cutting Tool Travel Path (CTTP) 184, 186,
     194

D

Danger Theory 199, 210, 212
Dantzing–Fulkerson–Johnson algorithm
     216
data mining 198, 202
Data mining and classification 198, 202
data validation 198
Defuzzification 2, 7, 8, 11, 12
degeneracy models 199
Device 37, 42, 46, 47, 70, 94, 96, 99, 101–103,
     133, 306
Diffusive process 133
Dirac Notation 311
Direction of rotation 173
disease 198, 199
Distributed system 201, 212
Diversity 34, 35, 169, 198, 200, 201, 212, 216,
     229, 233, 246, 269
domination 235, 245
Drilling 173, 174, 176, 177, 179–181,
     183–185, 192, 195
Dual vector 311, 313
DXF 177, 183–185, 188, 191, 194
dynamic and long lasting memory 198

E

EA 233
Efficiency 80, 122, 147, 166, 187, 191–193,
     233, 275, 302
elapsed time 60
Elitism 84, 233, 247
Elitist ant system 148, 283, 291, 293
Elitist AS 152, 158, 162
Ending point 173
engineering 3, 30, 34, 73, 75, 110, 197–199,
     201, 212, 216
Entanglement 306, 307, 319, 320
Euclidian distance 168, 218, 220
Evolutionary algorithm 83, 164, 211, 213,
     233, 255

Executable program 174
execution time 3, 49, 59, 60, 61, 63, 67, 69,
    83, 87, 90, 131, 191–193, 230, 231, 279,
    296, 300
Expansion 204, 206, 218–220, 225, 226, 230,
    314
extrapolation 197, 201

F

fast non-dominated sorting procedure 245
Fault detection and anomalies 202
fault tolerance 27, 198, 212, 214
Feed rate 173, 175, 177, 180, 181
Finite difference 139, 143, 147
fitness 218
fixed-size 67, 69
Fixed-time 67
Fixed-time speedup model 67
Fletcher-Reeves 110, 122, 123
Flynn classification 52, 56
Folk theorem 191
foreign agents 198, 203
FPGA 3, 168
FPS 108
fundamental algorithms 197
Fuzzification 2, 7, 8, 11, 12
Fuzzy logic 1–3, 6, 7, 11, 27, 167
Fuzzy rule 6–8, 11
Fuzzy set 2–5, 8–12, 28
Fuzzy system 2, 3, 9, 12, 207

G

G commands 172, 184, 188
Garbage 309, 310
Genetic Algorithm (GA) 74, 76, 87, 90, 216,
    219, 225, 229, 230, 237, 238, 246, 255,
    283
genetic code 203
Genetic operator 232, 244, 255, 267
genetic programming 95, 198
geometrical Center 223, 224, 226
GFLOPS 95
Global memory 99–101, 187, 190, 248
Global Optima 75
Global search 73, 83, 84, 86
Goals non reachable problem 266
Gradient 24–27, 29, 30, 72, 75–79, 83, 85, 90,
    110–113, 115, 120, 122–124, 131, 260,
    262–265, 271, 272
Gradient descent Method 27, 30, 76, 77

Graphics Processing Unit (GPU) 3, 32, 34,
    42, 58, 70, 73, 88, 89, 94–102, 106, 108,
    110, 111, 126–132, 146, 147, 168, 169
Graphics User Interface (GUI) 105, 185
Greedy Search 217, 218
Grid 97–99, 102–105, 125, 130, 168

H

HAAS Automation™ 193
Hamiltonian cycle 218, 221
Harvard model 44, 45
Hermitian conjugate 316
Hermitian matrix 316, 317
Hessian 78
Heuristic Crossover 239
Heuristic optimization 83
High-performance 31, 33, 34, 56, 69, 73, 142,
    195, 258, 259, 270, 271, 275, 277–279
Hole-cutting operation 173, 184
Holes 85, 86, 90, 172, 177, 179, 180, 190–192,
    194
Host 96, 97, 99, 101–103, 210
Human Immune Systems 198
Hybrid Learning 27, 29
Hybrid models 199
hybridization 207
Hyper-Cube AS 165
hypermutation 201, 204, 206, 211
Hyperthreading 39, 278
Hyper-threading or Simultaneous
    Multithreading (SMT) 52

I

Idiotopes 207
IGES 177
Immune Networks (IN) 222
immune system (IS) 198–204, 207, 210, 212
immunological computation 199
immunology 198–201, 210, 212
improve performance 30, 52, 60, 168
Incremental programming 178, 179
Individuals 25, 84, 86, 88, 134, 148, 149,
    199, 233, 244, 246–247, 249–251, 254,
    268, 269
Inference 1, 2, 7, 8, 11, 12, 27, 75
Innate Immune System 199
Interval 4, 10–14, 31, 126, 162, 165, 240, 249,
    275, 293
Iterative algorithms 14, 72, 74, 75, 77, 83

J

Jobs 189
Join 20, 166, 188, 190, 247

K

Kernel functions 97, 98
Ket 311–314, 316, 325
Ket vector 312–314, 316
KM algorithm 14
Kronecker product 317

L

L2 Cache 46, 47
Laplace Crossover 241, 251
Laplace crossover operator (LX) 241, 249,
    251–253, 255
Learning 16, 23, 27, 29, 158, 159, 200, 266
Learning paradigms 20, 23
Learning process 17, 20, 22, 24, 28, 160, 308
Levenberg-Marquardt 25
Linear crossover 238, 239, 251
linear crossover operator (LinX) 249,
    251–253, 255
Linearly dependent 312
Linearly independent 119, 313–315
LMS algorithm 24
Local minimum problem 261, 265
Local optima 73, 75, 292
Local search 73, 75, 83–86, 166, 187
Lymphocyte 202, 204

M

machine learning 96, 198, 199, 202, 266
Machine tools 172, 173
MADALINE 16
Mäkinen 243
Mäkinen Periaux and Toivanen Mutation
    (MPTM) 243, 249, 250, 252, 253, 255
Mamdani 1, 8, 9, 13, 15
Mandelbrot fractal 94, 96, 108
Manufacturing time 172
MasterCAM™ 193, 194
Mathematics 1, 4, 197, 199, 306, 311
Matlab 35, 72, 81, 82, 91, 92, 133, 142, 143,
    146, 206, 209, 275–276, 294
Max-Min 152, 161, 166, 187, 291, 293
Max-Min ant System 166, 291, 293
McCulloch-Pitt 19
Membership Function 2, 5, 10, 11
Memetic Algorithm 72, 74, 83, 84
Memetic operator 84

memory 22, 23, 33–38, 41, 44–48, 50, 51,
    54–59, 67, 68, 95, 96, 99–101, 103, 127,
    128, 160, 168, 169, 174, 186–190, 198,
    204, 205, 207, 208, 248, 254
memory-bounded 67
Metaheuristics 148, 149, 153, 182, 195
Methodology 3, 52, 95, 96, 195, 211,
    215–217, 219, 224, 229, 231, 233, 286,
    289
MEX files 72, 277
Microprocessor 33, 36, 40, 47, 50
Microprocessor without Interlocked
    Pipeline Stages (MIPS) 50, 61, 62
Milling machine 173, 200
Minimize 3, 27, 75–78, 99, 100, 118, 172, 173,
    184, 234
Mobile robots 258, 259, 265, 279, 308
Modified height type reduction 13
molecules 198, 200, 201, 203, 204, 210
Multi Ant Parallel Colonies Metaheuristic
    172
Multicore 33, 39, 41–44, 49–51, 56, 57,
    63–67, 69, 72, 73, 86–88, 90, 95
Multicore processors 39, 41, 43, 56, 73
Multilayer feedforward 20, 21
Multi-Objective Optimization Problem
    (MOOP) 233, 234, 236
Multiple Instructions Multiple Data
    (MIMD) 52, 56
Multiple Instructions Single Data (MISD)
    52, 55
Multi-qubit 324
Multitasking 186
Mutate 164, 203, 216
Mutation 84, 164, 165, 201–204, 206, 219,
    229, 233, 237, 238, 242–245, 248–251,
    254, 269, 270, 274
Mutation of the B-cells 204

N

natural system 201
Navigation system 3, 258, 259, 270–275,
    277–280
NC Programming 173
Negative Selection Algorithm (NSA) 202,
    203
network intrusion detection 198
Neural network 15, 16, 18, 23, 27, 95, 307
Neuro-fuzzy 1, 27, 75
Newton Method 131
Non-deterministic 149
non-dominated set 235, 236

Non Dominated Sorting Genetic Algorithm-
      II 234, 244
non-self antigens 204
Non uniform mutation 243, 249, 250, 251,
      254
NSGA 232–234, 244, 246, 255
NSGA-II 232, 234, 244–246, 255
NVIDIA 94–98, 100, 108, 127, 130

O

Objective value 86, 87
object-oriented C# language 217
Offspring 219, 237, 239–245, 248, 269
Optimal value 75, 76, 90, 236
Optimization 26, 27, 39, 72–77, 79, 81, 83,
      90, 95, 112, 122, 148–150, 152, 153,
      156, 166, 168, 172, 182–188, 190, 192,
      193, 197, 198, 202, 204–207, 215–219,
      224–226, 230, 232–236, 244, 253, 266,
      267, 282–284, 291, 292
Optimization algorithms 72, 152, 168, 218
organs 149, 198, 201

P

Parallel ant colony optimization (P-ACO)
      185, 186, 192, 193, 195, 284, 285,
      287–290
Parallel computing 35, 49, 79, 94, 96, 100,
      133, 137, 215, 258, 259, 266, 271, 275,
      277, 279, 300
Parallel implementation 72, 88, 90, 127, 128,
      166, 183, 186–188, 190, 192, 232, 246,
      291, 292
Parallel processing 67, 80, 191, 275, 292,
      296–298, 302–304
Paratopes 207
Parent 219, 239, 242–244, 246, 247
Pareto Optimal set 232–236
Pareto optimal value 236
Partial Differential Equation 125, 133
Path planning 167, 168, 259, 265, 271, 279
Pathogens 198, 199, 201
Pattern recognition 15, 198, 201, 202, 204,
      212
Pauli Gates 327
PCB 183–185, 191, 192, 194, 195
Perceptron 15, 16, 19, 24
performance 1–330
Periaux 243
Phase Field 110, 123
Pheromone 150–158, 160–166, 182, 183,
      187–189, 283–285, 292, 293, 302

Population 72, 84, 86–90, 110, 169, 205,
      206, 229, 233, 237, 242, 244–251, 254,
      267–270, 274, 275, 278, 279
Printed circuit board 172, 173
probabilistic 18, 19, 149, 152–154, 157, 182,
      199, 307, 308
Probabilistic Turing Machine 307
Proliferation of B-cells 204
Proportional gains 259, 265, 271–275, 278
pseudocode 26, 76–79, 83, 84, 172, 156, 203,
      205, 207, 244, 246

Q

Q-Learning 158–160, 307
Quantum artificial neural network 307
Quantum Circuits 311, 328
Quantum computation 34, 307
Quantum computer 34, 306, 307
Quantum computing 306–308, 311, 320, 329
Quantum gate 326
Quantum-Inspired Immune Clonal
      Algorithm (QICA) 199
Quantum measurement (QM) 306, 307, 311,
      315, 318, 319, 325
Quantum Mechanics 199
Quantum register 307, 324
Quantum Turing Machine 306, 307
Qubit 199, 320–322, 324–328
qubit antibody 199
Qubyte 324
Qutrit 324

R

Random Walk 133–135
randomly 151, 154, 156, 182, 203, 219, 222,
      223, 237, 238, 240, 244, 248, 268–270,
      284, 303, 318
Rank based ant system 291, 293
Rank-based AS 12
Recurrent network 16, 20–22
Reduce-Optimize-Expand (ROE) 215–217,
      229–231
Reduce-Optimize-Expand (ROE)
      methodology 215
Reduction 12, 13, 51, 67, 216, 217, 227, 230,
      284
Repulsive potential 259–264, 271
Repulsive potential field 259, 261, 264, 271
Repulsive potential function 262, 263
response time 59, 60
Reversible 306, 309–311, 326

robotic control 198
robustness 198, 266

S

SACO 148, 152, 153, 167, 168
sanity checks 101, 219
scalability 64, 66, 67, 96, 97, 198
Scheduling 59, 63, 202
Schematic 183, 244, 310
SDRAM 46
search 2, 30, 73–76, 83–86, 113, 117–123,
   155, 161–163, 165, 166, 187, 189, 198,
   201–203, 205, 217, 218, 232, 236, 239,
   240, 242, 260, 266, 267, 270, 283, 293
Selection 5, 27, 84, 153, 194, 199, 200,
   202–206, 212, 224, 229, 233, 234, 237,
   238, 245, 254, 263, 266, 267–269
self and non-self classification 203
self-antigens 204
Self-organization 16, 149, 200, 212
Sequence 33, 52, 77, 107, 137, 172, 174, 176,
   183, 184, 186, 195, 328
Sequential 24, 25, 43, 48, 49, 51, 53, 63–67,
   76, 80, 83, 86–90, 94, 96, 97, 106, 128,
   166, 167, 187–189, 258, 275, 279, 280,
   295, 297, 299, 301–304
Sequential and batch trainning 24
Sequential processing 67, 295, 304
Simulated Annealing 149
Simulated Binary Crossover 240, 254
Single Instruction Single Data (SISD) 43,
   52, 53
Single Instruction, Multiple Data (SIMD)
   39, 52–55
Solidworks™ 184, 185
Somatic hypermutation 201, 204, 206
Speed 3, 15, 34, 37, 39, 44, 46, 47, 59, 63–69,
   73, 79, 80, 83, 90, 94, 136, 151, 166–168,
   173, 175–177, 181, 187, 189, 191, 192,
   194, 246, 263, 266, 271, 276, 278, 301,
   302, 304, 307, 308
Speed up 34, 63–69, 83, 90, 94, 166–168,
   187, 189, 191, 192, 266, 276, 301, 302,
   304, 308
SRAM 46, 47
Stochastic 18, 19, 24, 25, 73, 74, 83, 138,
   149–151, 156, 236, 255, 269
Stochastic global search 83
Superposition 138, 264, 306, 307, 319, 320,
   324
Superscalar 37, 49, 51

Swarm Intelligence 149
swarm systems 198
system CPU time 60

T

Target position 259–262, 265, 271, 274, 275,
   277, 278
T-cells 202, 203
Telegrapher's equation 138, 144, 146
Thread 41, 82, 88, 92, 93, 96–99, 102–105,
   108, 128, 137, 168, 175, 176, 186, 187,
   190, 232, 246, 275, 279
Threading 41, 52, 186
Time-lagged feed-forward 21
Toffoli gate 311, 326
Toivanen 243
Tool path 172, 183–186, 188, 192–194
Total potential 260, 264, 265
Tournament 238, 245, 248, 269
training pattern 23, 209
Trajectory prediction problem 265
Travel time 173
Travelers Salesman Problem (TSP) 153–157,
   161, 166, 173, 182, 185, 188, 194, 206,
   215–219, 224, 226–231, 282–288,
   290–292
TSP solving algorithm 216, 218, 226, 229
Turing Machine 306–308
Type-1 1–5, 7, 9, 11–13
Type-2 1–4, 9–13
Type reduction 12, 13

U

Uncertainty 2, 3, 6, 9, 10, 27, 149, 319, 320
Uniform Memory Access 57
Uniform mutation 242–244, 249–251, 254
Universal gate 309
Universal Turing Machine (UTM) 308, 309
user CPU time 60

V

vaccination 199, 210, 211, 217, 219, 226, 229,
   230, 231
Vaccine by Elitist Selector (VES) 216, 217,
   223, 224, 227, 230, 231
Vaccine by Random Selector (VRS) 216,
   217, 222–224, 226–230
vaccines 210, 216, 217, 219, 222, 223–228,
   231
vaccinia 198

Variants of Ant Colony 291, 292
VDA 177
Vector Space 311–313, 320
VHDL 3
VLSI 3, 16
Volatile memory 46, 47
Von Neumann 15, 42–44, 48, 51, 53, 266

W

wall-clock-time 60
Wave equation 138

Worker 80, 92, 101, 102, 149, 167, 189, 247, 248
Workpiece 173, 177

Z

Zig-zag trajectory 76

Color Plate Section

Chapter 2

Fig. 2.1 i4004 processor.

Fig. 2.2 Pentium processor.

Fig. 2.3 PowerPC 601 processor.

Fig. 2.4 Pentium 4 processor.

Fig. 2.6 Pentium D processor.

Chapter 3

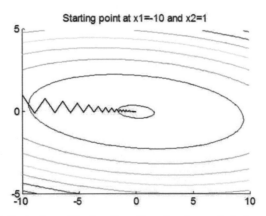

Fig. 3.3 Optimization of the function . Here the zig-zag trajectory is shown.

F1: Sphere function

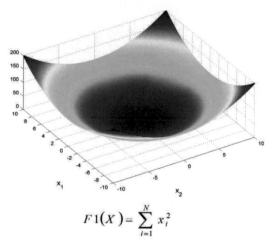

$$F1(X) = \sum_{i=1}^{N} x_i^2$$

Fig. 3.8 Sphere Function (F1).

F2: Rosenbrock function

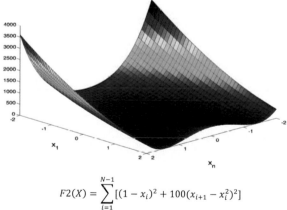

$$F2(X) = \sum_{i=1}^{N-1} [(1 - x_i)^2 + 100(x_{i+1} - x_i^2)^2]$$

Fig. 3.9 Rosenbrock Function (F2).

F5: Foxholes function

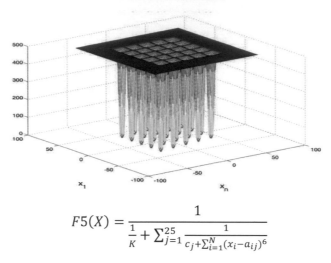

$$F5(X) = \cfrac{1}{\cfrac{1}{K} + \sum_{j=1}^{25} \cfrac{1}{c_j + \sum_{i=1}^{N}(x_i - a_{ij})^6}}$$

Fig. 3.10 Fox holes Function (F5).

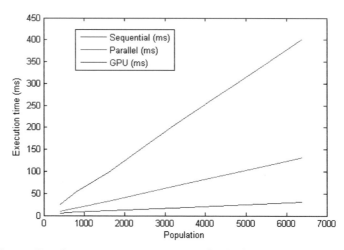

Fig. 3.11 Population size vs. Execution times for the function F5 (Fox holes).

Chapter 4

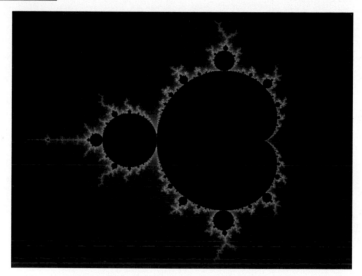

Fig. 4.4 Mandelbrot set representation.

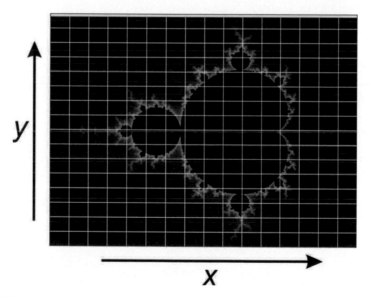

Fig. 4.5 2D Fractal Image workload division. Each grid section represents a block of data that will be processed by each worker thread. If the image size is 800x600 pixels, this will generate a grid of blocks of 50x37.5 pixels, rounded up to 38 for a total of 11900 blocks. The excess pixels are denoted with yellow and are not calculated as we are out of bounds.

Chapter 5

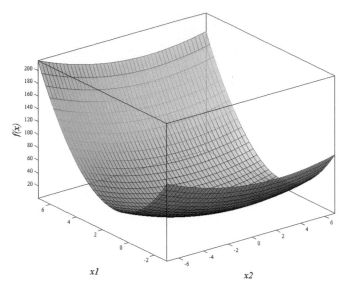

Fig. 5.1 Graph of the quadratic form $f(x)$ with a positive definite matrix.

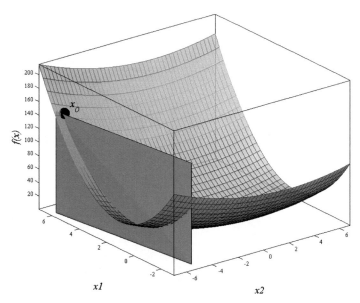

Fig. 5.2 The α value finds the point of the intersection of the surfaces that minimizes $f(x)$.

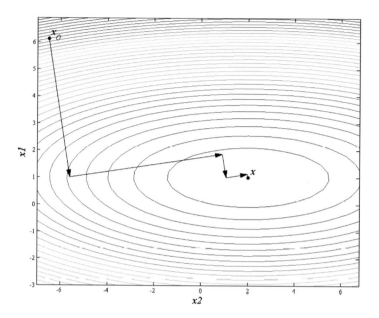

Fig. 5.3 Zigzag path produced by the SD method.

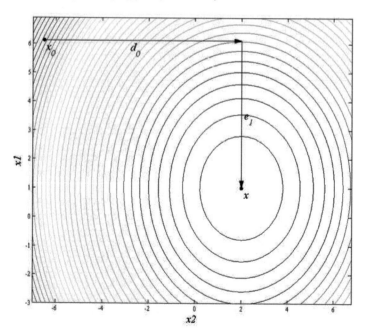

Fig. 5.4 Method of one-step per orthogonal direction.

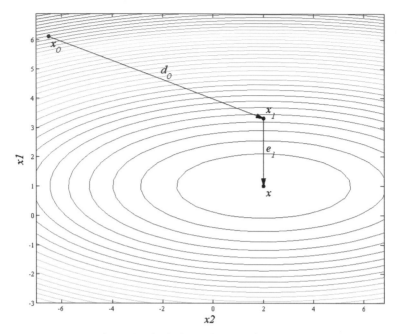

Fig. 5.5 Method of A-orthogonal directions.

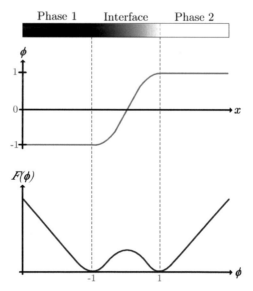

Fig. 5.7 Phase field concept.

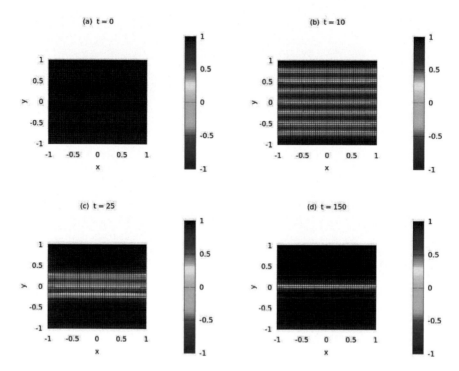

Fig. 5.9 Process of phase separation in 2-D.

Chapter 6

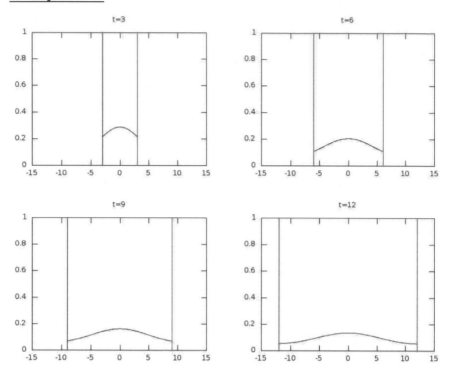

Fig. 6.1 Mesoscopic diffusion for the problem (6.4.2) under conditions (6.4.3). As far as time increases the general trend is the classical description.

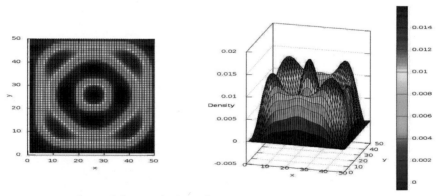

Fig. 6.3 Solution of Telegrapher's equation in two dimensions.

Chapter 8

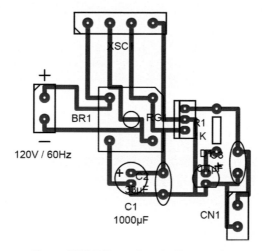

Fig. 8.7 5V DC Power Supply PCB Board.

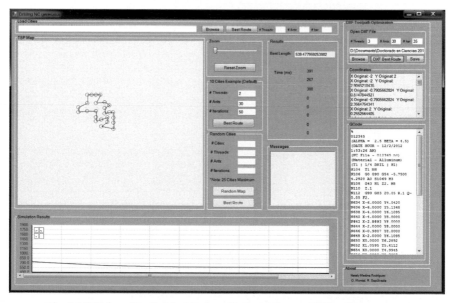

Fig. 8.9 GUI for a Parallel Ant Colony Optimization and NC code generator.

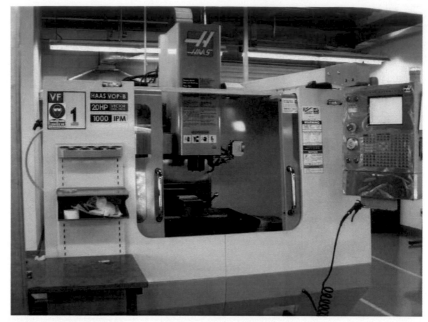

Fig. 8.15 HAAS Automation CNC machine used in this research.

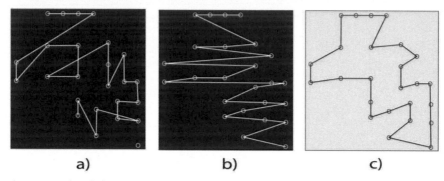

a)   b)   c)

Fig. 8.16 Tool path for a 25 holes PCB design. a) Option 1 from MasterCAM™, b) Option 2 from MasterCAM™ and c) Tool path using our Parallel ACO algorithm.

Chapter 9

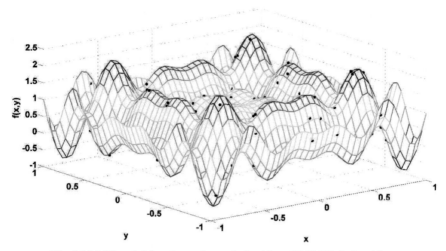

Fig. 9.1 Multimodal function to be optimized by a CLONALG algorithm.

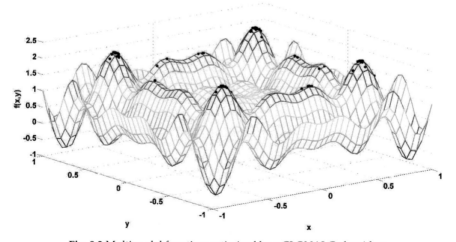

Fig. 9.2 Multimodal function optimized by a CLONALG algorithm.

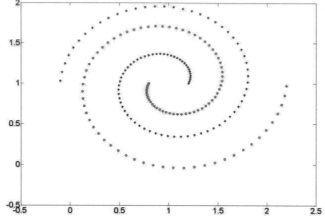

Fig. 9.3 Training patterns for the aiNet algorithm.

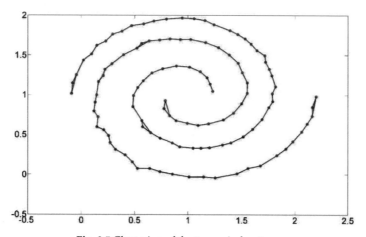

Fig. 9.5 Clustering of the two spiral patterns.

Chapter 10

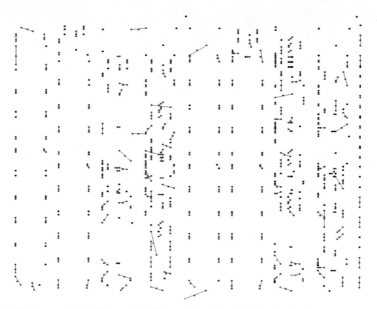

Fig. 10.5 TSP Instance RBX711 from TSPLIB vaccinated by VRS Method with NV=ONLSize×0.4, and NNV=2. Blue dots represent the Start and End Nodes and the red dot represents the Geometrical Central that represents the vaccine.

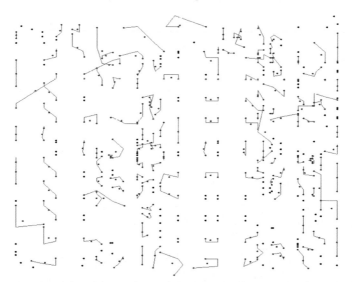

Fig. 10.6 TSP Instance RBX711 from TSPLIB vaccinated by VRS Method with NV=ONLSize×0.20, and NNV=4. Blue dots represent the Start and End Nodes and the red dot represents the Geometrical Central that represents the vaccine. Green lines are the vaccines.

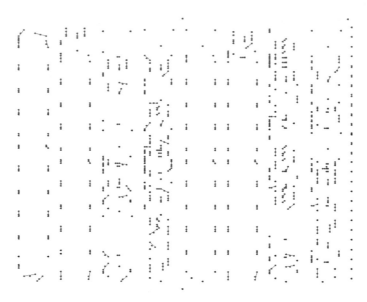

Fig. 10.7 TSP Instance RBX711 from TSPLIB vaccinated by VRS Method with NV=ONLSize×0.4, and NNV=2. Blue dots represent the Start and End Nodes and the red dot represents the Geometrical Central that represents the vaccine.

Chapter 11

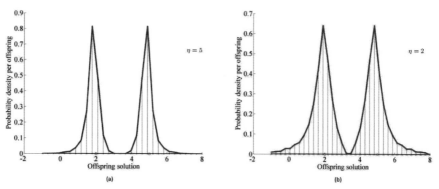

Fig. 11.7 Probability distribution for creating children solutions of continuous variables using the SBX crossover operator.

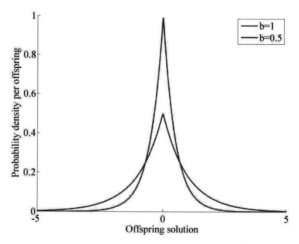

Fig. 11.8 Density function of Laplace distribution ($a = 0$, $b = 0.5$ and $b = 1$).

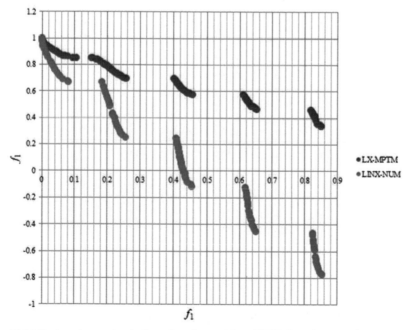

Fig. 11.16 Best and worst solutions for the problem ZDT3 found using the PNSGA-II algorithm.

Chapter 12

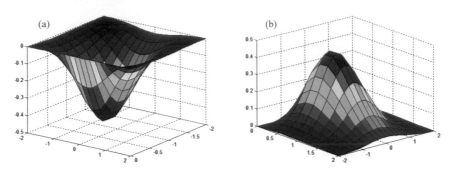

Fig. 12.1 Simulation of (a) an attractive potential surface (b) a repulsive potential surface.

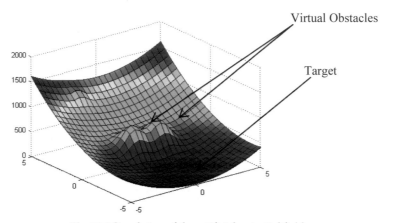

Fig. 12.2 Simulation of the artificial potential field.

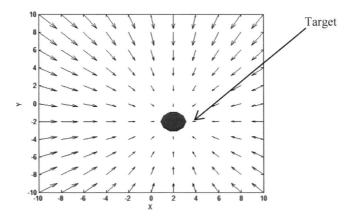

Fig. 12.3 Simulation representing the gradient of an attractive potential field.

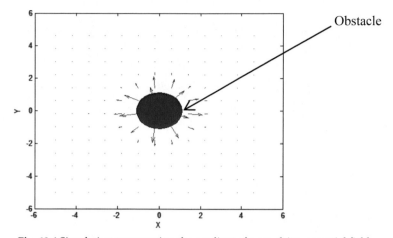

Fig. 12.4 Simulation representing the gradient of a repulsive potential field.

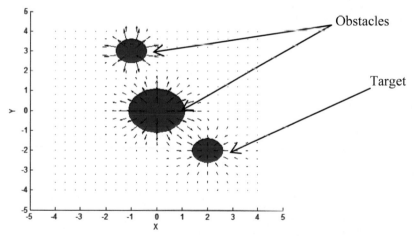

Fig. 12.5 Simulation representing the negative gradient of a total potential field.

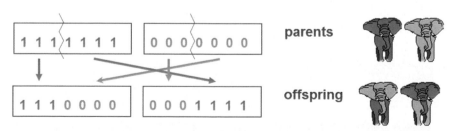

Fig. 12.8 Example of crossover (Rodríguez 2007).

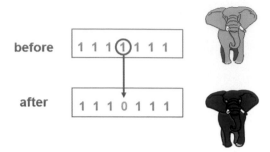

Fig. 12.9 Example of mutation (Rodríguez 2007).

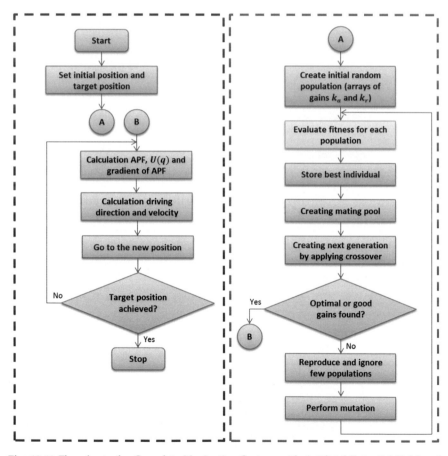

Fig. 12.11 Flowchart of a Complete Navigation System with Artificial Potential Field and Genetic Algorithms.

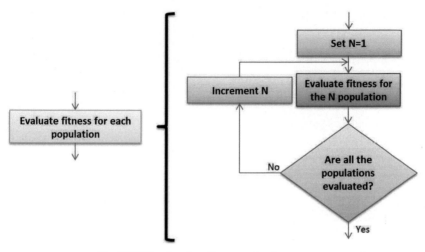

Fig. 12.12 Step by step fitness evaluation process.

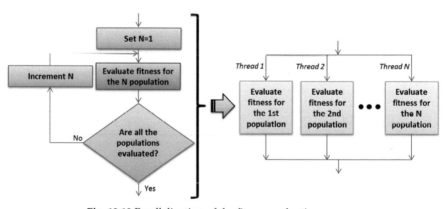

Fig. 12.13 Parallelization of the fitness evaluation process.

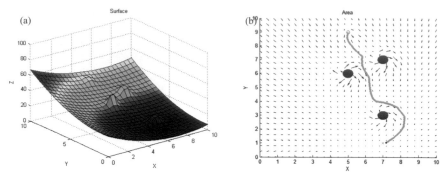

Fig. 12.15 (a) Surface generated by artificial potential field. (b) Path generated by the high-performance navigation system.

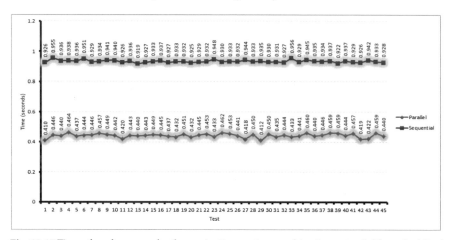

Fig. 12.17 Time of performance for the navigation system working in sequential form (red line) and parallel form (blue line). The vertical axis represents time in seconds and the horizontal axis represents the number of tests.

Chapter 13

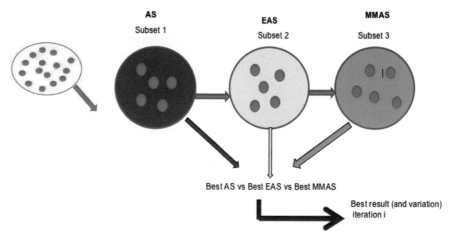

Fig. 13.4 Picture representing how the partition is made.

Chapter 14

Fig. 14.1 Main Interface.

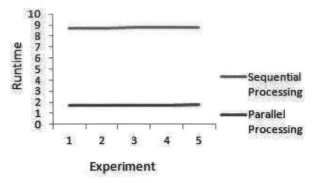

Fig. 14.7 Runtime sequential and parallel.

Chapter 15

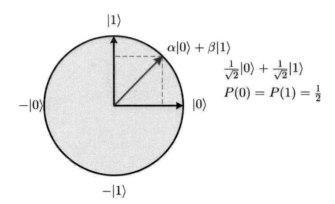

Fig. 15.6 2D representation of the state of a qubit. The sign in the middle of the two values changes the internal evolution of the qubit, but not the outcome of the measurement. Therefore, $\frac{1}{\sqrt{2}}|0\rangle + \frac{1}{\sqrt{2}}|1\rangle$ and $\frac{1}{\sqrt{2}}|0\rangle - \frac{1}{\sqrt{2}}|1\rangle$ have the same output values and probabilities, but different computational phase.